Studies in Systems, Decision and Control

Volume 317

Series Editor

Janusz Kacprzyk, Systems Research Institute, Polish Academy of Sciences, Warsaw, Poland

The series "Studies in Systems, Decision and Control" (SSDC) covers both new developments and advances, as well as the state of the art, in the various areas of broadly perceived systems, decision making and control–quickly, up to date and with a high quality. The intent is to cover the theory, applications, and perspectives on the state of the art and future developments relevant to systems, decision making, control, complex processes and related areas, as embedded in the fields of engineering, computer science, physics, economics, social and life sciences, as well as the paradigms and methodologies behind them. The series contains monographs, textbooks, lecture notes and edited volumes in systems, decision making and control spanning the areas of Cyber-Physical Systems, Autonomous Systems, Sensor Networks, Control Systems, Energy Systems, Automotive Systems, Biological Systems, Vehicular Networking and Connected Vehicles, Aerospace Systems, Automation, Manufacturing, Smart Grids, Nonlinear Systems, Power Systems, Robotics, Social Systems, Economic Systems and other. Of particular value to both the contributors and the readership are the short publication timeframe and the world-wide distribution and exposure which enable both a wide and rapid dissemination of research output.

** Indexing: The books of this series are submitted to ISI, SCOPUS, DBLP, Ulrichs, MathSciNet, Current Mathematical Publications, Mathematical Reviews, Zentralblatt Math: MetaPress and Springerlink.

More information about this series at http://www.springer.com/series/13304

Shuyi Shao · Mou Chen · Peng Shi

Robust Discrete-Time Flight Control of UAV with External Disturbances

Springer

Shuyi Shao
College of Automation Engineering
Nanjing University of Aeronautics
and Astronautics
Jiangsu, Nanjing, China

Mou Chen
College of Automation Engineering
Nanjing University of Aeronautics
and Astronautics
Jiangsu, Nanjing, China

Peng Shi
School of Electrical and Electronic
Engineering
University of Adelaide
Adelaide, SA, Australia

ISSN 2198-4182 ISSN 2198-4190 (electronic)
Studies in Systems, Decision and Control
ISBN 978-3-030-57956-2 ISBN 978-3-030-57957-9 (eBook)
https://doi.org/10.1007/978-3-030-57957-9

This Springer imprint is published by the registered company Springer Nature Switzerland AG
The registered company address is: Gewerbestrasse 11, 6330 Cham, Switzerland

To our families for their love and support

Preface

This book is devoted to studying some discrete-time flight control schemes for the fixed-wing UAV system in the presence of system uncertainties, external disturbances, and the input saturation. On the basis of lemmas of the stability analysis of discrete-time nonlinear systems, approximation methods of system uncertainties, design technologies of nonlinear discrete-time disturbance observers (DTDOs), and tackling methods of input saturation, the main research motives of this book are given as follows:

(1) With the development of modern industrial production technology, the system mathematical model is more and more complex. In order to meet the high demand for practical control, the high performance computers have been widely used in the control field. Since computers can only process discrete-time digital signals for the data storage and calculation, the continuous-time signals need to be converted to discrete-time signals when the UAV system is controlled by computers. In addition, the actual control laws of the UAV system need to be realized by the digital controllers, so the studies of discrete-time flight control schemes are particularly important. On the other hand, the control performance of the digital controller based on the approximated discrete-time UAV model may be better than that of the digital controller obtained by discretizing the continuous-time controller. Therefore, in order to facilitate the digital realization of the flight control schemes, it is of practical significance to study discrete-time flight control methods.

(2) In practice, a large number of systems possess modeling errors and other uncertainties. System uncertainties may not only degrade the performance of the control plant but also even lead to instability of dynamics systems. Due to the coupling between each channel of the fixed-wing UAV and the nonlinear actuator, the model of the fixed-wing UAV cannot be accurately modeled, so there will be modeling errors between the accurate model and the constructed model. Thus, the control issue considering system uncertainties should be solved to improve the performance of the UAV system. Due to their inherent ability of universal approximation, the neural network (NN) is usually applied

to model the unknown nonlinear system functions. The existing research works demonstrate that uncertain nonlinear systems can be efficiently controlled by using adaptive NN control schemes.

(3) It is generally known that practical systems are often subject to external unknown disturbances, which can cause the performance degradation and instability of systems. Due to the variable flight environment of fixed-wing UAVs during flight missions, UAVs are often affected by wind disturbances. However, the Kalman filter cannot be used to tackle external disturbances if the external disturbance cannot be measured. It is well known that external disturbances are very difficult or even impossible to be measured physically by sensors. Disturbance observers (DOs) can well estimate external disturbances using the known information of controlled plants and the control law can include the outputs of disturbance observers. As a result, the disturbance rejection ability is ensured to improve the performance of the closed-loop system. Therefore, the development of DTDO-based control scheme is significant for the control of uncertain UAV systems with external disturbances.

(4) On the other hand, the system control performance of the UAV can be degraded under the input nonlinearities such as input saturation, dead-zone, and hysteresis, and the input nonlinearities even can lead to the close-loop system instability. Therefore, considering the input nonlinearities problem is very significant for the control of uncertain fixed-wing UAV system. Moreover, it is a challenging problem to design the discrete-time control scheme for the control of the uncertain fixed-wing UAV system in the presence of external disturbances and input saturation. Therefore, DTDO-based control schemes need to be further studied for uncertain UAV systems with external wind disturbances and input saturation.

(5) In previous studies, most of the control methods focused on how to guarantee that the tracking error converges to a bounded region or asymptotically converges to origin, which belongs to steady-state performance research. The study on transient performance including overshoot and convergence rate needs to be further considered. Actually, the transient performance plays an important role in improving the performance of the actual UAV system. The bigger overshoot may lead the actuators to exceed the physical limitation, consequently causing the instability of the closed-loop system. Therefore, the transient and steady-state performance should be further improved for the control of UAV systems in the presence of system uncertainties, external disturbances, and input saturation.

Based on the above research motivation, the main contributions of this book are shown as follows:

Firstly, on the basis of the previous research works, the continuous-time fixed-wing UAV attitude dynamics model under external wind disturbances is given. By using the Euler approximation method, the fixed-wing UAV attitude dynamics model with continuous-time form is transformed into an approximate discrete-time one. A discrete-time neural network (NN) control scheme based on the DTDO is proposed for the discrete-time fixed-wing UAV attitude dynamics model with external disturbances and system uncertainties. This scheme combines the DTDO, the NN approximation method, and the backstepping control (BC) technique to realize the discrete-time tracking control of the fixed-wing UAV, and the robustness of the resulted UAV control system is also enhanced. The feasibility of the discrete-time flight control scheme is demonstrated by the numerical simulation of the fixed-wing UAV attitude dynamics system.

Secondly, since the actuator is prone to occurring saturation, the input saturation problem is further considered in the design of the robust discrete-time flight control scheme, and a discrete-time robust anti-saturation control method based on the auxiliary system and the nonlinear DTDO is proposed to achieve the stable flight control of the fixed-wing UAV with external wind disturbances, the input saturation, and system uncertainties. On the basis of the BC technique, the stability of the closed-loop system is proven by using Lyapunov stability theory. The simulation results of the uncertain fixed-wing UAV attitude dynamics model show that the discrete-time-NN-based control scheme is effective.

Thirdly, for the trajectory control model under external wind disturbances and the uncertain discrete-time fixed-wing UAV attitude nonlinear model under wind disturbances and input saturation, a discrete-time robust control scheme is proposed for the uncertain fixed-wing UAV system based on an auxiliary system and the discrete-time sliding mode disturbance observer (SMDO). The scheme introduces an auxiliary system to suppress or eliminate the adverse effects of input saturation for the closed-loop flight control of the fixed-wing UAV, and the NN approximation is used to deal with the uncertainties in the fixed-wing UAV system, and a NN-based discrete-time SMDO is employed to compensate for the negative effects of external disturbances, and a discrete-time robust controller is designed by utilizing the BC technique. The numerical simulation results verify that the flight control performance is satisfactory under the proposed NN-based discrete-time control method.

Fourthly, the system uncertainties and the problem of the prescribed performance are considered in the design of discrete-time fractional-order (DTFO) control scheme for the fixed-wing UAV attitude system with external disturbances, and the control problems of the steady state and the transient performance for the fixed-wing UAV attitude closed-loop system are comprehensively studied, and then by combining the prescribed performance control method, the NN-based DTDO and the discrete-time fractional-order calculus (FOC) theory, the DTFO tracking controller based on the DTDO and the NN was designed with the BC technique. On the basis of the designed DTFO tracking controller, the flight control of the fixed-wing UAV system in the presence of external disturbances and system uncertainties is realized. The simulation results show that the designed DTFO tracking controller can ensure that the fixed-wing UAV closed-loop system satisfies the steady state and the transient performance, and the designed DTDO can guarantee that the disturbance estimation errors are bounded.

Finally, aiming at the DTFO control problems for the trajectory control system and the attitude system of the fixed-wing UAV with external wind disturbances, a DTFO control method is proposed for the fixed-wing UAV system. In this method, the DTFO control method based on the designed DTDO is proposed by using the BC technique, and then the DTFO tracking control of the fixed-wing UAV system is realized. Simulation results show the effectiveness of the proposed DTFO control scheme based on DTDO.

This book intends to provide the readers a good understanding on how to achieve the discrete-time tracking control schemes of the fixed-wing UAV system with system uncertainties, external wind disturbances, and input saturation. The book can be used as a reference for the academic research on uncertain UAV systems or used in Ph.D. study of control theory and engineering.

We would like to acknowledge Prof. Bin Jiang, Prof. Youmin Zhang, Prof. Qingxian Wu, Dr. Kenan Yong, and Dr. Yankai Li for their help and support. We would also like to thank the support of research grants, including National Science Fund for Distinguished Young Scholars (No. 61825302), Natural Science Foundation of Jiangsu Province for Young Scholars (N0. SBK2020042328), and Aeronautical Science Foundation of China (No. 201957052001).

Nanjing, China
June 2020

Shuyi Shao
Mou Chen
Peng Shi

Contents

Acronyms

$\Re$	The field of real numbers
$\Re^r$	The r-dimensional real vector space
$\lvert\cdot\rvert$	The absolute value
$\lvert\lvert\cdot\rvert\rvert$	The 2-norm
$\lambda_{\max}(\cdot)$	The maximal eigenvalue
$\lambda_{\min}(\cdot)$	The minimal eigenvalue
sign	The signum function
sat	The saturation function
diag	The diagonal matrix
sec	The second
deg	The degree
M	The quality of UAV
T	The transposition
$\overline{V}_A$	The flight velocity of UAV
χ	The track azimuth
γ	The track inclination
β, α, μ	The sideslip angle, attack angle, and roll angle
$\overline{T}$	The engine thrust
p_b, q_b, r_b	The roll, pitch, and yaw angular velocity
$\overline{x}_g, \overline{y}_g, \overline{z}_g$	The position of UAV in the ground coordinate system
I_{xx}, I_{yy}, I_{zz}	The rotational inertia
I_{xz}	The product of inertia
g_0	The gravitational constant
g	The vertical gravity component
$\overline{R}$	The equatorial radius of earth
$\overline{D}$	The resistance
$\overline{L}$	The lift force
$\overline{X}, \overline{Y}, \overline{Z}$	The aerodynamic
$\overline{q}$	The aerodynamic pressure
S_r	The wing area of UAV

$\overline{\rho}$	The current density
ρ_0	The sea level air density
$\overline{\overline{T}}$	The temperature
$C_{X_{\overline{T}}}$	The axial force coefficient
h_e	The engine angular momentum
$C_{Y_{\overline{T}}}$	The lateral force coefficient
$C_{Z_{\overline{T}}}$	The normal force coefficient
$\overline{b}$	The wing length of UAV
$\overline{c}$	The average aerodynamic chord length
$\delta_e, \delta_a, \delta_r$	The pneumatic rudder surface deflection angle
$C_{l_{\overline{T}}}, C_{m_{\overline{T}}}$	The total rolling and pitching moment coefficients
$C_{n_{\overline{T}}}$	The total yaw moment coefficient
$C_{n\delta_r}$	The moment coefficient
$C_{l\delta_a}, C_{l\delta_r}$	The moment coefficients
$C_{m\delta_e}, C_{n\delta_a}$	The moment coefficients
UAVs	Unmanned aerial vehicles
DO	Disturbance observer
DTDO	Discrete-time disturbance observer
NN	Neural network
BC	Backstepping control
SM	Sliding mode
SMC	Sliding mode control
SMDO	Sliding mode disturbance observer
FO	Fractional-order
DTFO	Discrete-time fractional-order
DTDO	Discrete-time disturbance observer
FOC	Fractional-order calculus
FBL	Feedback linearization
DI	Dynamic inversion
ABC	Adaptive backstepping control
MIMO	Multiple-input-multiple-output
CTDO	Continuous-time disturbance observer
FL	Fuzzy logic
FOPID	Fractional-order proportion integration differentiation
RBFNNs	Radial basis function neural networks

Chapter 1
Introduction

1.1 Background

UAVs can be defined as vehicles without carrying pilots, and are unmanned aircrafts that can operate remotely or autonomously during the flights mission [1]. Due to the advantages of low cost, strong flexibility and wide application range, UAVs have developed rapidly in recent years and been increasingly applied in civil and military fields [1]. For example, aerial photography, crop monitoring and pesticide spraying, sheep monitoring, coast guard search and rescue, coastline and waterway monitoring, pollution and land monitoring, forest fire detection, reconnaissance, enemy activity monitoring, mine location and destruction, and so on [1–7]. Therefore, on the basis of the wide application of UAVs, researchers have designed many types of UAVs in recent decades, and the researches on flight control systems of UAVs have received high attention.

According to the size, weight, endurance, range and altitude of UAVs, the international organization for unmanned vehicle systems classifies UAVs into three categories [8]: (i) *Tactical UAVs*, including micro, close, short, medium range, low altitude deep penetration, low altitude long endurance and medium altitude long duration systems. The mass varies from a few kilograms to 1000 kg, the range varies from a few kilometers to 500 km, the flight height varies from a few hundred meters to 5 km, the flight time varies from a few minutes to 2–3 days; (ii) *Strategical UAVs*, including high-altitude long-endurance, stratospheric and extra-stratospheric systems, which fly at a height of more than 20,000 m and last for 2–4 days; (iii) *Special mission UAVs*, such as unmanned combat autonomous aircraft and decoys systems. UAVs have been widely studied in military field. The US military introduced UAVs during the first world war, which promoted the development of UAVs technologies. So far, a variety of advanced military UAVs have been produced [9], for example, the global hawk, pterosaurs, predators, sword, X47-B and fire scout, and some of these UAVs can land, take off and finish the flight trajectory tracking tasks independently. Moreover, some UAVs can adjust their flying attitude automatically to adapt to changes in the environment. In addition, due to the mission requirements of UAVs, the control technologies applied to military UAVs are more advanced than those of civilian

S. Shao et al., *Robust Discrete-Time Flight Control of UAV with External Disturbances*, Studies in Systems, Decision and Control 317,
https://doi.org/10.1007/978-3-030-57957-9_1

UAVs. Based on the characteristics of the easy operation and the slow flight speed, rotary-wing UAVs are more widely used in civil fields than fixed-wing UAVs, for example, applications such as power patrol, express delivery and so on [10]. Whether it is military UAVs or civil UAVs, flight control system is the main component of UAV system and the premise of performing various flight control tasks. Therefore, research on flight control systems with low manual intervention and strong autonomy is an important direction of UAV technology development in the future [9].

With the continuous development of modern industrial production technologies, the abstract mathematical model systems based on practical engineering are becoming more and more complex. In order to meet the high demand of modern industrial control, high-performance computers have been widely used in the field of control. Since computers can only process discrete-time digital signals during data storage and calculation, continuous time signals need to be converted into discrete-time signals when the controlled system is controlled by computers [11]. In addition, the actual nonlinear control law is realized by the digital controller, so the research on the control problem of discrete nonlinear systems is particularly important. Furthermore, the control performance of the digital controller designed based on the approximate discrete-time controlled object model may be better than that of the digital controller designed by discretization methods based on the continuous time controlled object model [12]. Therefore, in order to facilitate the digital implementation, it is of practical significance to further study the discrete-time flight control methods for UAVs systems in the discrete-time framework.

Most of the flight control methods studied at present are designed based on the continuous time UAVs systems. Therefore, based on the continuous time control schemes, this book takes the design of discrete-time flight control schemes as the main research direction. Combining with the approximated discrete-time fixed-wing UAV dynamics system, the robust discrete flight control methods with system uncertainties and external wind disturbances in the flight process of fixed-wing UAV are emphatically studied to ensure that the fixed-wing UAV could complete the flight control mission under the influence of system uncertainties, input saturation and external wind disturbances.

1.2 Significance and Purpose

The quality of flight control system not only affects the ability of UAVs to perform tasks, but also affects the flight safety of UAVs. Therefore, it is of great significance to study the flight control methods of UAVs. The flight environment of fixed-wing UAVs is changeable, and the flight tasks that UAVs need to perform are special, and the design of the flight control systems of the fixed-wing UAVs is more complicated than that of the rotary-wing UAVs. Therefore, the design of high-precision and high-efficiency flight control scheme plays an important role in improving the control performance of the fixed-wing UAVs.

First of all, in the process of modeling a fixed-wing UAV system, it is often difficult to obtain accurate aerodynamic parameters and body structure model, so there exist modeling errors between the accurate model and the constructed model, which results in system uncertainties in the UAV system. In addition, due to the variable flight environment of fixed-wing UAVs during the flight missions, UAVs are often affected by wind disturbances. However, system uncertainties and external disturbances will not only reduce the control performance of the flight control systems, but also lead to the instability of the UAVs systems and make the flight of UAVs unsafe. Therefore, the study of flight control methods with strong robustness to suppress the adverse impact of system uncertainties and external disturbances on the flight control performance of UAVs, so as to improve the closed-loop performance of UAVs systems, has become a key issue to be considered in the studies of discrete-time flight control schemes of UAVs systems.

Secondly, according to the physical characteristics of the engine and aerodynamic rudder surface of the fixed-wing UAVs, the rudder surface deflection angle, angular velocity and engine thrust of the fixed-wing UAVs are limited to a bounded range [13]. For the traditional design of flight control schemes, the problem of actuator limitation (i.e. input saturation) is ignored. However, if the designed control system does not consider the influence of input saturation, when the deflection angle of the aerodynamic rudder surface reaches the limit boundary, it will no longer change, which will affect the control performance of UAVs systems and even destroy the stability of UAVs systems. Therefore, for the research of discrete flight control method of UAVs, the robust restricted control is also one of the key problems in the flight control design of UAVs.

Finally, in the existing tracking control methods of fixed-wing UAVs, most of the flight control methods focus on how to ensure that the tracking error converges to the bounded region or asymptotically converges to zero, which belongs to steady-state performance research. So far, there are few researches on the transient performance of flight control for fixed-wing UAVs. In fact, studying transient performance plays an important role in improving the performance of UAVs. For example, excessive overshoot may cause the actuator to exceed the physical limit, resulting in instability of the closed-loop UAV systems. Therefore, when designing a robust discrete-time flight controller, the transient and steady-state performance of UAVs systems should be further considered simultaneously.

Based on the analysis motivation above, for a fixed-wing UAV system with system uncertainties, wind disturbances and input saturation, the design of a discrete-time flight controller with better flight control performance has practical significance for the flight control of a fixed-wing UAV. So far, the studies on the discrete-time flight control problems of the fixed-wing UAV systems are less, so this book in view of the external disturbance suppression and limited actuator control problem, and some robust discrete-time flight control technologies are proposed for the uncertain fixed-wing UAVs systems.

1.3 Advanced Flight Control Methods

Research on the flight control is of great significance to improve the maneuverability and stability of UAVs, and the ability of the UAVs to complete tasks in different environments can be improved by advanced flight control methods [14]. In recent years, with the development of control science and computer technology and the increasingly complex flight control requirements of UAVs, some advanced control theories have been gradually introduced into the design of flight control method. The following section mainly summarizes four advanced flight control methods for UAVs control, including the flight control methods based on H_∞ control, the flight control methods based on feedback linearization (FBL), the flight control method based on sliding mode control (SMC) and adaptive backstepping control (ABC) method.

Flight control methods based on H_∞ control: Due to the strong robustness of H_∞ control, it can effectively control the controlled system with the large parameter variation range, the dynamic uncertain model and serious nonlinear characteristics, so H_∞ control has been widely applied in the design of flight control scheme. In [15], the H_∞ control method was used to realize the control of an unmanned helicopter. For a small fixed-wing UAV system, a trajectory tracking controller was designed based on H_∞ robust control in [16]. In [17], an extended H_∞ controller was studied based on the multi-objective differential evolution algorithm, and the flight control of an autonomous helicopter was realized. For the CE-15 helicopter, an H_∞ control scheme was proposed in [18]. In [19], the H_2/H_∞ control technology was used to achieve landing control of aircraft. Based on the two-degree-of-freedom H_∞ robust control method, a multivariable controller based on observer is designed in [20], and the flight control for helicopter was realized. For the formation flight system of short-range UAV with unknown actuator fault, an adaptive fault-tolerant H_∞ output feedback control scheme was proposed in [21]. In [22], the game theory was employed to design H_∞ control scheme, and the longitudinal flight control of hypersonic vehicle was realized. For the attitude stability control of the rocket, a two-degree-of-freedom H_∞ robust control scheme was proposed in [23]. The above research results shown that the H_∞ control can be effectively applied to the design of flight control schemes. However, due to the large calculation amount of H_∞ control and the complexity of the designed controller, it is difficult to realize H_∞ control in practical engineering.

Flight control methods based on FBL: The FBL method is a common method to study the control of nonlinear systems, and it includes two important branches: the dynamic inversion (DI) control method and the differential geometric control method [14]. For the above two methods, the differential geometric control method is relatively abstract, which is not conducive to the application in the design of flight control schemes of UAVs. The DI control method can realize differential transformation of nonlinear system, so as to complete decoupling between channels in flight control system [14]. Therefore, the DI control method is more suitable for the design of flight control method. Up to now, many FBL-based flight control methods have been reported by researchers. For example, the FBL method was used in [24] to realize the tracking control of a four-rotor UAVs with faults. In [25], an UAV attitude tracking control scheme was proposed based on the FBL method. For a four rotor UAV system, the FBL control method was studied based on the DO in [26]. In [27], a DI controller was designed and the stability control of the tilt-turn missile system was realized. For the tilt-rotor UAV system, a flight control method was studied based on the incremental nonlinear DI technology in [28]. For the unmanned combat aircraft during aerial refueling in [29], the NN control method was investigated based on the DI control technology. In [30], an adaptive incremental nonlinear DI control method was explored and the attitude control of the micro air vehicle was studied.

The above research results show that the DI control technology can be effectively applied in the design of flight control methods. However, when the DI method is directly applied to the design of flight control method, the system model of the aircraft needs to be very accurate, which is difficult to exist in the real world. Due to the system modeling error, unmodeled dynamics, sensor noise, unknown environment and other uncertain factors [31], when studying the design of flight control method, the robustness of the controller should be fully considered.

Flight control method based on SMC: The design idea of SMC was to use high-frequency switch control signal to drive the controlled signal to arrive and stay near sliding mode (SM) surface in finited time [32]. In addition, the SMC method provides a method for maintaining stability and consistency in the case of low modeling accuracy, and SMC has strong robustness against internal and external disturbances and parameter uncertainties. Therefore, the main advantage of SMC is that the controlled signal response of the system was insensitive to uncertainties and disturbances of the model [33]. As a widely used nonlinear control method, SMC had good performance and robustness for a class of nonlinear control problems, and it was also used to solve practical problems in some fields [32, 34].

It is well known that SMC method has been widely used in the design of flight control schemes and some meaningful conclusions have been obtained. For example, aiming at the trajectory tracking problem of quadrotor aircraft, an adaptive SMC strategy was proposed based on the shape theory [35]. In [36], an adaptive SM relative motion control was studied for the landing of a fixed-wing UAV. Five control laws was designed based on the backstepping control (BC) and the SMC, and the flight control of a fixed-wing UAV was realized in [37]. In [38], a SMC scheme based on tangent vector field was proposed, and multiple UAVs were realized to track the ground moving target collaboratively. An integral SMC method was proposed, and the path tracking control was studied for the quadrotor aircraft under model uncertainties and external disturbances in [39]. In [40], an extended observer based on the adaptive second-order SMC method was designed and the tracking control of the fixed-wing UAV was realized. A control scheme was proposed based on the global fast dynamic terminal SMC method in [41], the designed controller can ensure that the position and attitude of the quad-rotor UAV was controlled in a limited time. For the fixed-wing UAV tracking the ground moving target, a fast terminal SMC method was proposed in [42]. In [43], a non-singular fast terminal SM attitude control method was studied based on a tracking differentiator and an extended state observer, and the flight control of the quad-rotor UAV was realized. In the above conclusions, the flight control schemes based on the SMC were studied and the control performance of the systems was improved by using SMC methods. While, due to the high frequency switching control signal in the designed SM controller, there will be chattering in the output signal of the controller and the controlled system. In order to suppress the influence of chattering on the system control performance, researchers proposed a variety of improved SMC methods, such as terminal SMC method [41–43], high-order SMC method [44] and super-distorted SMC method [45]. However, the influence of chattering on system control was not completely eliminated.

Adaptive BC method: Adaptive control has emerged since the early 1950s and has been a research field of control with theoretical and practical significance [46]. Since adaptive control can provide an adaptive mechanism to regulate the controller so that the system with uncertain parameters, structure and environment can achieve the desired system control performance [47]. Therefore, the adaptive control has been widely studied in the control of actual systems [47–49]. In addition, in the early 1990s, a BC method was proposed for the design of adaptive controller [50]. The BC is a control method based on the recursive Lyapunov function for a class of strict feedback systems. In fact, the BC method can ensure global or local tuning and tracking performance of the system when the controlled object belongs to a class of systems that can be converted into strict feedback form. An important advantage of the BC method was that it provides a systematic process to design the controller and follows the step-by-step solution method [50]. Therefore, the BC method can construct the feedback control law and the Lyapunov function systematically. In addition, another advantage of BC method was that it can flexibly avoid canceling useful nonlinearities and guarantee the stability of the tracking control system [50]. Therefore, some conclusions were obtained by using the BC method [50–52].

With the development of the adaptive control method and the BC method, researchers proposed ABC methods and applied ABC methods to the design of flight control scheme. In [53], an ABC based flight control method was proposed for the quadrotor UAV system with uncertain parameters. For small fixed-wing UAV systems, an ABC-based attitude control scheme was designed in [54]. In [55], the tracking control of a quadrotor UAV system was realized by using the ABC-based SMC technology. A flight control method was proposed based on the ABC for a six-degree-of-freedom unmanned helicopter system with unknown parameters, and the tracking of predetermined position and yaw track was realized in [56]. In [57], a robust flight control method was studied based on fuzzy logic (FL) system and the ABC for an uncertain unmanned helicopter system. Although the above research results show that the ABC method can effectively realize the flight control of the aircrafts, it is still rare to consider simultaneously the system uncertainties, external disturbances and input constraints in the design of flight control methods. Therefore, the issues mentioned above need to be further studied.

The above mentioned work mainly introduces the application results of the H_∞ control method, the FBL control method, the SMC control method and the ABC method in flight control systems. In addition, with the development of advanced flight control method research, in recent years, the researchers combined with intelligent control methods and flight control systems, and some flight control schemes were proposed based on the intelligent control, such as the flight control method based on the NN [58, 59], the flight control method based on the FL [60, 61], the flight control based on the intelligent optimization method [62, 63], the flight control method based on the reinforcement learning [64, 65], the flight control method based on the adaptive dynamic programming [66, 67], and so on.

1.4 Discrete-Time Flight Control Methods

In the past ten years, due to the application of digital controller in the actual control system, the design of discrete-time control scheme has been widely studied. In practice, the design of digital controller mainly included the continuous time controller design and the discrete-time controller design [68–70]. In the case of the continuous time design, a continuous time controller was first designed based on the continuous time controlled object model, and then the continuous time controller was discretized and applied by means of the sampler and holding device [71]. In the case of the discrete-time design, the discrete-time controller was designed by using the discrete-time controlled object model. Up to now, many scholars have proposed a variety of discrete-time control methods for the discrete-time controlled object model. In [72, 73], the model predictive control methods were respectively proposed for fuzzy discrete-time systems and discrete-time linear systems with unknown disturbances. For a discrete-time hydraulic tank system, the observer design problem with error limitation was studied in [74]. For the discrete-time offshore steel casing platform model, a discrete-time feedforward and feedback optimal tracking control scheme was proposed in [75]. In [76], the output integral discrete-time SMC of the piezoelectric micro/nano positioning system was studied. The above research results mainly studied the control of discrete-time linear system models. In addition, the research on the discrete-time nonlinear controlled object models has also attracted extensive attention. In [77], a robust adaptive fuzzy SMC method was proposed for a class of discrete-time nonlinear systems. For a class of discrete-time nonlinear affine systems with control constraints, an NN-based optimal control scheme was studied in [78]. In [79], a robust inverse optimal control for discrete-time nonlinear systems is studied. For unknown discrete-time non-affine nonlinear systems, an optimal control scheme based on adaptive dynamic programming method was explored in [80]. In [81], an adaptive minimum quadratic control of the discrete-time manipulator model was realized. Using the reinforcement learning control method, the adaptive NN control of an autonomous underwater vehicle with nonlinear control input was investigated in [82]. In [83], an NN-based adaptive predictive control scheme was proposed for discrete-time purely feedback nonlinear systems. For a class of multiple-input-multiple-output (MIMO) nonlinear discrete-time systems, a data-driven model-free adaptive control method was presented in [84]. In [85], an H_∞ finite time control scheme was studied for switched nonlinear discrete-time systems with disturbances. For the manipulator model, a discrete-time SMC method based on the NN was proposed in [86], and the control of the manipulator was realized.

According to the above research conclusions, not all the mentioned discrete-time control methods can be directly applied to the design of discrete-time flight control. In addition, the continuous time advanced flight control method in Sect. 1.2 cannot be directly applied to the design of discrete-time flight control.

Some significant research results on discrete-time flight control have been reported in recent years. In [87], the discrete-time adaptive flight control of an aircraft with a controlled layout was studied. For a quadrotor UAV system, a discrete-time SMC scheme was proposed by Euler approximation method in [88]. In [89], a fault estimation and regulation method was explored for the discrete-time linear MIMO system, and the vertical flight control of helicopter was realized. The discrete-time hypersonic aircraft model was obtained through Euler approximation method in [90–92], and a flight control scheme was presented based on the NN. According to the fixed-wing aircraft model and the Euler approximation method, a nonlinear discrete-time reconstruction flight control scheme was proposed in [93]. The above works have given some design techniques of discrete-time flight control methods, but the external wind disturbances, system uncertainties and input constraints are not considered in the studies of discrete-time flight control. However, UAVs are susceptible to external disturbances due to the volatile work environment. At the same time, the structure of UAV is relatively complex, and there are often uncertainties in the control system, so it is difficult to establish an accurate mathematical model. In addition, the aerodynamic rudder surface and engine thrust of UAVs are restricted to a certain range. If the influence is not considered in the design of the control system, when the aerodynamic rudder surface and engine thrust reach the limit boundary, which will affect the control performance and even destroy the stability of the system. Furthermore, in previous studies, external disturbances, system uncertainties and input constraints were considered less in the design of discrete-time flight control schemes. Therefore, the discrete-time flight control methods need to be further studied for uncertain UAVs with external disturbances and constrained inputs.

1.5 Flight Control Methods Based on DOs

A variety of disturbances exist widely in practical control systems, and they will not only affect the control performance of the systems, but even cause instability of the control systems [94, 95]. Therefore, the study of disturbances suppression is of great significance to improve the control performance of the actual system. As is known to all, if the disturbances in the control system can be measured directly, the influence of disturbances can be reduced or eliminated by introducing feedforward control method [95, 96]. However, in general, it is difficult to measure the disturbance information in the control system [95, 96]. To address the above mentioned problems, the researchers designed DO with the known information of the control system and used the output of the DO to suppress the adverse effects of disturbances. Because the DO can effectively deal with disturbances in the control system, a lot of research results of control methods based on the DOs have been reported. In [96], some design forms and application scopes of DOs were introduced. For manipulator systems with external disturbances, a control method based on a nonlinear DO was proposed in [97]. In [98], an integral SMC method based on the DO was studied for continuous time linear systems with mismatched disturbances. A DO was used to implement the NN learning control for a class of strict feedback systems in [99]. For uncertain fractional-order (FO) mechanical systems, a DO-based adaptive NN control scheme was proposed in [100]. The research works mentioned above were to study the control methods of general linear systems, nonlinear systems and the FO mechanical system, and the control performance was improved for the systems by using DOs.

In recent years, the research on DOs and flight control methods has attracted extensive attention due to the fact that the DOs can enhance the disturbances suppression ability and robustness of the control system, and some significant research results have been reported. The DO-based DI control methods were proposed for the missile longitudinal dynamics model with external disturbance in [101] and the transport aircraft with continuous heavy cargo airdrop in [102]. In view of the flapping wing micro air vehicle model with external disturbances and system uncertainties, an adaptive flight control scheme was studied based on the DO and the NN in [103]. Based on the SMC technology and the DO design method, the flight control was realized for the hypersonic vehicle system model with input saturation [104]. For a quadrotor UAV system with external disturbances, a nonlinear elastic control scheme was designed by using the BC technology and a nonlinear DO in [105]. In [106], the flight control of a quadrotor UAV was realized by using a sliding model controller and a SMDO. In [107, 108], a robust attitude controller was designed for a DO-based quadrotor UAV system. For the aspirated hypersonic vehicle system with external disturbance, a DO-based flight control scheme was proposed in [109–111]. In [112], an adaptive SMC method based on the DO was designed for the near-space vehicle system. With the BC technology based on a DO, the flight control of small helicopter was realized in [113]. In [114, 115], the robust flight control of the fixed-wing UAV systems based on DOs was studied. The above research works mainly for continuous time disturbance observer (CTDO) design and the design of flight control scheme based on CTDO.

In addition, for the research achievements in the past, some reports on the design and application of DTDOs. The design methods of DTDOs were given for general linear systems with disturbances in [116, 117]. For a discrete-time stochastic linear system, a reduced order DTDO was designed in reference [118]. Combined with the switching control method and the DTDO design method, the control of a discrete-time delay system with actuator failure was studied in [119]. By combining the discrete-time SMC method and the DTDO design method, the current control was realized for the induction motor in [120]. In [121], a robust deadbeat predictive power control method based on a DTDO was proposed for the pulse-width modulated rectifier. Although the above works have studied the control methods based on DTDOs, there is few research on combining DTDOs with discrete-time flight control schemes. Therefore, the discrete-time flight control methods based on DTDOs need to be further studied for UAVs with external disturbances.

1.6 Flight Control Methods Under Input Nonlinearities and System Uncertainties

In the UAV control systems, the relationship between the input signal and the output signal of the controller is often nonlinear. Previous studies on the nonlinear characteristics of controllers in the UAV control systems mainly include the input saturation, the dead-zone nonlinearity and the nonlinearity satisfying asymmetric saturation and dead-zone input. In addition, there will be modeling errors in the modeling process of the UAV control systems, which will lead to system uncertainties in the UAV control systems. The input nonlinearities and system uncertainties mentioned above will not only reduce the control performance of the systems but also cause the instability of the UAV control systems. Therefore, the control performance and robustness of the systems can be enhanced by considering the input nonlinearities and system uncertainties in the control studies of the actual systems. The following is an introduction of the system control methods for the problems of input nonlinearities and system uncertainties.

Input nonlinearities: As is known to all, actuators in actual control systems can only provide limited control power. Therefore, researchers have proposed some control schemes for systems with input saturation. In [122], the closed-loop quadratic stability and the $\mathscr{L}_2$ performance characteristics were studied for the linear control system with input saturation. For a linear system with input saturation, a compound nonlinear feedback control method was proposed in [123]. In [124], the stability was analyzed for a linear system with the existential state delay and the input saturation. For the manipulator model with input saturation, an adaptive NN impedance control scheme was investigated in [125]. In [126], a robust adaptive control was explored for the uncertain nonlinear system with external disturbances and the input saturation. For general linear and nonlinear systems with input saturation, some control schemes have been proposed in the above works. In addition, in recent years, some research works have been reported with the consideration of input saturation in the flight control. In [127–129], adaptive SMC methods were studied for spacecraft models with input saturation. For the spacecraft model with input saturation, fault-tolerant control schemes were proposed in [130, 131]. In [132], the robust flight tracking control was presented for the spacecraft with input saturation. For the spacecraft with external disturbances and input saturation, a DO-based adaptive DI control scheme was proposed in [133]. The control of systems with input saturation has been studied in the above mentioned research works. On the other hand, the dead-zone nonlinear phenomenon often occurs in the actual control system, that is, when the input control signal is less than a certain value, the control output signal of the actuator is zero. In order to solve the problem of dead-zone nonlinearity in the systems, scholars have studied the control methods of dead-zone nonlinearity. In [134], a robust adaptive control for the nonlinear system with unknown dead-zone nonlinearities was studied. For a nonlinear system with uncertain dead-zone nonlinearity, an adaptive output control method was proposed in [135]. In [136], an adaptive dynamic surface control scheme was studied for nonlinear pure feedback systems with unknown dead-zone nonlinearity. In [137], an adaptive tracking control method was presented for the nonlinear system with asymmetric dead-zone nonlinearity. For the MIMO system with dead-zone nonlinearity, an adaptive fuzzy output feedback control scheme was investigated in [138]. In addition, the dead-zone nonlinear problem has also received extensive attention in the studies of flight control methods. For the hypersonic aircraft model with dead-zone nonlinearity, the ABC method was designed in [139], the BC scheme based on Lyapunov function was proposed in [140], the prescribed performance control method was studied in [141], and a robust adaptive NN control scheme was analyzed in [142]. For the near-space vehicle with dead-zone nonlinearity, an adaptive NN prescribed performance tracking control scheme was proposed in [143]. In the above mentioned research works, the proposed control methods can effectively suppress the negative effects of input saturation and dead zone nonlinearities. Moreover, in view of the control problem for the asymmetric saturation and dead zone-input nonlinearity, the hysteresis input and the gap nonlinearity, a robust adaptive NN control scheme was proposed for the MIMO system with the asymmetric saturation and dead zone input nonlinearity in [144, 145], the adaptive control was studied for uncertain nonlinear systems with the hysteresis input in [146–148], the nonlinear tracking control problems of uncertain nonlinear systems were explored in [149, 150]. The above research results proposed some control schemes for continuous time systems with input nonlinearities, and the designed methods can effectively suppress the influence of input nonlinearities on the system control performance.

So far, researchers have done a lot of research on the continuous time systems with input nonlinearities. In addition, some research results have also been reported for discrete-time systems with input nonlinearities. In [151], an adaptive fuzzy control scheme was proposed for the nonlinear discrete-time system with the gap. For a discrete-time system with unknown control direction and input hysteresis nonlinearity, an output feedback adaptive control method was studied in [152]. In [153], the problem of adaptive control design was analyzed for the purely feedback discrete-time nonlinear MIMO system with unknown hysteresis. The modeling and identification was investigated for a discrete-time nonlinear dynamic cascade system with input hysteresis in [154]. In [155–157], fuzzy adaptive control schemes were studied for uncertain nonlinear discrete-time systems with dead-zone nonlinearity. An NN-based adaptive reinforcement learning control method was explored for the uncertain nonlinear discrete-time system with unknown dead zone nonlinearity in [158]. In [159], the adaptive NN control problem was analyzed for an uncertain nonlinear discrete-time system with dead-zone nonlinearity. By designing the Lyapunov function, the stability of a discrete-time system with actuator saturation was studied in [160]. In [161], the global consistency was investigated for the discrete-time multi-intelligence system with input saturation. For the discrete-time switched network system with external disturbance and input saturation, the problem of semi-global consistency was studied in [162]. In [163], an adaptive NN control method was presented for an uncertain discrete-time feedback system in the presence of input saturation. The tracking control problem was explored for the nonlinear discrete-time system with input saturation in [164], and the proposed control method can guarantee the fast response of the tracking control and the performance of small overshoot. In [165], a robust output tracking controller was designed for the uncertain discrete-time system with input saturation. In [166], the exponential synchronization control of the discrete-time networked oscillator with partial input saturation is realized. Although the above results have studied the control of continuous time system with input nonlinearities, there is few study considering input nonlinearities in flight control design and realize discrete-time flight control. Since the energy provided by the actuators in the UAVs is limited, and the computer can only process the discrete-time digital signals in data storage and calculation. Therefore, the discrete-time flight control methods of UAVs with input nonlinearities need to be further investigated.

System uncertainties: In the actual control systems, due to modeling errors, measurement error, component aging and other irresistible factors, there are uncertainties in the system modeling. In addition, the system uncertainties will not only destroy the performance of the system, but also may lead to the instability of the system. In order to deal with the uncertainties in the control systems, the NN and the fuzzy logic (FL) system can be used to compensate for the uncertainties of the systems. In recent years, some control methods based on the NN and the FL have been proposed to control nonlinear systems with system uncertainties. In [167], a NN-based adaptive control scheme was proposed for the uncertain pure feedback system with hysteresis input. DO-based adaptive NN control methods were studied for uncertain nonlinear systems in [168, 169] respectively. In [170, 171], the adaptive NN control method and the adaptive NN output feedback control scheme were studied respectively based on the reduced order observers. The NN-based dynamic surface control methods were proposed for uncertain nonlinear systems in [172, 173]. In [174], an adaptive $\mathscr{L}_2$ control scheme was explored based on the BC technology for an uncertain nonlinear system. An adaptive NN control of the uncertain manipulator system with full state constraints was studied in [175]. In [176], the NN-based adaptive tracking control method was proposed for an uncertain MIMO system with unknown dead-zone nonlinearity and control direction. For the consistency problem of multi-intelligent systems, the NN-based decentralized robust adaptive control was implemented in [177]. For chaotic systems with uncertain functions, the NN-based synchronization control was presented in [178]. The above works mainly use the NN to deal with the uncertainties of the systems. Moreover, the design of control methods based on the FL system has been widely concerned for uncertain systems. In [179], an adaptive FL tracking control was implemented for the uncertain nonlinear MIMO system based on the DO. Using SMC control method based on FL system, the adaptive control of an uncertain nonlinear system and the microelectromechanical system were studied in [180, 181] respectively. In [182, 183], the observer-based FL control schemes were proposed for the switched uncertain nonlinear large systems and the uncertain MIMO system with dead-zone nonlinearity. The FL control of electro-hydraulic active suspension system with input constraint was explored in [184]. In the above works, the NN and the FL system were effectively utilized to suppress the influence of system uncertainties. Therefore, some researchers applied the NN and the FL system to the design of flight control, and realized the effective control of aircrafts. In [185, 186], the flight control of the unstable UAV system and the F-8 fighter were realized respectively by the direct adaptive NN control method. Based on the BC technology, a nonlinear adaptive flight controller based on the NN was designed in [187]. In [188, 189], the adaptive NN control methods for the small unmanned rotorcraft and the micro air vehicle were studied, respectively. For the hypersonic vehicle with uncertain parameters and input constraints, a NN-based robust adaptive control scheme was proposed in [190]. In [191], a flight control of a small rotor UAV based on the NN was investigated. The longitudinal motion control was analyzed for the aircraft based on the FL system in [192]. In [193], an adaptive control method based on the FL system was presented for the uncertain tilt-turn missile system. Based on the above research results, continuous time control schemes based on the NN and the FL system were proposed for the flight control under uncertainties, and the negative influence of system uncertainties on the flight control performance was effectively offset.

On the other hand, according to the approximation methods of the NN and the FL system, the system uncertainties have been also widely studied in the control of discrete-time systems. In [194, 195], the adaptive dynamic programming control method was used to realize the optimal control of the uncertain nonlinear discrete-time system based on the NN. For an uncertain nonlinear discrete-time system under disturbance environment, a NN control scheme was proposed based on the approximate internal model in [196]. In [197], an NN-based adaptive decentralized discrete-time control method was studied for the power distribution system. For the uncertain nonlinear discrete-time non-strict feedback system, the NN-based state feedback control method was proposed in [198]. In [199], a NN-based constraint optimal control scheme was presented for the discrete-time switching system by using the adaptive dynamic programming method. For the uncertain nonlinear discrete-time system with noise, a SM controller based on the FL system was designed in [200]. In [201], an adaptive control scheme based on the FL system was proposed for the discrete-time switched nonlinear system with system uncertainties. For the uncertain nonlinear discrete-time system with time delay, a FL tracking control scheme based on the model reference output feedback was studied in [202]. In [203], an adaptive SMC scheme based on the FL system was explored for the uncertain discrete-time chaotic system. For the nonlinear discrete-time system with existing state and input delay, an output feedback H_∞ tracking control method based on the FL system was investigated in [204]. In the above studies, the control of uncertain discrete-time systems based on NN and FL system have been investigated and the designed control method can effectively inhibit the negative influence of system uncertainties on the control performance of discrete-time systems. In addition, some researches on the discrete-time discrete flight control with uncertainties were shown in [90–93]. However, so far, few studies have taken the NN approximation method, the design of DTDO and the input constraint into the discrete-time flight control. Therefore, for UAVs systems with external disturbances, system uncertainties and input constraints, discrete-time flight control methods need to be further studied.

1.7 Flight Control Methods Based on FO Control

As is well known to all, the development of fractional-order calculus (FOC) has a long history, and the FOC can be seen as an extension of the classical integer order differential and integral [205, 206]. In addition, one of the most important advantages of using the FOC is its non-local nature, that is, the FOC has historical dependency and long-range correlation characteristics [205, 206]. This means that the next state of the control system depends not only on its current state, but on all its previous states. Therefore, the FOC is a powerful way to describe the data memory and inheritance [207]. In the past few decades, researchers have found that the FOC can describe some non-classical phenomena in natural science and engineering applications, and the FOC theory has been successfully applied in many practical fields. For example, the fractional-order (FO) circuit models in the biology and the biomedicine were studied in [208], and the FOC applications in biomedicine and biology were reviewed in [209]. In [210], the application of FOC in pharmacokinetics was introduced and the clinical significance of FOC was discussed. The application of FOC in physics was studied in [211]. The nonlinear dynamics and chaos phenomena were analyzed for the FO financial system and the FO financial system with time delay respectively in [212, 213]. In [214], the adaptive control and the circuit simulation of the FO financial system were implemented. A macro-economic modeling method based on the FOC theory was studied in [215]. The dynamic simulation and optimization of chemical treatment systems modeled by using the FOC were discussed in [216]. In [217], the stability and chaos control of the FO chaotic chemical reactor system were analyzed, and the function projection synchronization control of the FO chaotic chemical reactor system with uncertain parameters was realized. The FO system modeling and control were investigated for the speed servo system of permanent magnet synchronous motor in [218]. In [219], the application of FOC in engineering was introduced. The solving of the theoretical problem of the FO thermoelasticity was analyzed for a class of functionally graded materials in [220]. Experimental studies on the fractional inductance and the FO bandpass filter were implemented in [221]. In [222], the projectile motion was analyzed by using the FOC theory. In the above mentioned results, the applications of FOC in biological, physical, chemical, economic, material science, engineering and other fields ware studied.

In the actual control systems, the actual closed-loop control system consists of four forms [223]: (I) the integer order control object with integer order controller; (ii) the FO control object with integer order controller; (iii) the integer order control object with FO controller; (iv) the FO control object with FO controllers. For the case of an integer order control object with an integer order controller, a large number of research results have been given in the above research results. For the case of FO control objects with integer order controllers, some research conclusions have been reported. For example, the integer order controller was used to realize the synchronization of FO Chua's system in [224], and the scheme of secure communication was studied based on the synchronization control. In [225], the chaotic phenomenon was analyzed for the FO Chen system and a linear feedback control method was proposed. The chaotic dynamics of FO Lü system were studied and the master slave synchronization control of FO Lü system was realized in [226]. In [227], the synchronization control of FO hyperchaotic Lorenz system was studied by using the feedback control method. An active SM controller of integer order was designed in [228], and the synchronization control of FO chaotic system was realized. In the studies of the above, the control of FO systems has been realized by using integer order controllers. In addition, as an important tool that can improve control performance, the FOC has been combined with many traditional control methods [207]. Furthermore, compared with traditional integer order controllers, the controller performance was improved based on the extra degree of freedom for the FO integrator and differentiator [229]. Therefore, the FOC was introduced in [229] to reduce the shaking phenomenon of traditional SMC. Since the advantage of FO adaptive controller was that it has more freedom and flexibility than the integer order control law [230], the design of FO controller and the research on the control of FO system and integer order systems based on FO controllers were received extensive attention, and some meaningful conclusions were obtained [231]. In [232], a fractional-order proportional integral-differential (FOPID) controller form was studied that includes FO integral and differential, and the control of FO linear system was realized. The development of FOPID controller design was reviewed in [233]. In [234], the automatic generation control of power system was realized by using the FOPID controller. The stability control of the FO time delay system was studied in [235]. In [236], the trajectory tracking control of three-degree-of-freedom parallel manipulator was explored. For a class of second-order control objects, a debugging method of PD controller in FO form was presented in [237]. In [238], the application of PI controller with FO form in temperature and motor control was implemented. An adaptive feedback control and the synchronization were investigated for different chaotic FO systems based on FO controller in [239]. In [240], a FO integral SMC method based on the FL system was proposed for the FO dynamic system. For the control of FO economic system, a FO control scheme based on SMC was developed in [241]. In [242], the control of nonequilibrium integer order chaotic system based on a FO controller was studied. In addition, with the development of FO control method, FO controllers were applied in flight control. In [243], an attitude control scheme based on the FO controller was proposed for flexible spacecraft. The FOPID controller was used to control the flight path of the six-degree of freedom flying object model in [244]. In [245], a plane phase edge FOPID controller was designed and the UAV roll angle control experiment was tested. A control scheme for a small fixed-wing UAV based on a roll channel FO controller was studied in [246]. In [247], the flight control performance and the robustness were improved for the small fixed-wing UAV through a FO controller. The pitch angle control of vertical take-off and landing UAV was studied by using a FO controller in [248]. In the above mentioned research works, the FO controller with continuous time form has been successfully applied in the control of FO chaotic system, temperature control, motor control, spacecraft attitude control and UAV system control.

On the other hand, the FOC with the form of DT has also attracted extensive attention from researchers [249], and some design methods of discrete-time fractional-order (DTFO) controllers and discrete-time control schemes based on DTFO controllers have been studied. For example, the design method of DTFO controller was discussed and the control performance of DTFO controller was studied in [250]. In [251], the time scale trapezoidal integral rule of DTFO controller was analyzed. The DTFO controller based on the least square approximation method was implemented in [252]. In [253], the continuous time and discrete-time linear quadratic regulator theories were adopted to design the optimal analog PID controller and the discrete-time PID controller respectively. A discrete-time FOPID controller was designed, along with the performance and the application of FO controller were studied in [254]. In [255], the synchronization control of DTFO chaotic system was realized by using the variable structure discrete-time SMC method. A tracking control scheme based on the DTFO terminal SMC was proposed for the linear motor system in [256]. Although the above works have studied the design methods and applications of DTFO controllers, flight control methods based on DTFO controller have not been fully investigated so far. Therefore, for the UAV system with external disturbances and system uncertainties, discrete-time flight control methods based on DTFO controllers can be further investigated.

1.8 The Structure of the Book

For the attitude dynamics model of a fixed-wing UAV, the discrete-time robust control problem with external disturbances, system uncertainties and input saturation will be studied in this book. Figure 1.1 shows the relations among chapters.

The rest of this book is organized as follows: In Chap. 2, the six-degree-of-freedom dynamic mathematical model of a fixed-wing UAV with wind disturbance is firstly derived, and the nonlinear affine model of UAV attitude dynamics is presented, which provide a model basis for the subsequent design of discrete-time flight control scheme for the UAV attitude system. Secondly, the input saturation, the dead-zone nonlinearity and the non-symmetric saturation and dead-zone nonlinearity are introduced, which provides a basis for further research on the UAV input constraint control. Finally, some lemmas and definitions are given, including the definition of the Grunwald-Letnikov difference operator in discrete-time fractional-order (DTFO) form and the approximation theory of radial basis function neural networks in discrete-time form, which provide a theoretical basis for the design of robust DTFO control scheme for UAV attitude nonlinear model under external disturbances.

In Chap. 3, the discrete-time BC and the discrete-time SMC are introduced for the fixed-wing UAV system without wind disturbances. According to the BC technology, the discrete-time controllers are designed, and the stability theory of Lyapunov in the form of discrete-time is used to prove that the discrete-time controller designed can ensure the boundedness of the closed-loop system signals. Finally, the simulation results are given to show the effectiveness of the discrete-time control schemes.

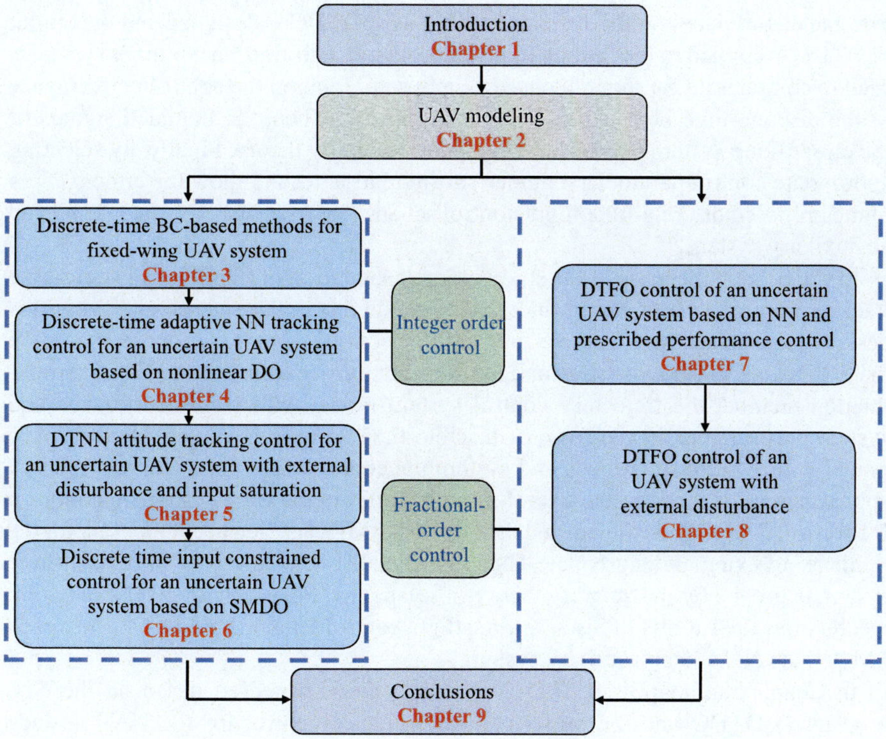

Fig. 1.1 A block diagram of this book

In Chap. 4, aiming at the UAV attitude dynamics model with external wind disturbances and system uncertainties, a discrete-time flight control scheme based on the discrete-time disturbance observer (DTDO) and the neural network (NN) is proposed. Firstly, the NN is used to approach the system uncertainties in the UAV attitude dynamics model. Secondly, an NN-based DTDO is designed, and the designed DTDO is used to estimate the external disturbance. Then, according to the NN-based nonlinear DTDO and the BC technology, an NN-based discrete-time controller is designed, and the stability theory of Lyapunov in discrete-time form is used to prove that the designed discrete-time controller can ensure the boundedness of closed-loop system signals. Finally, the UAV attitude dynamics model with external wind disturbances and system uncertainties is simulated and analyzed, and the simulation results are used to further illustrate the effectiveness of the proposed discrete-time flight control scheme based on the DTDO and the NN.

Based on the research in Chap. 4, a NN tracking control scheme based on the DTDO and an auxiliary system is proposed for the uncertain UAV attitude dynamic model with external wind disturbances and input saturation in Chap. 5. Firstly, the system uncertainties in the attitude dynamics model of UAV is dealt with by using the NN approximation principle. Secondly, in order to suppress the adverse influence of external disturbances on the system, an NN-based DTDO is designed and the output of DTDO is applied to the design of the controller. In addition, an auxiliary system is used to compensate for the influence of input saturation on the control performance of the discrete-time UAV attitude dynamics model. Then, the bounded signals of the closed-loop system is proven by Lyapunov stability theory. Finally, by selecting appropriate control parameters, numerical simulation results show the effectiveness of the proposed discrete-time flight control scheme based on the NN, the DTDO and an auxiliary system.

In Chap. 6, on the basis of the study of input saturation in Chap. 5, a discrete-time tracking control scheme based on a sliding mode disturbance observer (SMDO) is designed and an adaptive discrete-time attitude control scheme is proposed based on the SMDO, an auxiliary system and the NN. Firstly, according to the Euler approximation method, the trajectory control system model with wind disturbance and system uncertainties in the form of discrete-time and the UAV attitude dynamics model with wind disturbances and system uncertainties are obtained, and the NN approximation is used to deal with the uncertainties in the UAV system. Secondly, a DT form of SMDO is designed, and the output of SMDO is used to compensate for the influence of external disturbance. Then, the signals boundedness of the closed-loop system is proven by the stability theory of Lyapunov. Finally, numerical simulation results show the validity of discrete-time flight control scheme based on the designed SMDO, the NN and an auxiliary system.

In Chap. 7, an adaptive DTFO control scheme is proposed based on the NN, a designed DTDO and the prescribed performance control for the UAV attitude dynamics model with external disturbance and system uncertainties. Firstly, the NN is used to approximate the system uncertainties in the UAV attitude dynamics model. Secondly, a NN-based DTDO is designed to compensate the negative influence of external disturbance on the control performance. Then, based on the NN, the designed DTDO, the prescribed performance control method and the BC technology, a DTFO controller is designed, and the stability theory of Lyapunov is used to prove the bounded signal of the closed-loop system. Finally, numerical simulation results show that the proposed NN-based prescribed performance adaptive DTFO control scheme is available.

According to the research in Chap. 7, a DTDO-based DTFO control scheme is proposed for the trajectory control system model with external wind disturbances and the UAV attitude dynamic model with wind disturbances and input saturation in Chap. 8. Firstly, a DTDO is designed to compensate for the adverse effects of external disturbance on the UAV attitude dynamics system. Secondly, a DTFO control method based on the DTDO is designed by using the output of DTDO, the DTFO theory and the BC technology, and the boundedness of closed-loop system signals is proven by Lyapunov stability theory. Finally, the UAV trajectory control system model and attitude dynamics model with external disturbances are simulated and analyzed, and the simulation results show the effectiveness of the proposed DTDO-based DTFO control scheme.

In Chap. 9, several future research directions are predicated.

References

1. Şöhret, Y., Dinç, A., Karakoç, T.H.: Exergy analysis of a turbofan engine for an unmanned aerial vehicle during a surveillance mission. Energy **93**, 716–729 (2015)
2. Merino, L., Caballero, F., Martínez-De-Dios, J.R., et al.: An unmanned aircraft system for automatic forest fire monitoring and measurement. J. Intell. Robot. Syst. **65**(1–4), 533–548 (2012)
3. Kabamba, P.T., Meerkov, S.M., Zeitz, F.H.: Optimal path planning for unmanned combat aerial vehicles to defeat radar tracking. J. Guid. Control Dyn. **29**(2), 279–288 (2006)
4. Kingston, D., Beard, R.W., Holt, R.S.: Decentralized perimeter surveillance using a team of UAVs. IEEE Trans. Robot. **24**(6), 1394–1404 (2008)
5. Mahmud, I., Cho, Y.-Z.: Detection avoidance and priority-aware target tracking for UAV group reconnaissance operations. J. Intell. Robot. Syst. **92**(2), 381–392 (2018)
6. Honkavaara, E., Saari, H., Kaivosoja, J., et al.: Processing and assessment of spectrometric, stereoscopic imagery collected using a lightweight UAV spectral camera for precision agriculture. Remote Sens. **5**(10), 5006–5039 (2013)
7. Berni, J.A.J., Zarco-Tejada, P.J., Suarez, L., et al.: Thermal and narrowband multispectral remote sensing for vegetation monitoring from an unmanned aerial vehicle. IEEE Trans. Geosci. Remote Sens. **47**(3), 722–738 (2009)
8. Nex, F., Remondino, F.: UAV for 3D mapping applications: a review. Appl. Geomat. **6**(1), 1–15 (2014)
9. Zhang, T., Li, Q., Zhang, C., et al.: Current trends in the development of intelligent unmanned autonomous systems. Unmanned Syst. Technol. **1**, 11–22 (2017)
10. Li, Z., Liu, Y., Walker, R., et al.: Towards automatic power line detection for a UAV surveillance system using pulse coupled neural filter and an improved hough transform[J]. Mach. Vis. Appl. **21**(5), 677–686 (2010)
11. Guo, Z.: The stability analysis and observer design for classes of discrete systems with time-delay. Ocean University of China, Qingdao (2013)
12. Ping, Z., Hu, H., Huang, Y. et al.: Discrete-time neural network approach for tracking control of spherical inverted pendulum. In: IEEE Transactions on Systems, Man, and Cybernetics: Systems (2018). https://doi.org/10.1109/TSMC.2018.2834560
13. Sonneveldt, L.: Nonlinear F-16 model description. Technical report, Delft University of Technology, Netherlands (2006)
14. Li, Y., Li, Z., Zhang, X.: Prediction of ice accretions based on the neural net. Flight Dyn. **29**(2), 1–5 (2011)
15. Gadewadikar, J., Lewis, F., Subbarao, K., et al.: Structured H_∞ command and control-loop design for unmanned helicopters. J. Guid. Control Dyn. **31**(4), 1093–1102 (2008)
16. Muniraj, D., Palframan, M.C., Guthrie, K.T., et al.: Path-following control of small fixed-wing unmanned aircraft systems with H_∞ type performance. Control Eng. Pract. **67**, 76–91 (2017)
17. Tijani, I.B., Akmeliawati, R., Legowo, A., et al.: Optimization of an extended H_∞ controller for unmanned helicopter control using Multiobjective Differential Evolution (MODE). Aircr. Eng. Aerosp. Technol. Int. J. **87**(4), 330–344 (2015)
18. Kafi, M.R., Chaoui, H., Miah, S., et al.: Local model networks based mixed-sensitivity H_∞ control of CE-150 helicopters. Control Theory Technol. **15**(1), 34–44 (2017)
19. Lungu, M., Lungu, R.: Application of H_2/H_∞ technique to aircraft landing control. Asian J. Control **17**(6), 2153–2164 (2015)
20. Walker, D., Postlethwaite, I.: Advanced helicopter flight control using two-degree-of-freedom H_∞ optimization. J. Guid. Control Dyn. **19**(2), 461–468 (1996)
21. Liu, C., Jiang, B., Zhang, K.: Adaptive fault-tolerant H_∞ output feedback control for lead-wing close formation flight. In: IEEE Transactions on Systems, Man, and Cybernetics: Systems (2018). https://doi.org/10.1109/TSMC.2018.2830511
22. Yuan, Y.: A dynamic games approach to H^∞ control design of dos with application to longitudinal flight control. Sci. China Inf. Sci. **58**(9), 1–10 (2015)

23. Kanade, S.P., Mathew, A.T.: 2 DOF H-infinity loop shaping robust control for rocket attitude stabilization. Int. J. Aerosp. Sci. **2**(3), 71–91 (2013)
24. Ghandour, J., Aberkane, S., Ponsart, J.-C.: Feedback linearization approach for standard and fault tolerant control: application to a quadrotor UAV testbed. J. Phys. Conf. Ser. **570**, 082003 (2014)
25. Zhou, W., Yin, K., Wang, R. et al.: Design of attitude control system for UAV based on feedback linearization and adaptive control. In: Mathematical Problems in Engineering, Article ID 492680, 8 pp (2014)
26. Aboudonia, A., El-Badawy, A., Rashad, R.: Disturbance observer-based feedback linearization control of an unmanned quadrotor helicopter. Proc. Inst. Mech. Eng. Part I: J. Syst. Control Eng. **230**(9), 877–891 (2016)
27. Schumacher, C., Khargonekar, P.P.: Stability analysis of a missile control system with a dynamic inversion controller. J. Guid. Control Dyn. **21**(3), 508–515 (1998)
28. Di Francesco, G., Mattei, M.: Modeling and incremental nonlinear dynamic inversion control of a novel unmanned tiltrotor. J. Aircr. **53**(1), 73–86 (2015)
29. Pedro, J.O., Panday, A., Dala, L.: A nonlinear dynamic inversion-based neurocontroller for unmanned combat aerial vehicles during aerial refuelling. Int. J. Appl. Math. Comput. Sci. **23**(1), 75–90 (2013)
30. Smeur, E.J., Chu, Q., de Croon, G.C.: Adaptive incremental nonlinear dynamic inversion for attitude control of micro air vehicles. J. Guid. Control Dyn. **38**(12), 450–461 (2015)
31. Harris, J.A.: Nonlinear Adaptive Dynamic Inversion Control for Variable Stability Small Unmanned Aircraft Systems. Texas A & M University, United States (2017)
32. Chen, F., Jiang, R., Zhang, K., et al.: Robust backstepping sliding-mode control and observer-based fault estimation for a quadrotor UAV. IEEE Trans. Ind. Electron. **63**(8), 5044–5056 (2016)
33. Xu, H., Mirmirani, M.D., Ioannou, P.A.: Adaptive sliding mode control design for a hypersonic flight vehicle. J. Guid. Control Dyn. **27**(5), 829–838 (2004)
34. Liu, J., Laghrouche, S., Harmouche, M., et al.: Adaptive-gain second-order sliding mode observer design for switching power converters. Control Eng. Pract. **30**, 124–131 (2014)
35. Ma, D., Xia, Y., Shen, G., et al.: Flatness-based adaptive sliding mode tracking control for a quadrotor with disturbances. J. Frankl. Inst. **355**(14), 6300–6322 (2018)
36. Zheng, Z., Jin, Z., Sun, L., et al.: Adaptive sliding mode relative motion control for autonomous carrier landing of fixed-wing unmanned aerial vehicles. IEEE Access **5**, 5556–5565 (2017)
37. Espinoza, T., Dzul, A., Lozano, R., et al.: Backstepping-sliding mode controllers applied to a fixed-wing UAV. J. Intell. Robot. Syst. **73**(1–4), 67–79 (2014)
38. Oh, H., Kim, S., Tsourdos, A., et al.: Decentralised standoff tracking of moving targets using adaptive sliding mode control for UAVs. J. Intell. Robot. Syst. **76**(1), 169–183 (2014)
39. Mu, B., Zhang, K., Shi, Y.: Integral sliding mode flight controller design for a quadrotor and the application in a heterogeneous multi-agent system. IEEE Trans. Ind. Electron. **64**(12), 9389–9398 (2017)
40. Castañeda, H., Salas-Peña, O.S., de León-Morales, J.: Extended observer based on adaptive second order sliding mode control for a fixed wing UAV. ISA Trans. **66**, 226–232 (2017)
41. Xiong, J.-J., Zhang, G.-B.: Global fast dynamic terminal sliding mode control for a quadrotor UAV. ISA Trans. **66**, 233–240 (2017)
42. Wu, K., Cai, Z., Zhao, J., et al.: Target tracking based on a nonsingular fast terminal sliding mode guidance law by fixed-wing UAV. Appl. Sci. **7**(4), 333–1–18 (2017)
43. Hua, C.-C., Wang, K., Chen, J.-N., et al.: Tracking differentiator and extended state observer-based nonsingular fast terminal sliding mode attitude control for a quadrotor. Nonlinear Dyn. **94**(1), 343–354 (2018)
44. Yamasaki, T., Balakrishnan, S., Takano, H.: Integrated guidance and autopilot design for a chasing UAV via high-order sliding modes. J. Frankl. Inst. **349**(2), 531–558 (2012)
45. Babaei, A.-R., Malekzadeh, M., Madhkhan, D.: Adaptive super-twisting sliding mode control of 6-DOF nonlinear and uncertain air vehicle. In: Aerospace Science and Technology (2018). https://doi.org/10.1016/j.ast.2018.09.013

46. Wang, W., Wen, C., Zhou, J.: Adaptive Backstepping Control of Uncertain Systems with Actuator Failures, Subsystem Interactions, and Nonsmooth Nonlinearities. CRC Press, USA (2017)
47. Tao, G.: Adaptive Control Design and Analysis. Wiley, USA (2003)
48. Craig, J.J., Hsu, P., Sastry, S.S.: Adaptive control of mechanical manipulators. Int. J. Robot. Res. **6**(2), 16–28 (1987)
49. Fukao, T., Nakagawa, H., Adachi, N.: Adaptive tracking control of a nonholonomic mobile robot. IEEE Trans. Robot. Autom. **16**(5), 609–615 (2000)
50. Zhou, J., Wen, C.: Adaptive Backstepping Control of Uncertain Systems: Nonsmooth Nonlinearities, Interactions or Time-variations. Springer, Berlin (2008)
51. Wen, C., Soh, Y.C.: Decentralized adaptive control using integrator backstepping. Automatica **33**(9), 1719–1724 (1997)
52. Zhang, Y., Wen, C., Soh, Y.C.: Adaptive backstepping control design for systems with unknown high-frequency gain. IEEE Trans. Autom. Control **45**(12), 2350–2354 (2000)
53. Choi, I.-H., Bang, H.-C.: Adaptive command filtered backstepping tracking controller design for quadrotor unmanned aerial vehicle. Proc. Inst. Mech. Eng. Part G: J. Aerosp. Eng. **226**(5), 483–497 (2012)
54. Lungu, M., Lungu, R.: Adaptive backstepping flight control for a mini-UAV. Int. J. Adapt. Control Signal Process. **27**(8), 635–650 (2013)
55. Gong, X., Hou, Z.-C., Zhao, C.-J., et al.: Adaptive backstepping sliding mode trajectory tracking control for a quad-rotor. Int. J. Autom. Comput. **9**(5), 555–560 (2012)
56. Xian, B., Guo, J., Zhang, Y.: Adaptive backstepping tracking control of a 6-DOF unmanned helicopter. IEEE/CAA J. Autom. Sin. **2**(1), 19–24 (2015)
57. Choi, Y.H., Yoo, S.J.: A simple fuzzy-approximation-based adaptive control of uncertain unmanned helicopters. Int. J. Control Autom. Syst. **14**(1), 340–349 (2016)
58. Lin, C.-M., Tai, C.-F., Chung, C.-C.: Intelligent control system design for UAV using a recurrent wavelet neural network. Neural Comput. Appl. **24**(2), 487–496 (2014)
59. Xu, B., Fan, Y., Zhang, S.: Minimal-learning-parameter technique based adaptive neural control of hypersonic flight dynamics without back-stepping. Neurocomputing **164**, 201–209 (2015)
60. Cetin, O., Kurnaz, S., Kaynak, O.: Fuzzy logic based approach to design of autonomous landing system for unmanned aerial vehicles. J. Intell. Robot. Syst. **61**(1–4), 239–250 (2011)
61. Kurnaz, S., Cetin, O., Kaynak, O.: Adaptive neuro-fuzzy inference system based autonomous flight control of unmanned air vehicles. Expert Syst. Appl. **37**(2), 1229–1234 (2010)
62. Nodland, D., Zargarzadeh, H., Jagannathan, S., et al.: Neural network-based optimal adaptive output feedback control of a helicopter UAV. IEEE Trans. Neural Netw. Learn. Syst. **24**(7), 1061–1073 (2013)
63. Horn, J.F., Schmidt, E.M., Geiger, B.R., et al.: Neural network-based trajectory optimization for unmanned aerial vehicles. J. Guid. Control Dyn. **35**(2), 548–562 (2012)
64. Wang, T., Qin, R., Chen, Y. et al.: A reinforcement learning approach for UAV target searching and tracking. In: Multimedia Tools and Applications (2018). https://doi.org/10.1007/s11042-018-5739-5
65. Kim, D., Oh, G., Seo, Y., et al.: Reinforcement learning-based optimal flat spin recovery for unmanned aerial vehicle. J. Guid. Control Dyn. **40**(4), 1076–1084 (2016)
66. Krishnamoorthy, K., Pachter, M., Darbha, S., et al.: Approximate dynamic programming with state aggregation applied to UAV perimeter patrol. Int. J. Robust Nonlinear Control **21**(12), 1396–1409 (2011)
67. Quintero, S.A., Ludkovski, M., Hespanha, J.P.: Stochastic optimal coordination of small UAVs for target tracking using regression-based dynamic programming. J. Intell. Robot. Syst. **82**(1), 135–162 (2016)
68. Nešić, D., Teel, A.R.: Stabilization of sampled-data nonlinear systems via backstepping on their Euler approximate model. Automatica **42**(10), 1801–1808 (2006)
69. Üstüntürk, A.: Output feedback stabilization of nonlinear dual-rate sampled-data systems via an approximate discrete-time model. Automatica **48**(8), 1796–1802 (2012)

70. Nešić, D., Teel, A.R., Kokotović, P.V.: Sufficient conditions for stabilization of sampled-data nonlinear systems via discrete-time approximations. Syst. Control Lett. **38**(4–5), 259–270 (1999)
71. Laila, D.S., Nešić, D., Teel, A.R.: Open-and closed-loop dissipation inequalities under sampling and controller emulation. Eur. J. Control **8**(2), 109–125 (2002)
72. Xia, Y., Yang, H., Shi, P., et al.: Constrained infinite-horizon model predictive control for fuzzy-discrete-time systems. IEEE Trans. Fuzzy Syst. **18**(2), 429–436 (2010)
73. Gao, Y., Chong, K.T.: The explicit constrained min-max model predictive control of a discrete-time linear system with uncertain disturbances. IEEE Trans. Autom. Control **57**(9), 2373–2378 (2012)
74. Araújo, J.M., Barros, P.R., Dorea, C.E.: Design of observers with error limitation in discrete-time descriptor systems: a case study of a hydraulic tank system. IEEE Trans. Control Syst. Technol. **20**(4), 1041–1047 (2012)
75. Zhang, B.-L., Liu, Y.-J., Ma, H., et al.: Discrete feedforward and feedback optimal tracking control for offshore steel jacket platforms. Ocean Eng. **91**, 371–378 (2014)
76. Xu, Q., Li, Y.: Micro-/nanopositioning using model predictive output integral discrete sliding mode control. IEEE Trans. Ind. Electron. **59**(2), 1161–1170 (2012)
77. Lin, T.-C., Chang, S.-W., Hsu, C.-H., et al.: Robust adaptive fuzzy sliding mode control for a class of uncertain discrete-time nonlinear systems. Int. J. Innov. Comput. Inf. Control **8**(1), 347–359 (2012)
78. Zhang, H., Luo, Y., Liu, D.: Neural-network-based near-optimal control for a class of discrete-time affine nonlinear systems with control constraints. IEEE Trans. Neural Netw. **20**(9), 1490–1503 (2009)
79. Ornelas-Tellez, F., Sanchez, E.N., Loukianov, A.G., et al.: Robust inverse optimal control for discrete-time nonlinear system stabilization. Eur. J. Control **20**(1), 38–44 (2014)
80. Wang, D., Liu, D., Wei, Q., et al.: Optimal control of unknown nonaffine nonlinear discrete-time systems based on adaptive dynamic programming. Automatica **48**(8), 1825–1832 (2012)
81. de Jesús Rubio, J.: Adaptive least square control in discrete time of robotic arms. Soft Comput. **19**(12), 3665–3676 (2015)
82. Cui, R., Yang, C., Li, Y., et al.: Adaptive neural network control of AUVs with control input nonlinearities using reinforcement learning. IEEE Trans. Syst. Man Cybern. Syst. **47**(6), 1019–1029 (2017)
83. Ge, S.S., Yang, C., Lee, T.H.: Adaptive predictive control using neural network for a class of pure-feedback systems in discrete time. IEEE Trans. Neural Netw. **19**(9), 1599–1614 (2008)
84. Hou, Z., Jin, S.: Data-driven model-free adaptive control for a class of mimo nonlinear discrete-time systems. IEEE Trans. Neural Netw. **22**(12), 2173–2188 (2011)
85. Xiang, W., Xiao, J.: H_∞ finite-time control for switched nonlinear discrete-time systems with norm-bounded disturbance. J. Frankl. Inst. **348**(2), 331–352 (2011)
86. Corradini, M.L., Fossi, V., Giantomassi, A., et al.: Minimal resource allocating networks for discrete time sliding mode control of robotic manipulators. IEEE Trans. Ind. Inform. **8**(4), 733–745 (2012)
87. Osa, Y., Mabuchi, T., Uchikado, S.: Synthesis of discrete time adaptive flight control system using nonlinear model matching. IEEE Int. Symp. Ind. Electron. **1**, 58–63 (2001)
88. Xiong, J.-J., Zhang, G.: Discrete-time sliding mode control for a quadrotor UAV. Optik **127**(8), 3718–3722 (2016)
89. Jiang, B., Chowdhury, F.N.: Fault estimation and accommodation for linear MIMO discrete-time systems. IEEE Trans. Control Syst. Technol. **13**(3), 493–499 (2005)
90. Xu, B., Sun, F., Yang, C., et al.: Adaptive discrete-time controller design with neural network for hypersonic flight vehicle via back-stepping. Int. J. Control **84**(9), 1543–1552 (2011)
91. Xu, B., Wang, D., Sun, F., et al.: Direct neural discrete control of hypersonic flight vehicle. Nonlinear Dyn. **70**(1), 269–278 (2012)
92. Xu, B., Zhang, Y.: Neural discrete back-stepping control of hypersonic flight vehicle with equivalent prediction model. Neurocomputing **154**, 337–346 (2015)

93. Shin, D.-H., Kim, Y.: Nonlinear discrete-time reconfigurable flight control law using neural networks. IEEE Trans. Control Syst. Technol. **14**(3), 408–422 (2006)
94. Gao, Z.: On the centrality of disturbance rejection in automatic control. ISA Trans. **53**(4), 850–857 (2014)
95. Li, S., Yang, J., Chen, W.-H., et al.: Disturbance Observer-based Control: Methods and Applications. CRC Press, USA (2016)
96. Chen, W.-H., Yang, J., Guo, L., et al.: Disturbance-observer-based control and related methods-an overview. IEEE Trans. Ind. Electron. **63**(2), 1083–1095 (2016)
97. Chen, W.-H., Ballance, D.J., Gawthrop, P.J., et al.: A nonlinear disturbance observer for robotic manipulators. IEEE Trans. Ind. Electron. **47**(4), 932–938 (2000)
98. Zhang, J., Liu, X., Xia, Y., et al.: Disturbance observer-based integral sliding-mode control for systems with mismatched disturbances. IEEE Trans. Ind. Electron. **63**(11), 7040–7048 (2016)
99. Xu, B., Shou, Y., Luo, J. et al.: Neural learning control of strict-feedback systems using disturbance observer. In: IEEE transactions on neural networks and learning systems (2018). https://doi.org/10.1109/TNNLS.2018.2862907
100. Shao, S., Chen, M., Chen, S., et al.: Adaptive neural control for an uncertain fractional-order rotational mechanical system using disturbance observer. IET Control Theory Appl. **10**(16), 1972–1980 (2016)
101. Chen, W.-H.: Nonlinear disturbance observer-enhanced dynamic inversion control of missiles. J. Guid. Control Dyn. **26**(1), 161–166 (2003)
102. Xu, B.: Disturbance observer-based dynamic surface control of transport aircraft with continuous heavy cargo airdrop. IEEE Trans. Syst. Man Cyber. Syst. **47**(1), 161–170 (2017)
103. He, W., Yan, Z., Sun, C., et al.: Adaptive neural network control of a flapping wing micro aerial vehicle with disturbance observer. IEEE Trans. Cybern. **47**(10), 3452–3465 (2017)
104. Chen, M., Ren, B., Wu, Q., et al.: Anti-disturbance control of hypersonic flight vehicles with input saturation using disturbance observer. Sci. China Inf. Sci. **58**(7), 1–12 (2015)
105. Chen, F., Lei, W., Zhang, K., et al.: A novel nonlinear resilient control for a quadrotor uav via backstepping control and nonlinear disturbance observer. Nonlinear Dyn. **85**(2), 1281–1295 (2016)
106. Besnard, L., Shtessel, Y.B., Landrum, B.: Quadrotor vehicle control via sliding mode controller driven by sliding mode disturbance observer. J. Frankl. Inst. **349**(2), 658–684 (2012)
107. Lee, K., Back, J., Choy, I.: Nonlinear disturbance observer based robust attitude tracking controller for quadrotor UAVs. Int. J. Control Autom. Syst. **12**(6), 1266–1275 (2014)
108. Wang, H., Chen, M.: Trajectory tracking control for an indoor quadrotor UAV based on the disturbance observer. Trans. Inst. Meas. Control **38**(6), 675–692 (2016)
109. Yang, J., Li, S., Sun, C., et al.: Nonlinear-disturbance-observer-based robust flight control for airbreathing hypersonic vehicles. IEEE Trans. Aerosp. Electron. Syst. **49**(2), 1263–1275 (2013)
110. Wu, G., Meng, X.: Nonlinear disturbance observer based robust backstepping control for a flexible air-breathing hypersonic vehicle. Aerosp. Sci. Technol. **54**, 174–182 (2016)
111. Sun, H., Li, S., Yang, J., et al.: Non-linear disturbance observer-based back-stepping control for airbreathing hypersonic vehicles with mismatched disturbances. IET Control Theory Appl. **8**(17), 1852–1865 (2014)
112. Chen, M., Yu, J.: Disturbance observer-based adaptive sliding mode control for near-space vehicles. Nonlinear Dyn. **82**(4), 1671–1682 (2015)
113. Lu, H., Liu, C., Guo, L., et al.: Flight control design for small-scale helicopter using disturbance-observer-based backstepping. J. Guid. Control Dyn. **38**(11), 2235–2240 (2015)
114. Han, Y., Li, P., Zheng, Z.: A non-decoupled backstepping control for fixed-wing UAVs with multivariable fixed-time sliding mode disturbance observer. In: Transactions of the Institute of Measurement and Control (2018). https://doi.org/10.1177/0142331218793178
115. Smith, J., Su, J., Liu, C., et al.: Disturbance observer based control with anti-windup applied to a small fixed wing UAV for disturbance rejection. J. Intell. Robot. Syst. **88**(2–4), 329–346 (2017)

116. Kim, K.-S., Rew, K.-H.: Reduced order disturbance observer for discrete-time linear systems. Automatica **49**(4), 968–975 (2013)
117. Su, J., Chen, W.-H.: Further results on "reduced order disturbance observer for discrete-time linear systems". Automatica **93**, 550–553 (2018)
118. Su, J., Li, B., Chen, W.-H., et al.: Reduced-order disturbance observer design for discrete-time linear stochastic systems. Trans. Inst. Meas. Control **38**(6), 657–664 (2016)
119. Sun, H., Hou, L.: Composite anti-disturbance control for a discrete-time time-varying delay system with actuator failures based on a switching method and a disturbance observer. Nonlinear Anal. Hybrid Syst. **14**, 126–138 (2014)
120. Vieira, R.P., Gabbi, T.S., Gründling, H.A.: Combined discrete-time sliding mode and disturbance observer for current control of induction motors. J. Control Autom. Electr. Syst. **28**(3), 380–388 (2017)
121. Yang, H., Zhang, Y., Liang, J. et al.: Robust deadbeat predictive power control with a discrete-time disturbance observer for PWM rectifiers under unbalanced grid conditions. In: IEEE Transactions on Power Electronics (2018). https://doi.org/10.1109/TPEL.2018.2816742
122. Grimm, G., Hatfield, J., Postlethwaite, I., et al.: Antiwindup for stable linear systems with input saturation: an LMI-based synthesis. IEEE Trans. Autom. Control **48**(9), 1509–1525 (2003)
123. Chen, B.M., Lee, T.H., Peng, K., et al.: Composite nonlinear feedback control for linear systems with input saturation: theory and an application. IEEE Trans. Autom. Control **48**(3), 427–439 (2003)
124. Cao, Y.-Y., Lin, Z., Hu, T.: Stability analysis of linear time-delay systems subject to input saturation. IEEE Trans. Circuits Syst. I: Fundam. Theory Appl. **49**(2), 233–240 (2002)
125. He, W., Dong, Y., Sun, C.: Adaptive neural impedance control of a robotic manipulator with input saturation. IEEE Trans. Syst. Man Cybern. Syst. **46**(3), 334–344 (2016)
126. Wen, C., Zhou, J., Liu, Z., et al.: Robust adaptive control of uncertain nonlinear systems in the presence of input saturation and external disturbance. IEEE Trans. Autom. Control **56**(7), 1672–1678 (2011)
127. Hu, Q.: Adaptive output feedback sliding-mode manoeuvring and vibration control of flexible spacecraft with input saturation. IET Control Theory Appl. **2**(6), 467–478 (2008)
128. Zhu, Z., Xia, Y., Fu, M., et al.: Adaptive sliding mode control for attitude stabilization with actuator saturation. IEEE Trans. Ind. Electron. **58**(10), 4898–4907 (2011)
129. Bošković, J.D., Li, S.-M., Mehra, R.K.: Robust adaptive variable structure control of spacecraft under control input saturation. J. Guid. Control Dyn. **24**(1), 14–22 (2001)
130. Hu, Q., Xiao, B., Friswell, M.: Robust fault-tolerant control for spacecraft attitude stabilisation subject to input saturation. IET Control Theory Appl. **5**(2), 271–282 (2011)
131. Bustan, D., Pariz, N., Sani, S.K.H.: Robust fault-tolerant tracking control design for spacecraft under control input saturation. ISA Trans. **53**(4), 1073–1080 (2014)
132. Boskovic, J.D., Li, S.-M., Mehra, R.K.: Robust tracking control design for spacecraft under control input saturation. J. Guid. Control Dyn. **27**(4), 627–633 (2004)
133. Chen, M., Yu, J.: Adaptive dynamic surface control of NSVs with input saturation using a disturbance observer. Chin. J. Aeronaut. **28**(3), 853–864 (2015)
134. Wang, X.-S., Su, C.-Y., Hong, H.: Robust adaptive control of a class of nonlinear systems with unknown dead-zone. Automatica **40**(3), 407–413 (2004)
135. Zhou, J., Wen, C., Zhang, Y.: Adaptive output control of nonlinear systems with uncertain dead-zone nonlinearity. IEEE Trans. Autom. Control **51**(3), 504–511 (2006)
136. Zhang, T.-P., Ge, S.S.: Adaptive dynamic surface control of nonlinear systems with unknown dead zone in pure feedback form. Automatica **44**(7), 1895–1903 (2008)
137. Ibrir, S., Xie, W.F., Su, C.-Y.: Adaptive tracking of nonlinear systems with non-symmetric dead-zone input. Automatica **43**(3), 522–530 (2007)
138. Tong, S., Li, Y.: Adaptive fuzzy output feedback control of MIMO nonlinear systems with unknown dead-zone inputs. IEEE Trans. Fuzzy Syst. **21**(1), 134–146 (2013)
139. Hu, Q., Meng, Y., Wang, C., et al.: Adaptive backstepping control for air-breathing hypersonic vehicles with input nonlinearities. Aerosp. Sci. Technol. **73**, 289–299 (2018)

140. An, H., Xia, H., Wang, C.: Barrier lyapunov function-based adaptive control for hypersonic flight vehicles. Nonlinear Dyn. **88**(3), 1833–1853 (2017)
141. Wang, Y., Hu, J.: Improved prescribed performance control for air-breathing hypersonic vehicles with unknown deadzone input nonlinearity. ISA Trans. **79**, 95–107 (2018)
142. Xu, B.: Robust adaptive neural control of flexible hypersonic flight vehicle with dead-zone input nonlinearity. Nonlinear Dyn. **80**(3), 1509–1520 (2015)
143. Yang, Q., Chen, M.: Adaptive neural prescribed performance tracking control for near space vehicles with input nonlinearity. Neurocomputing **174**, 780–789 (2016)
144. Chen, M., Ge, S.S., Ren, B.: Adaptive tracking control of uncertain MIMO nonlinear systems with input constraints. Automatica **47**(3), 452–465 (2011)
145. Chen, M., Ge, S.S., How, B.V.E.: Robust adaptive neural network control for a class of uncertain MIMO nonlinear systems with input nonlinearities. IEEE Trans. Neural Netw. **21**(5), 796–812 (2010)
146. Chen, M., Ge, S.S.: Adaptive neural output feedback control of uncertain nonlinear systems with unknown hysteresis using disturbance observer. IEEE Trans. Ind. Electron. **62**(12), 7706–7716 (2015)
147. Liu, Y.-J., Tong, S., Chen, C.P., et al.: Neural controller design-based adaptive control for nonlinear MIMO systems with unknown hysteresis inputs. IEEE Trans. Cybern. **46**(1), 9–19 (2016)
148. Yu, T., Ma, L., Zhang, H.: Prescribed performance for bipartite tracking control of nonlinear multiagent systems with hysteresis input uncertainties. In: IEEE Transactions on Cybernetics (2018). https://doi.org/10.1109/TCYB.2018.2800297
149. Lai, G., Liu, Z., Zhang, Y., et al.: Adaptive fuzzy tracking control of nonlinear systems with asymmetric actuator backlash based on a new smooth inverse. IEEE Trans. Cybern. **46**(6), 1250–1262 (2016)
150. Lai, G., Liu, Z., Zhang, Y., et al.: Asymmetric actuator backlash compensation in quantized adaptive control of uncertain networked nonlinear systems. IEEE Trans. Neural Netw. Learn. Syst. **28**(2), 294–307 (2017)
151. Liu, Y.-J., Tong, S.: Adaptive fuzzy control for a class of nonlinear discrete-time systems with backlash. IEEE Trans. Fuzzy Syst. **22**(5), 1359–1365 (2014)
152. Hosseini-Ardali, S.M., Ghaderi, A.: Output feedback adaptive control of a class of nonlinear discrete-time systems with unknown control directions and preceded by hysteresis. Int. J. Control **88**(11), 2412–2422 (2015)
153. Tang, L., Liu, Y.-J., Chen, C.P.: Adaptive critic design for pure-feedback discrete-time MIMO systems preceded by unknown backlashlike hysteresis. In: IEEE Transactions on Neural Networks and Learning Systems (2018). https://doi.org/10.1109/TNNLS.2018.2805689
154. Vörös, J.: Modeling and identification of discrete-time nonlinear dynamic cascade systems with input hysteresis. In: Mathematical Problems in Engineering, Article ID 393572, 8 pp (2015)
155. Tang, L., Zhao, J.: Adaptive tracking control for discrete-time switched nonlinear systems with dead-zone inputs. Fuzzy Sets Syst. **344**, 51–69 (2018)
156. Liu, Y.-J., Tong, S., Li, D.-J., et al.: Fuzzy adaptive control with state observer for a class of nonlinear discrete-time systems with input constraint. IEEE Trans. Fuzzy Syst. **24**(5), 1147–1158 (2016)
157. Liu, Y.-J., Gao, Y., Tong, S., et al.: Fuzzy approximation-based adaptive backstepping optimal control for a class of nonlinear discrete-time systems with dead-zone. IEEE Trans. Fuzzy Syst. **24**(1), 16–28 (2016)
158. Liu, Y.-J., Li, S., Tong, S. et al.: Adaptive reinforcement learning control based on neural approximation for nonlinear discrete-time systems with unknown nonaffine dead-zone input. IN: IEEE Transactions on Neural Networks and Learning Systems (2018). https://doi.org/10.1109/TNNLS.2018.2844165
159. Liu, Y.-J., Gao, Y., Tong, S., et al.: A unified approach to adaptive neural control for nonlinear discrete-time systems with nonlinear dead-zone input. IEEE Trans. Neural Netw. Learn. Syst. **27**(1), 139–150 (2016)

160. Cao, Y.-Y., Lin, Z.: Stability analysis of discrete-time systems with actuator saturation by a saturation-dependent lyapunov function. Automatica **39**(7), 1235–1241 (2003)
161. Yang, T., Meng, Z., Dimarogonas, D.V., et al.: Global consensus for discrete-time multi-agent systems with input saturation constraints. Automatica **50**(2), 499–506 (2014)
162. Su, H., Ye, Y., Qiu, Y. et al.: Semi-global output consensus for discrete-time switching networked systems subject to input saturation and external disturbances. In: IEEE Transactions on Cybernetics (2018). https://doi.org/10.1109/TCYB.2018.2859436
163. Wang, X., Liu, Z., Cai, Y.: Adaptive single neural network control for a class of uncertain discrete-time nonlinear strict-feedback systems with input saturation. Nonlinear Dyn. **82**(4), 2021–2030 (2015)
164. He, Y., Chen, B.M., Lan, W.: On improving transient performance in tracking control for a class of nonlinear discrete-time systems with input saturation. IEEE Trans. Autom. Control **52**(7), 1307–1313 (2007)
165. Binazadeh, T., Bahmani, M.: Design of robust controller for a class of uncertain discrete-time systems subject to actuator saturation. IEEE Trans. Autom. Control **62**(3), 1505–1510 (2017)
166. Li, Z., Fang, J.-A., Huang, T., et al.: Impulsive synchronization of discrete-time networked oscillators with partial input saturation. Inf. Sci. **422**, 531–541 (2018)
167. Ren, B., Ge, S.S., Su, C.-Y., et al.: Adaptive neural control for a class of uncertain nonlinear systems in pure-feedback form with hysteresis input. IEEE Trans. Syst. Man Cybern. Part B (Cybern.) **39**, 431–443 (2009)
168. Chen, M., Shao, S.-Y., Jiang, B.: Adaptive neural control of uncertain nonlinear systems using disturbance observer. IEEE Trans. Cybern. **47**(10), 3110–3123 (2017)
169. Chen, M., Ge, S.S.: Direct adaptive neural control for a class of uncertain nonaffine nonlinear systems based on disturbance observer. IEEE Trans. Cybern. **43**(4), 1213–1225 (2013)
170. Park, J.-H., Kim, S.-H., Moon, C.-J.: Adaptive neural control for strict-feedback nonlinear systems without backstepping. IEEE Trans. Neural Netw. **20**(7), 1204–1209 (2009)
171. Liu, Y.-J., Tong, S.-C., Wang, D., et al.: Adaptive neural output feedback controller design with reduced-order observer for a class of uncertain nonlinear siso systems. IEEE Trans. Neural Netw. **22**(8), 1328–1334 (2011)
172. Xu, B., Shi, Z., Yang, C., et al.: Composite neural dynamic surface control of a class of uncertain nonlinear systems in strict-feedback form. IEEE Trans. Cybern. **44**(12), 2626–2634 (2014)
173. Yan, X., Chen, M., Wu, Q., et al.: Dynamic surface control for a class of stochastic non-linear systems with input saturation. IET Control Theory Appl. **10**(1), 35–43 (2016)
174. Chen, M., Jiang, C., Wu, Q.: Backstepping control for a class of uncertain nonlinear systems with neural network. Int. J. Nonlinear Sci. **3**(2), 137–143 (2007)
175. He, W., Chen, Y., Yin, Z.: Adaptive neural network control of an uncertain robot with full-state constraints. IEEE Trans. Cybern. **46**(3), 620–629 (2016)
176. Zhang, T., Ge, S.S.: Adaptive neural network tracking control of mimo nonlinear systems with unknown dead zones and control directions. IEEE Trans. Neural Netw. **20**(3), 483–497 (2009)
177. Hou, Z.-G., Cheng, L., Tan, M.: Decentralized robust adaptive control for the multiagent system consensus problem using neural networks. IEEE Trans. Syst. Man Cybern. Part B: Cybern. **39**(3), 636–647 (2009)
178. Shao, S., Chen, M., Yan, X.: Prescribed performance synchronization for uncertain chaotic systems with input saturation based on neural networks. Neural Comput. Appl. **29**(12), 1349–1361 (2018)
179. Chen, M., Chen, W., Wu, Q.: Adaptive fuzzy tracking control for a class of uncertain MIMO nonlinear systems using disturbance observer. Sci. China Inform. Sci. **57**(1), 1–13 (2014)
180. Roopaei, M., Zolghadri, M., Meshksar, S.: Enhanced adaptive fuzzy sliding mode control for uncertain nonlinear systems. Commun. Nonlinear Sci. Numer. Simul. **14**(9–10), 3670–3681 (2009)
181. Yau, H.-T., Wang, C.-C., Hsieh, C.-T., et al.: Nonlinear analysis and control of the uncertain micro-electro-mechanical system by using a fuzzy sliding mode control design. Comput. Math. Appl. **61**(8), 1912–1916 (2011)

182. Tong, S., Zhang, L., Li, Y.: Observed-based adaptive fuzzy decentralized tracking control for switched uncertain nonlinear large-scale systems with dead zones. IEEE Trans. Syst. Man Cybern. Syst. **46**(1), 37–47 (2016)
183. Liu, Y.-J., Tong, S.-C., Li, T.-S.: Observer-based adaptive fuzzy tracking control for a class of uncertain nonlinear MIMO systems. Fuzzy Sets Syst. **164**(1), 25–44 (2011)
184. Du, H., Zhang, N.: Fuzzy control for nonlinear uncertain electrohydraulic active suspensions with input constraint. IEEE Trans. Fuzzy Syst. **17**(2), 343–356 (2009)
185. Suresh, S., Kannan, N.: Direct adaptive neural flight control system for an unstable unmanned aircraft. Appl. Soft Comput. **8**(2), 937–948 (2008)
186. Suresh, S., Omkar, S.N., Mani, V., et al.: Direct adaptive neural flight controller for F-8 fighter aircraft. J. Guid. Control Dyn. **29**(2), 454–464 (2006)
187. Lee, T., Kim, Y.: Nonlinear adaptive flight control using backstepping and neural networks controller. J. Guid. Control Dyn. **24**(4), 675–682 (2001)
188. Lei, X., Ge, S.S., Fang, J.: Adaptive neural network control of small unmanned aerial rotorcraft. J. Intell. Robot. Syst. **75**(2), 331–341 (2014)
189. Zhang, C., Hu, H., Wang, J.: An adaptive neural network approach to the tracking control of micro aerial vehicles in constrained space. Int. J. Syst. Sci. **48**(1), 84–94 (2017)
190. Bu, X., Wu, X., Wei, D., et al.: Neural-approximation-based robust adaptive control of flexible air-breathing hypersonic vehicles with parametric uncertainties and control input constraints. Inf. Sci. **346**, 29–43 (2016)
191. Lei, X., Lu, P.: The adaptive radial basis function neural network for small rotary-wing unmanned aircraft. IEEE Trans. Ind. Electron. **61**(9), 4808–4815 (2014)
192. Hušek, P., Narenathreyas, K.: Aircraft longitudinal motion control based on takagi-sugeno fuzzy model. Appl. Soft Comput. **49**, 269–278 (2016)
193. Rong, H.-J., Yang, Z.-X., Wong, P.K., et al.: Adaptive self-learning fuzzy autopilot design for uncertain bank-to-turn missiles. J. Dyn. Syst. Meas. Control **139**(4), 041002 (2017)
194. Liu, D., Wang, D., Zhao, D., et al.: Neural-network-based optimal control for a class of unknown discrete-time nonlinear systems using globalized dual heuristic programming. IEEE Trans. Autom. Sci. Eng. **9**(3), 628 (2012)
195. Huang, Y., Liu, D.: Neural-network-based optimal tracking control scheme for a class of unknown discrete-time nonlinear systems using iterative adp algorithm. Neurocomputing **125**, 46–56 (2014)
196. Li, H.-X., Deng, H.: An approximate internal model-based neural control for unknown nonlinear discrete processes. IEEE Trans. Neural Netw. **17**(3), 659–670 (2006)
197. Kazemlou, S., Mehraeen, S.: Decentralized discrete-time adaptive neural network control of interconnected DC distribution system. IEEE Trans. Smart Grid **5**(5), 2496–2507 (2014)
198. Jagannathan, S., He, P.: Neural-network-based state feedback control of a nonlinear discrete-time system in nonstrict feedback form. IEEE Trans. Neural Netw. **19**(12), 2073–2087 (2008)
199. Zhang, H., Qin, C., Luo, Y.: Neural-network-based constrained optimal control scheme for discrete-time switched nonlinear system using dual heuristic programming. IEEE Trans. Autom. Sci. Eng. **11**(3), 839–849 (2014)
200. Yoshimura, T.: Design of an adaptive fuzzy sliding mode control for uncertain discrete-time nonlinear systems based on noisy measurements. Int. J. Syst. Sci. **47**(3), 617–630 (2016)
201. Wang, H., Wang, Z., Liu, Y.-J., et al.: Fuzzy tracking adaptive control of discrete-time switched nonlinear systems. Fuzzy and Syst. **316**, 35–48 (2017)
202. Tseng, C.-S.: Model reference output feedback fuzzy tracking control design for nonlinear discrete-time systems with time-delay. IEEE Trans. Fuzzy Syst. **14**(1), 58–70 (2006)
203. Medhaffar, H., Feki, M., Derbel, N.: Adaptive discrete-time fuzzy sliding mode control for a class of chaotic systems. Adv. Sci. Technol. Eng. Syst. J. **2**(3), 395–400 (2017)
204. Esfahani, S.H.: Improvement on the problem of output feedback fuzzy H_∞-tracking control design for non-linear discrete-time systems with state and input delay. IET Control Theory Appl. **10**(1), 24–34 (2016)
205. Miller, K.S., Ross, B.: An Introduction to the Fractional Calculus and Fractional Differential Equations. Wiley, New York (1993)

206. Kilbas, A.A., Srivastava, H.M., Trujillo, J.J.: Theory and Applications of Fractional Differential Equations. Elsevier, Netherlands (2006)
207. Wang, J., Shao, C., Chen, Y.-Q.: Fractional order sliding mode control via disturbance observer for a class of fractional order systems with mismatched disturbance. Mechatronics **53**, 8–19 (2018)
208. Freeborn, T.J.: A survey of fractional-order circuit models for biology and biomedicine. IEEE J. Emerg. Sel. Top. Circuits Syst. **3**(3), 416–424 (2013)
209. Ionescu, C., Lopes, A., Copot, D., et al.: The role of fractional calculus in modeling biological phenomena: a review. Commun. Nonlinear Sci. Numer. Simul. **51**, 141–159 (2017)
210. Sopasakis, P., Sarimveis, H., Macheras, P., et al.: Fractional calculus in pharmacokinetics. J. Pharmacokinet. Pharmacodyn. **45**(1), 107–125 (2018)
211. Shivanian, E., Jafarabadi, A.: Applications of Fractional Calculus in Physics. World Scientific, New Jersey (2000)
212. Chen, W.-C.: Nonlinear dynamics and chaos in a fractional-order financial system. Chaos Solitons Fractals **36**(5), 1305–1314 (2008)
213. Wang, Z., Huang, X., Shi, G.: Analysis of nonlinear dynamics and chaos in a fractional order financial system with time delay. Comput. Math. Appl. **62**(3), 1531–1539 (2011)
214. Tacha, O., Volos, C.K., Kyprianidis, I.M., et al.: Analysis, adaptive control and circuit simulation of a novel nonlinear finance system. Appl. Math. Comput. **276**, 200–217 (2016)
215. Škovránek, T., Podlubny, I., Petráš, I.: Modeling of the national economies in state-space: a fractional calculus approach. Econ. Model. **29**(4), 1322–1327 (2012)
216. Flores-Tlacuahuac, A., Biegler, L.T.: Optimization of fractional order dynamic chemical processing systems. Ind. Eng. Chem. Res. **53**(13), 5110–5127 (2014)
217. Yadav, V.K., Das, S., Bhadauria, B.S., et al.: Stability analysis, chaos control of a fractional order chaotic chemical reactor system and its function projective synchronization with parametric uncertainties. Chin. J. Phys. **55**(3), 594–605 (2017)
218. Yu, W., Luo, Y., Chen, Y., et al.: Frequency domain modelling and control of fractional-order system for permanent magnet synchronous motor velocity servo system. IET Control Theory Appl. **10**(2), 136–143 (2016)
219. Gutierrez, R.E., Rosario, J.M., Tenreiro Machado, J.: Fractional order calculus: Basic concepts and engineering applications. In: Mathematical Problems in Engineering, Article ID 375858, 19 pp (2010)
220. Abbas, I.A.: A problem on functional graded material under fractional order theory of thermoelasticity. Theor. Appl. Fract. Mech. **74**, 18–22 (2014)
221. Tripathy, M.C., Mondal, D., Biswas, K., et al.: Experimental studies on realization of fractional inductors and fractional-order bandpass filters. Int. J. Circuit Theory Appl. **43**(9), 1183–1196 (2015)
222. Ebaid, A.: Analysis of projectile motion in view of fractional calculus. Appl. Math. Model. **35**(3), 1231–1239 (2011)
223. Chen, Y.: Ubiquitous fractional order controls? IFAC Proc. Vol. **39**, 481–492 (2006)
224. Shao, S.-Y., Min, F.-H., Ma, M.-L., Wang, E.R.: Non-inductive modular circuit of dislocated synchronization of fractional-order Chuas system and its application. Acta Physica Sinica **62**(13), 130504–1–8 (2013)
225. Li, C., Chen, G.: Chaos in the fractional order chen system and its control. Chaos Solitons Fractals **22**(3), 549–554 (2004)
226. Lu, J.G.: Chaotic dynamics of the fractional-order lü system and its synchronization. Phys. Lett. A **354**(4), 305–311 (2006)
227. Wang, X.-Y., Song, J.-M.: Synchronization of the fractional order hyperchaos lorenz systems with activation feedback control. Commun. Nonlinear Sci. Numer. Simul. **14**(8), 3351–3357 (2009)
228. Tavazoei, M.S., Haeri, M.: Synchronization of chaotic fractional-order systems via active sliding mode controller. Phys. A: Stat. Mech. Appl. **387**(1), 57–70 (2008)
229. Zhang, B., Pi, Y., Luo, Y.: Fractional order sliding-mode control based on parameters autotuning for velocity control of permanent magnet synchronous motor. ISA Trans. **51**(5), 649–656 (2012)

230. Ullah, N., Ali, M.A., Ahmad, R., et al.: Fractional order control of static series synchronous compensator with parametric uncertainty. IET Gener. Transm. Distrib. **11**(1), 289–302 (2017)
231. Chen, M., Shao, S., Shi, P.: Robust Adaptive Control for Fractional-Order Systems with Disturbance and Saturation. ASME Press and Wiley, UK (2017)
232. Podlubny, I.: Fractional-order systems and $PI^{\lambda}D^{\mu}$-controllers. IEEE Trans. Autom. Control **44**(1), 208–214 (1999)
233. Shah, P., Agashe, S.: Review of fractional PID controller. Mechatronics **38**, 29–41 (2016)
234. Debbarma, S., Saikia, L.C., Sinha, N.: Automatic generation control using two degree of freedom fractional order PID controller. Int. J. Electr. Power Energy Syst. **58**, 120–129 (2014)
235. Hamamci, S.E.: An algorithm for stabilization of fractional-order time delay systems using fractional-order PID controllers. IEEE Trans. Autom. Control **52**(10), 1964–1969 (2007)
236. Dumlu, A., Erenturk, K.: Trajectory tracking control for a 3-DOF parallel manipulator using fractional-order $PI^{\lambda}D^{\mu}$ control. IEEE Trans. Ind. Electron. **61**(7), 3417–3426 (2014)
237. Li, H., Luo, Y., Chen, Y.: A fractional order proportional and derivative (FOPD) motion controller: tuning rule and experiments. IEEE Trans. Control Syst. Technol. **18**(2), 516–520 (2010)
238. Bohannan, G.W.: Analog fractional order controller in temperature and motor control applications. J. Vib. Control **14**(9–10), 1487–1498 (2008)
239. Odibat, Z.M.: Adaptive feedback control and synchronization of non-identical chaotic fractional order systems. Nonlinear Dyn. **60**(4), 479–487 (2010)
240. Balasubramaniam, P., Muthukumar, P., Ratnavelu, K.: Theoretical and practical applications of fuzzy fractional integral sliding mode control for fractional-order dynamical system. Nonlinear Dyn. **80**(1–2), 249–267 (2015)
241. Dadras, S., Momeni, H.R.: Control of a fractional-order economical system via sliding mode[J]. Physica A **389**(12), 2434–2442 (2010)
242. Shao, S.-Y., Chen, M.: Fractional-order control for a novel chaotic system without equilibrium. In: IEEE/CAA Journal of Automatica Sinica (2016). https://doi.org/10.1109/JAS.2016.7510124
243. Manabe, S.: A suggestion of fractional-order controller for flexible spacecraft attitude control. Nonlinear Dyn. **29**(1–4), 251–268 (2002)
244. Aboelela, M.A., Ahmed, M.F., Dorrah, H.T.: Design of aerospace control systems using fractional PID controller. J. Adv. Res. **3**(3), 225–232 (2012)
245. Seyedtabaii, S.: New flat phase margin fractional order PID design: perturbed UAV roll control study. Robot. Auton. Syst. **96**, 58–64 (2017)
246. Chao, H., Luo, Y., Di, L., et al.: Roll-channel fractional order controller design for a small fixed-wing unmanned aerial vehicle. Control Eng. Pract. **18**(7), 761–772 (2010)
247. Luo, Y., Chao, H., Di, L., et al.: Lateral directional fractional order $(PI)^{\alpha}$ control of a small fixed-wing unmanned aerial vehicles: controller designs and flight tests. IET Control Theory Appl. **5**(18), 2156–2167 (2011)
248. Han, J., Di, L., Coopmans, C., et al.: Pitch loop control of a vtol uav using fractional order controller. J. Intell. Robot. Syst. **73**(1–4), 187–195 (2014)
249. Goodrich, C., Peterson, A.C.: Discrete Fractional Calculus. Springer, Berlin (2015)
250. Machado, J.: Discrete-time fractional-order controllers. Fract. Calc. Appl. Anal. **4**, 47–66 (2001)
251. Ma, C., Hori, Y.: The time-scaled trapezoidal integration rule for discrete fractional order controllers. Nonlinear Dyn. **38**(1–4), 171–180 (2004)
252. Barbosa, R.S., Machado, J.T.: Implementation of discrete-time fractional-order controllers based on LS approximations. Acta Polytechnica Hungarica **3**(4), 5–22 (2006)
253. Das, S., Pan, I., Halder, K., et al.: LQR based improved discrete PID controller design via optimum selection of weighting matrices using fractional order integral performance index. Appl. Math. Model. **37**(6), 4253–4268 (2013)
254. Merrikh-Bayat, F., Mirebrahimi, N., Khalili, M.R.: Discrete-time fractional-order PID controller: definition, tuning, digital realization and some applications. Int. J. Control Autom. Syst. **13**(1), 81–90 (2015)

255. Huang, L., Wang, L., Shi, D.: Discrete fractional order chaotic systems synchronization based on the variable structure control with a new discrete reaching-law. In: IEEE/CAA Journal of Automatica Sinica (2016). https://doi.org/10.1109/JAS.2016.7510148
256. Sun, G., Ma, Z., Yu, J.: Discrete-time fractional order terminal sliding mode tracking control for linear motor. IEEE Trans. Ind. Electron. **65**(4), 3386–3394 (2018)

Chapter 2
Modeling of UAV and Preliminaries

2.1 Mathematical Model of UAV System

In order to define the physical variables in the UAV model, this section the coordinate system is defined at first, and then the six-degree-of-freedom dynamic model of the fixed-wing UAV with external wind disturbance is established through derivation, which provide a model basis for the design and research of the subsequent flight control.

2.1.1 Definition of Coordinate System

To accurately describe the meaning of the physical variables in the UAV model, the following coordinate system is defined [1, 2]:

(i) *Ground coordinate system*: $\bar{O}_g\bar{x}_g\bar{y}_g\bar{z}_g(\bar{S}_g)$. The ground coordinate system is an inertial system. Select a point $\bar{O}_g$ on the horizontal ground as the origin of the coordinate. x_g axis and y_g axis are in the horizontal plane, and the direction can be customized. $\bar{z}_g$ axis is plumb downward, $\bar{S}_g$ denotes the ground coordinate system.

(ii) *Body coordinate system*: $\bar{O}_b\bar{x}_b\bar{y}_b\bar{z}_b(\bar{S}_b)$. The origin $\bar{O}_b$ is the center of mass of the fixed-wing UAV. $\bar{x}_b$ axis is in the UAV symmetric plane, and pointing forward along the longitudinal axis of the structure. $\bar{y}_b$ is perpendicular to the symmetric plane and points to the right. $\bar{z}_b$ axis is in the symmetric plane, perpendicular to the $\bar{x}_b$ axis and pointing down. $\bar{S}_b$ represents the body coordinate system.

(iii) *Airflow coordinate system*: $\bar{O}_a\bar{x}_a\bar{y}_a\bar{z}_a(\bar{S}_a)$. The airflow coordinate system is also known as the aerodynamics coordinate system. The origin $\bar{O}_a$ is taken as the center of mass of UAV. $\bar{x}_a$ axis along the flight speed vector V_A, and pointing forward. $\bar{z}_a$ axis is in the UAV symmetric plane and is perpendicular to the $\bar{x}_a$ axis and points down. $\bar{y}_a$ is perpendicular to the $\bar{x}_a$ axis and $\bar{z}_a$ axis, and pointing to the right. $\bar{S}_a$ represents the airflow coordinate system.

(iv) *Path coordinate system*: $\bar{O}_k\bar{x}_k\bar{y}_k\bar{z}_k(\bar{S}_k)$. The origin $\bar{O}_k$ is taken as the center of mass of UAV. $\bar{x}_k$ axis along track velocity vector V_K pointing forward $\bar{z}_k$ axis is in the plumb plane passing through the $\bar{x}_k$ axis and is perpendicular to the $\bar{x}_k$ axis

S. Shao et al., *Robust Discrete-Time Flight Control of UAV with External Disturbances*, Studies in Systems, Decision and Control 317,
https://doi.org/10.1007/978-3-030-57957-9_2

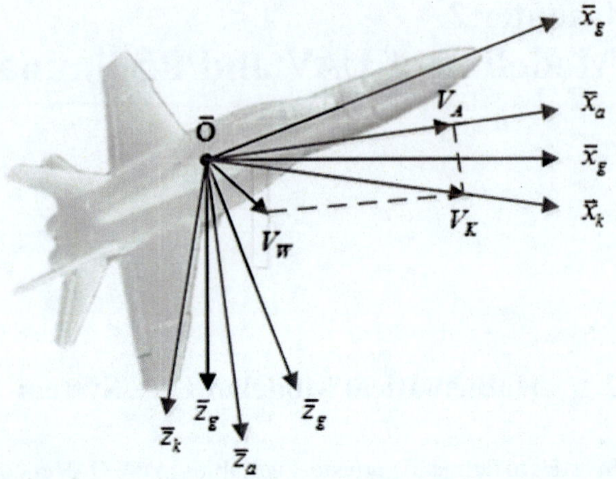

Fig. 2.1 The coordinate system relation of UAV

and points down. $\bar{y}_k$ axis is perpendicular to the plumb plane $\bar{x}_k\bar{z}_k$ and pointing to the right. $\bar{\text{S}}_k$ represents path coordinate system.

The definitions of ground coordinate system $\bar{\text{S}}_g$, the body coordinate system $\bar{\text{S}}_b$, the airflow coordinate system $\bar{\text{S}}_a$ and the path coordinate system $\bar{\text{S}}_k$ have been given in the above description, and the relationship between each coordinate system is shown in Fig. 2.1, where V_W represents the real-time wind speed vector.

2.1.2 *Dynamic Model of Six-Degree-of-Freedom UAV Under Wind Disturbance*

In order to establish the mathematical model of UAV under wind disturbances, the following expression is given based on Fig. 2.1 and the triangle relation of coordinate system [1, 2]:

$$V_K = V_A + V_W \tag{2.1}$$

where $V_K = [u_{Kb}, v_{Kb}, w_{Kb}]^{\mathrm{T}}$ is the UAV velocity vector to the ground, $V_A = [u_{Ab}, v_{Ab}, w_{Ab}]^{\mathrm{T}}$ is the UAV velocity vector to airspeed, $V_W = [u_{Wb}, v_{Wb}, w_{Wb}]^{\mathrm{T}}$ is the real-time wind speed vector, and the right subscript b of the variable is the decomposition of the corresponding coordinate system.

Based on the previous research results on aircraft modeling [3], the derivation process of the six-degree-of-freedom dynamic model of the fixed-wing UAV is given in detail below [1, 2]:

First of all, since the curvature and rotation of the earth are not considered, the rotation angular velocity of the ground coordinate system $\bar{S}_g$ is zero. Therefore, according to (2.1), the following equation can be obtained:

$$\begin{bmatrix} \dot{\bar{x}}_g \\ \dot{\bar{y}}_g \\ \dot{\bar{z}}_g \end{bmatrix} = \begin{bmatrix} u_{Kg} \\ v_{Kg} \\ w_{Kg} \end{bmatrix} = {}_b^gT \begin{bmatrix} u_{Kb} \\ v_{Kb} \\ w_{Kb} \end{bmatrix} = {}_b^gT \left(\begin{bmatrix} u_{Ab} \\ v_{Ab} \\ w_{Ab} \end{bmatrix} + \begin{bmatrix} u_{Wb} \\ v_{Wb} \\ w_{Wb} \end{bmatrix} \right)$$
$$= {}_b^gT \left({}_a^bT \begin{bmatrix} \bar{V}_A \\ 0 \\ 0 \end{bmatrix} + {}_g^bT \begin{bmatrix} u_{Wg} \\ v_{Wg} \\ w_{Wg} \end{bmatrix} \right) = {}_a^gT \begin{bmatrix} \bar{V}_A \\ 0 \\ 0 \end{bmatrix} + \begin{bmatrix} u_{Wg} \\ v_{Wg} \\ w_{Wg} \end{bmatrix} \tag{2.2}$$

where $[\bar{x}_g, \bar{y}_g, \bar{z}_g]^{\mathrm{T}}$ is the UAV position vector in the inertial coordinate system $\bar{\mathrm{S}}_g$, $[\bar{V}_A, 0, 0]^{\mathrm{T}}$ is airspeed vector V_A in the airflow coordinate system, and $\bar{V}_A$ is the module value of V_A. In addition, the variables ${}_b^gT$, ${}_a^bT$ and ${}_g^bT$ are transformation matrices between two coordinate systems, and the corresponding inverse matrices are ${}_g^bT$, ${}_b^aT$, and ${}_b^gT$. The right subscript g of the variable is the decomposition of the corresponding coordinate system, ${}_b^gT \cdot {}_a^bT = {}_a^gT$, and ${}_b^gT \cdot {}_g^bT$ is the unit matrix.

According to ${}_a^gT$, (2.2) can be written as

$$\begin{bmatrix} \dot{\bar{x}}_g \\ \dot{\bar{y}}_g \\ \dot{\bar{z}}_g \end{bmatrix} = \begin{bmatrix} \bar{V}_A \cos\gamma \cos\chi \\ \bar{V}_A \cos\gamma \sin\chi \\ -\bar{V}_A \sin\gamma \end{bmatrix} + \begin{bmatrix} u_{Wg} \\ v_{Wg} \\ w_{Wg} \end{bmatrix} \tag{2.3}$$

where γ and χ are the inclination and azimuth between the airflow coordinate system and the ground coordinate system, respectively.

Based on Newton's second law, the vector equation of mass center motion of UAV can be described as

$$M\frac{\mathrm{d}V_{Kg}}{\mathrm{d}t} = \mathscr{F}_{ag} + \mathscr{F}_{\bar{T}g} + \mathscr{F}_g \tag{2.4}$$

where M is the mass of the UAV, $\mathscr{F}_{ag}$ is the aerodynamic force of the UAV, $\mathscr{F}_{\bar{T}g}$ is the thrust of the UAV, and $\mathscr{F}_g$ is the gravity of the UAV.

From (2.4), the motion equation of the center of mass in the body coordinate system $\bar{\mathrm{S}}_b$ can be expressed as

$$\begin{bmatrix} \dot{u}_{Kb} \\ \dot{v}_{Kb} \\ \dot{w}_{Kb} \end{bmatrix} + \begin{bmatrix} p_b \\ q_b \\ r_b \end{bmatrix} \times \begin{bmatrix} u_{Kb} \\ v_{Kb} \\ w_{Kb} \end{bmatrix} = \frac{1}{M}\left({}_a^bT \begin{bmatrix} -\bar{D} \\ \bar{Y} \\ -\bar{L} \end{bmatrix} + \begin{bmatrix} \bar{T}_{\bar{x}} \\ \bar{T}_{\bar{y}} \\ \bar{T}_{\bar{z}} \end{bmatrix} + {}_g^bT \begin{bmatrix} 0 \\ 0 \\ Mg \end{bmatrix} \right) \tag{2.5}$$

where p_b, q_b and r_b are rotation angular velocities on the body coordinate system $\bar{\mathrm{S}}_b$; $\bar{T}_{\bar{x}}$, $\bar{T}_{\bar{y}}$ and $\bar{T}_{\bar{z}}$ are thrusts on the body coordinate system $\bar{\mathrm{S}}_b$; $\bar{D}$, $\bar{L}$ and $\bar{Y}$ are respectively drag force, lift force and side force, and the following expression is given by:

$$\begin{bmatrix} p_b \\ q_b \\ r_b \end{bmatrix} \times \begin{bmatrix} u_{Kb} \\ v_{Kb} \\ w_{Kb} \end{bmatrix} = \begin{bmatrix} 0 & -r_b & q_b \\ r_b & 0 & -p_b \\ -q_b & p_b & 0 \end{bmatrix} \begin{bmatrix} u_{Kb} \\ v_{Kb} \\ w_{Kb} \end{bmatrix} \tag{2.6}$$

According to (2.2), it yields

$$\begin{aligned} \begin{bmatrix} \dot{u}_{Kb} \\ \dot{v}_{Kb} \\ \dot{w}_{Kb} \end{bmatrix} &= \frac{\mathrm{d}}{\mathrm{d}t}\left({}_a^bT \begin{bmatrix} \bar{V}_A \\ 0 \\ 0 \end{bmatrix} + {}_g^bT \begin{bmatrix} u_{Wg} \\ v_{Wg} \\ w_{Wg} \end{bmatrix} \right) \\ &= {}_a^b\dot{T} \begin{bmatrix} \bar{V}_A \\ 0 \\ 0 \end{bmatrix} + {}_a^bT \begin{bmatrix} \dot{\bar{V}}_A \\ 0 \\ 0 \end{bmatrix} + {}_g^b\dot{T} \begin{bmatrix} u_{Wg} \\ v_{Wg} \\ w_{Wg} \end{bmatrix} + {}_g^bT \begin{bmatrix} \dot{u}_{Wg} \\ \dot{v}_{Wg} \\ \dot{w}_{Wg} \end{bmatrix} \end{aligned} \tag{2.7}$$

By substituting (2.6) and (2.7) into (2.5) and multiplying both sides of the equation by the coordinate transformation matrix ${}_b^aT$, the following expression can be obtained:

$$\begin{aligned} \begin{bmatrix} \dot{\bar{V}}_A \\ \bar{V}_A\dot{\beta} \\ \bar{V}_A \cos\beta\dot{\alpha} \end{bmatrix} &= -\begin{bmatrix} 0 \\ r_b\bar{V}_A \cos\alpha - p_b\bar{V}_A \sin\alpha \\ r_b\bar{V}_A \sin\alpha \sin\beta + p_b\bar{V}_A \sin\beta\cos\alpha - q_b\bar{V}_A \cos\beta \end{bmatrix} \\ &\quad + \frac{1}{M}\left({}_b^aT \begin{bmatrix} \bar{T}_{\bar{x}} \\ \bar{T}_{\bar{y}} \\ \bar{T}_{\bar{z}} \end{bmatrix} + \begin{bmatrix} -\bar{D} \\ \bar{Y} \\ -\bar{L} \end{bmatrix} + {}_g^aT \begin{bmatrix} 0 \\ 0 \\ Mg \end{bmatrix} \right) - {}_g^aT \begin{bmatrix} \dot{u}_{Wg} \\ \dot{v}_{Wg} \\ \dot{w}_{Wg} \end{bmatrix} \\ &= \begin{bmatrix} \frac{1}{M}\left(-\bar{D} + \bar{T}_{\bar{y}} \sin\beta + \bar{T}_{\bar{x}} \cos\beta\cos\alpha + \bar{T}_{\bar{z}} \sin\alpha\cos\beta\right) - g\sin\gamma \\ p_b\bar{V}_A \sin\alpha - r_b\bar{V}_A \cos\alpha + \frac{1}{M}\left(\bar{Y} - \bar{T}_{\bar{x}} \sin\beta\cos\alpha + \bar{T}_{\bar{y}} \cos\beta \right. \\ \left. -\bar{T}_{\bar{z}} \sin\alpha\sin\beta\right) + g\sin\mu\cos\gamma \\ q_b\bar{V}_A \cos\beta - r_b\bar{V}_A \sin\alpha\sin\beta - p_b\bar{V}_A \sin\beta\cos\alpha \\ +\frac{1}{M}\left(-\bar{L} - \bar{T}_{\bar{x}} \sin\alpha + \bar{T}_{\bar{z}} \cos\alpha\right) + g\sin\mu\cos\gamma \end{bmatrix} \\ &\quad - {}_g^aT \begin{bmatrix} \dot{u}_{Wg} \\ \dot{v}_{Wg} \\ \dot{w}_{Wg} \end{bmatrix} \end{aligned} \tag{2.8}$$

Then, based on (2.1), the following expression under the inertial system $\bar{\mathrm{S}}_g$ can be written as:

$$\frac{\mathrm{d}V_{Kg}}{\mathrm{d}t} = {}_a^gT \frac{\mathrm{d}V_A}{\mathrm{d}t} + \Omega_{Ag} \times V_{Kg} + \frac{\mathrm{d}V_W}{\mathrm{d}t} \tag{2.9}$$

where Ω_{Ag} is the angular velocity of the airflow coordinate system observed in the airflow coordinate system relative to the ground coordinate system, and we have

$$\Omega_{Ag} = \begin{bmatrix} -\dot{\chi}\sin\gamma \\ \dot{\gamma} \\ \dot{\chi}\cos\gamma \end{bmatrix} \tag{2.10}$$

Combining (2.4) and (2.8), one has

$$M\left({}^g_aT\frac{\mathrm{d}V_A}{\mathrm{d}t} + \Omega_{Ag}\times V_A + \frac{\mathrm{d}V_W}{\mathrm{d}t}\right) = {}^g_aT\mathscr{F}_a + {}^g_bT\mathscr{F}_{\bar{T}} + \mathscr{F}_g \tag{2.11}$$

By using the matrix ${}^a_gT$, (2.11) can be described as

$$M\left(\frac{\mathrm{d}V_A}{\mathrm{d}t} + \Omega_{Ag}\times V_A + {}^a_gT\frac{\mathrm{d}V_W}{\mathrm{d}t}\right) = \mathscr{F}_a + {}^a_bT\mathscr{F}_{\bar{T}} + {}^a_gT\mathscr{F}_g \tag{2.12}$$

Therefore, the equation about $\bar{V}_A$, χ and γ can be written as follows:

$$\begin{aligned} &\begin{bmatrix} \dot{\bar{V}}_A \\ 0 \\ 0 \end{bmatrix} + \begin{bmatrix} -\dot{\chi}\sin\gamma \\ \dot{\gamma} \\ \dot{\chi}\cos\gamma \end{bmatrix} \times \begin{bmatrix} \bar{V}_A \\ 0 \\ 0 \end{bmatrix} = \begin{bmatrix} \dot{\bar{V}}_A \\ \dot{\chi}\bar{V}_A\cos\gamma \\ -\dot{\gamma}\bar{V}_A \end{bmatrix} \\ &= \frac{1}{M}\left({}^a_bT\begin{bmatrix} \bar{T}_{\bar{x}} \\ \bar{T}_{\bar{y}} \\ \bar{T}_{\bar{z}} \end{bmatrix} + \begin{bmatrix} -\bar{D} \\ \bar{Y} \\ -\bar{L} \end{bmatrix} + {}^a_gT\begin{bmatrix} 0 \\ 0 \\ Mg \end{bmatrix}\right) - {}^a_gT\begin{bmatrix} \dot{u}_{Wg} \\ \dot{v}_{Wg} \\ \dot{w}_{Wg} \end{bmatrix} \end{aligned} \tag{2.13}$$

According to the results in [1, 2], the expression of roll angle μ is as follows:

$$\dot{\mu} = \dot{\chi}\sin\gamma - \dot{\alpha}\sin\beta + p_b\cos\beta\cos\alpha + r_b\cos\beta\sin\alpha + q_b\sin\beta \tag{2.14}$$

In addition, $\left[\dot{u}_{Wg}, \dot{v}_{Wg}, \dot{w}_{Wg}\right]^{\mathrm{T}}$ in (2.6) and (2.13) is the description of wind variation. Since wind change is not easy to measure in the actual situation, the function of wind vector and wind gradient can be used to describe wind variation. The description equation of wind variation is given as follows:

$$\begin{aligned} \begin{bmatrix} \dot{u}_{Wg} \\ \dot{v}_{Wg} \\ \dot{w}_{Wg} \end{bmatrix} &= \begin{bmatrix} \frac{\partial u_{Wg}}{\partial \bar{x}_g}\frac{\mathrm{d}\bar{x}_g}{\mathrm{d}t} + \frac{\partial u_{Wg}}{\partial \bar{y}_g}\frac{\mathrm{d}\bar{y}_g}{\mathrm{d}t} + \frac{\partial u_{Wg}}{\partial \bar{z}_g}\frac{\mathrm{d}\bar{z}_g}{\mathrm{d}t} \\ \frac{\partial v_{Wg}}{\partial \bar{x}_g}\frac{\mathrm{d}\bar{x}_g}{\mathrm{d}t} + \frac{\partial v_{Wg}}{\partial \bar{y}_g}\frac{\mathrm{d}\bar{y}_g}{\mathrm{d}t} + \frac{\partial v_{Wg}}{\partial \bar{z}_g}\frac{\mathrm{d}\bar{z}_g}{\mathrm{d}t} \\ \frac{\partial w_{Wg}}{\partial \bar{x}_g}\frac{\mathrm{d}\bar{x}_g}{\mathrm{d}t} + \frac{\partial w_{Wg}}{\partial \bar{y}_g}\frac{\mathrm{d}\bar{y}_g}{\mathrm{d}t} + \frac{\partial w_{Wg}}{\partial \bar{z}_g}\frac{\mathrm{d}\bar{z}_g}{\mathrm{d}t} \end{bmatrix} \\ &= \begin{bmatrix} u_{W\bar{x}_g}\dot{\bar{x}}_g + u_{W\bar{y}_g}\dot{\bar{y}}_g + u_{W\bar{z}_g}\dot{\bar{z}}_g \\ v_{W\bar{x}_g}\dot{\bar{x}}_g + v_{W\bar{y}_g}\dot{\bar{y}}_g + v_{W\bar{z}_g}\dot{\bar{z}}_g \\ w_{W\bar{x}_g}\dot{\bar{x}}_g + w_{W\bar{y}_g}\dot{\bar{y}}_g + w_{W\bar{z}_g}\dot{\bar{z}}_g \end{bmatrix} \end{aligned} \tag{2.15}$$

where $u_{W\bar{x}_g} = \partial u_{Wg}/\partial \bar{x}_g$, $v_{W\bar{y}_g} = \partial v_{Wg}/\partial \bar{y}_g$ and $w_{W\bar{z}_g} = \partial w_{Wg}/\partial \bar{z}_g$ are the wind gradients, and the expression of other wind gradients is similar. Then, (2.2) can also be written as

$$\begin{bmatrix} \dot{x}_g \\ \dot{y}_g \\ \dot{z}_g \end{bmatrix} = \begin{bmatrix} u_{Kg} \\ v_{Kg} \\ w_{Kg} \end{bmatrix} = {}^g_k T \begin{bmatrix} \bar{V}_K \\ 0 \\ 0 \end{bmatrix} \tag{2.16}$$

From (2.1), it yields

$$V_K = \begin{bmatrix} \bar{V}_A \\ 0 \\ 0 \end{bmatrix} + {}^a_g T \begin{bmatrix} u_{Wg} \\ v_{Wg} \\ w_{Wg} \end{bmatrix}, \qquad \bar{V}_K = \| V_K \| \tag{2.17}$$

where $\|\cdot\|$ denotes the 2-norm.

Finally, according to Euler law, the rotation vector equation of UAV in inertial vertical ground coordinate system is as follows:

$$\frac{\mathrm{d}\mathscr{L}_g}{\mathrm{d}t} = \mathscr{M}_g \tag{2.18}$$

where $\mathscr{L}_g$ is the moment of momentum of the UAV against the center of mass, $\mathscr{M}_g$ is the total torque vector of all external forces acting on the UAV against the center of mass. The moment of momentum vector, aerodynamic moment vector and aircraft angular velocity vector under the body coordinate system can be expressed as

$$\mathscr{M}_b = \begin{bmatrix} l_\Delta \\ m_\Delta \\ n_\Delta \end{bmatrix}, \quad \Omega_b^g = \omega_b = \begin{bmatrix} p_b \\ q_b \\ r_b \end{bmatrix} \tag{2.19}$$

where l_Δ,m_Δ and n_Δ are components of the aerodynamic moment.

Therefore, it is necessary to convert the angular velocity in the body coordinate system to the vertical ground coordinate system. According to the relationship of time derivative in different coordinate systems, it can be obtained

$$\frac{d\mathscr{L}_g}{dt} = {}^g_b T \frac{\mathrm{d}\mathscr{L}_b}{\mathrm{d}t} + \Omega_b^g \times \mathscr{L}_g \tag{2.20}$$

By substituting (2.20) into (2.18) and multiplying both sides of the equation by the coordinate transformation matrix ${}^b_g T$, one has

$$\frac{\mathrm{d}\mathscr{L}_b}{\mathrm{d}t} + \Omega_b^g \times \mathscr{L}_b = \mathscr{M}_b \tag{2.21}$$

According to $L_b = I_b \omega_b$, (2.21) can be written as

$$I_b \frac{\mathrm{d}\omega_b}{\mathrm{d}t} + \omega_b \frac{\mathrm{d}I_b}{\mathrm{d}t} + \Omega_b^g \times (I_b \omega_b) = \mathscr{M}_b \tag{2.22}$$

The mass of the vehicle and its distribution are constant, so the $\frac{\mathrm{d}I_b}{\mathrm{d}t} = 0$, and (2.22) can be transformed into

$$I_b \frac{\mathrm{d}\omega_b}{\mathrm{d}t} + \Omega_b^g \times (I_b \omega_b) = \mathscr{M}_b \tag{2.23}$$

The UAV on profile is symmetrical, so $I_{xy} = I_{yz} = 0$ in the moment of inertia matrix, then one has

$$I_b = \begin{bmatrix} I_{xx} & -I_{xy} & -I_{xz} \\ -I_{xy} & I_{yy} & -I_{yz} \\ -I_{xz} & -I_{yz} & I_{zz} \end{bmatrix} = \begin{bmatrix} I_{xx} & 0 & -I_{xz} \\ 0 & I_{yy} & 0 \\ -I_{xz} & 0 & I_{zz} \end{bmatrix} \tag{2.24}$$

Thus, the rotation vector equation in the body coordinate system $\bar{\mathrm{S}}_b$ is obtained as follows:

$$\begin{bmatrix} \dot{p}_b \\ \dot{q}_b \\ \dot{r}_b \end{bmatrix} = \begin{bmatrix} \frac{[I_{zz}(I_{yy}-I_{zz})-I_{xz}^2]q_b r_b + I_{xz}(I_{xx}-I_{yy}+I_{zz})p_b q_b + I_{zz} l_\Delta + I_{xz} n_\Delta}{I_{xx} I_{zz} - I_{xz}^2} \\ \frac{1}{I_{yy}}\left[(I_{zz} - I_{xx})\, p_b r_b + I_{xz}\left(r_b^2 - p_b^2\right) + m_\Delta\right] \\ \frac{[I_{xx}(I_{xx}-I_{yy})+I_{xz}^2]p_b q_b - I_{xz}(I_{xx}-I_{yy}+I_{zz})q_b r_b + I_{xz} l_\Delta + I_{xx} n_\Delta}{I_{xx} I_{zz} - I_{xz}^2} \end{bmatrix} \tag{2.25}$$

where l_Δ, m_Δ and n_Δ are the functions of p_a, q_a and r_a in (2.25), and $[p_a, q_a, r_a]^\mathrm{T}$ is the rotating angular velocity vector. In the case of no wind influence, it is that $[p_a, q_a, r_a]^\mathrm{T} = [p_b, q_b, r_b]^\mathrm{T}$. But in the case of wind, the wind field affects the surrounding air field, thus $[p_a, q_a, r_a]^\mathrm{T}$ and $[p_b, q_b, r_b]^\mathrm{T}$ are not equal. For existing measurement technologies, only $[p_b, q_b, r_b]^\mathrm{T}$ can be measured, and $[p_a, q_a, r_a]^\mathrm{T}$ still cannot be directly measured, it is necessary to convert l_Δ, m_Δ and n_Δ into the functions related to p_b, q_b and r_b. In addition, according to [1, 4], the following formula can be obtained:

$$\begin{bmatrix} p_a \\ q_a \\ r_a \end{bmatrix} = \begin{bmatrix} p_b \\ q_b \\ r_b \end{bmatrix} - \begin{bmatrix} p_W \\ q_W \\ r_W \end{bmatrix} = \begin{bmatrix} p_b \\ q_b \\ r_b \end{bmatrix} - \begin{bmatrix} w_{W\bar{y}_g} \\ -w_{W\bar{x}_g} \\ v_{W\bar{x}_g} \end{bmatrix} \tag{2.26}$$

where $[p_W, q_W, r_W]^\mathrm{T}$ is the rotation angular velocity vector of the air medium relative to the ground coordinate system $\bar{\mathrm{S}}_g$, $[p_W, q_W, r_W]^\mathrm{T}$ is caused by the wind gradient, and the wind gradients $w_{W\bar{y}_g} = \partial w_{Wg}/\partial \bar{y}_g$, $w_{W\bar{x}_g} = \partial w_{Wg}/\partial \bar{x}_g$ and $v_{W\bar{x}_g} = \partial v_{Wg}/\partial \bar{x}_g$. For the convenience of expression, the variables l_Δ, m_Δ and n_Δ are decomposed into can be measured variables $l_{\Delta 0}$, $m_{\Delta 0}$ and $n_{\Delta 0}$ and uncertainties caused by the gradient wind Δl_Δ, Δm_Δ and Δn_Δ. Therefore, (2.25) can be described as

$$\begin{bmatrix} \dot{p}_b \\ \dot{q}_b \\ \dot{r}_b \end{bmatrix} = \begin{bmatrix} \frac{[I_{zz}(I_{yy}-I_{zz})-I_{xz}^2]q_b r_b + I_{xz}(I_{xx}-I_{yy}+I_{zz})p_b q_b + I_{zz}(l_{\Delta 0}+\Delta l_\Delta) + I_{xz}(n_{\Delta 0}+\Delta n_\Delta)}{I_{xx} I_{zz} - I_{xz}^2} \\ \frac{1}{I_{yy}}\left[(I_{zz} - I_{xx})\, p_b r_b + I_{xz}\left(r_b^2 - p_b^2\right) + m_{\Delta 0} + \Delta m_\Delta\right] \\ \frac{[I_{xx}(I_{xx}-I_{yy})+I_{xz}^2]p_b q_b - I_{xz}(I_{xx}-I_{yy}+I_{zz})q_b r_b + I_{xz}(l_{\Delta 0}+\Delta l_\Delta) + I_{xx}(n_{\Delta 0}+\Delta n_\Delta)}{I_{xx} I_{zz} - I_{xz}^2} \end{bmatrix} \tag{2.27}$$

Based on the above analysis, the equation about variables $\bar{x}_g$, $\bar{y}_g$, $\bar{z}_g$, $\bar{V}_A$, χ, γ, α, β, μ, p_b, q_b and r_b (for convenience, the subscripts of p_b, q_b and r_b in the equation are omitted) can be obtained, and the expression can be described as follows:

$$\dot{\bar{x}}_g = \bar{V}_A \cos\gamma \cos\chi + u_{Wg} \tag{2.28}$$

$$\dot{\bar{y}}_g = \bar{V}_A \cos\gamma \sin\chi + v_{Wg} \tag{2.29}$$

$$\dot{\bar{z}}_g = -\bar{V}_A \sin\gamma + w_{Wg} \tag{2.30}$$

$$\begin{aligned} \dot{\bar{V}}_A = & \frac{1}{M}\left(-\bar{D} + \bar{T}_{\bar{y}} \sin\beta + \bar{T}_{\bar{x}} \cos\beta\cos\alpha + \bar{T}_{\bar{z}} \sin\alpha\cos\beta\right) - g\sin\gamma \\ & - \dot{u}_{Wg} \cos\gamma\cos\chi - \dot{v}_{Wg}\cos\gamma\sin\chi + \dot{w}_{Wg}\sin\gamma \end{aligned} \tag{2.31}$$

$$\begin{aligned} \dot{\chi} = & \frac{1}{M\bar{V}_A\cos\gamma}\left(\bar{Y} + Mg\sin\mu\cos\gamma + \bar{T}_{\bar{y}}\cos\beta - \bar{T}_{\bar{x}}\sin\beta\cos\alpha - \bar{T}_{\bar{z}}\sin\alpha\sin\beta\right) \\ & - \frac{1}{\bar{V}_A\cos\gamma}\left[\dot{u}_{Wg}\left(\sin\mu\sin\gamma\cos\chi - \cos\mu\sin\chi\right) + \dot{w}_{Wg}\sin\mu\cos\gamma \right. \\ & \left. + \dot{v}_{Wg}\left(\sin\mu\sin\gamma\sin\chi + \cos\mu\cos\chi\right)\right] \end{aligned} \tag{2.32}$$

$$\begin{aligned} \dot{\gamma} = & \frac{1}{M\bar{V}_A}\left[\bar{L} - Mg\cos\mu\cos\gamma + \bar{T}_{\bar{x}}\sin\alpha - \bar{T}_{\bar{z}}\cos\alpha\right] \\ & + \frac{1}{\bar{V}_A}\left[\dot{u}_{Wg}\left(\cos\mu\sin\gamma\cos\chi + \sin\mu\sin\chi\right) \right. \\ & \left. + \dot{v}_{Wg}\left(\cos\mu\sin\gamma\sin\chi - \sin\mu\cos\chi\right) + \dot{w}_{Wg}\cos\mu\cos\gamma\right] \end{aligned} \tag{2.33}$$

$$\begin{aligned} \dot{\alpha} = & q + \frac{1}{M\bar{V}_A\cos\beta}\left(-\bar{L} + Mg\cos\gamma\cos\mu - \bar{T}_{\bar{x}}\sin\alpha + \bar{T}_{\bar{z}}\cos\alpha\right) \\ & - \frac{1}{\bar{V}_A\cos\beta}\left[\dot{u}_{Wg}\left(\cos\mu\sin\gamma\cos\chi + \sin\mu\sin\chi\right) + \dot{w}_{Wg}\cos\mu\cos\gamma \right. \\ & \left. + \dot{v}_{Wg}\left(\cos\mu\sin\gamma\sin\chi - \sin\mu\cos\chi\right)\right] - \left(p\cos\alpha + r\sin\alpha\right)\tan\beta \end{aligned} \tag{2.34}$$

$$\begin{aligned} \dot{\beta} = & p\sin\alpha - r\cos\alpha + \frac{1}{M\bar{V}_A}\left(\bar{Y} + Mg\cos\gamma\cos\mu - \bar{T}_{\bar{x}}\sin\beta\cos\alpha + \bar{T}_{\bar{y}}\cos\beta \right. \\ & \left. - \bar{T}_{\bar{z}}\sin\beta\sin\alpha\right) - \frac{1}{\bar{V}_A}\left[\dot{u}_{Wg}\left(\sin\mu\sin\gamma\cos\chi - \cos\mu\sin\chi\right) \right. \\ & \left. + \dot{v}_{Wg}\left(\sin\mu\sin\gamma\sin\chi + \cos\mu\cos\chi\right) + \dot{w}_{Wg}\sin\mu\cos\gamma\right] \end{aligned} \tag{2.35}$$

$$\dot{\mu} = (p\cos\alpha + r\sin\alpha)\sec\beta + \frac{1}{M\bar{V}_A}\left[\bar{L}\,(\tan\gamma\sin\mu + \tan\beta) - Mg\cos\gamma\cos\mu\tan\beta + \right.$$
$$\left(\bar{T}_{\bar{x}}\sin\alpha - \bar{T}_{\bar{z}}\cos\alpha\right)(\tan\gamma\sin\mu + \tan\beta) - \left.\left(\bar{T}_{\bar{x}}\cos\alpha + \bar{T}_{\bar{z}}\sin\alpha\right)\tan\gamma\cos\mu\sin\beta\right]$$
$$+\frac{1}{M\bar{V}_A}\left(\bar{Y} + \bar{T}_{\bar{y}}\right)\tan\gamma\cos\mu + \frac{1}{\bar{V}_A}\left[{}^g\dot{u}_W\tan\gamma\,(\sin\mu\sin\gamma\cos\chi - \cos\mu\,\sin\,\chi)\right.$$
$$\left. + {}^g\dot{v}_W\tan\beta\,(\sin\mu\sin\gamma\sin\chi + \cos\mu\cos\chi) + {}^g\dot{w}_W\sin\mu\sin\gamma\right]$$
$$+\frac{1}{\bar{V}_A}\left[\dot{u}_{Wg}\tan\beta\,(\cos\mu\sin\gamma\cos\chi + \sin\mu\,\sin\,\chi) + \dot{v}_{Wg}\tan\beta\,(\cos\mu\sin\gamma\sin\chi\right.$$
$$\left. - \sin\mu\cos\chi) + \dot{w}_{Wg}\tan\beta\cos\mu\sin\gamma\right] \tag{2.36}$$

$$\dot{p} = \frac{\left[I_{zz}\left(I_{yy} - I_{zz}\right) - I_{xz}^2\right]qr + I_{xz}\left(I_{xx} - I_{yy} + I_{zz}\right)pq}{I_{xx}I_{zz} - I_{xz}^2}$$
$$\frac{+I_{zz}(l_{\Delta 0} + \Delta l_{\Delta}) + I_{xz}(n_{\Delta 0} + \Delta n_{\Delta})}{I_{xx}I_{zz} - I_{xz}^2} \tag{2.37}$$

$$\dot{q} = \frac{1}{I_{yy}}\left[(I_{zz} - I_{xx})\,pr + I_{xz}\left(r^2 - p^2\right) + m_{\Delta 0} + \Delta m_{\Delta}\right] \tag{2.38}$$

$$\dot{r} = \frac{\left[I_{xx}\left(I_{xx} - I_{yy}\right) + I_{xz}^2\right]pq - I_{xz}\left(I_{xx} - I_{yy} + I_{zz}\right)qr}{I_{xx}I_{zz} - I_{xz}^2}$$
$$\frac{+I_{xz}(l_{\Delta 0} + \Delta l_{\Delta}) + I_{xx}(n_{\Delta 0} + \Delta n_{\Delta})}{I_{xx}I_{zz} - I_{xz}^2} \tag{2.39}$$

The parameters involved in the above are expressed as [3]: $g = g_0(\bar{R}^2/(\bar{R} + V_e)^2)$ is the vertical gravity component, g_0 is the gravitational constant, $\bar{R}$ the equatorial radius of the earth and $V_e < 11000$m; $\bar{D} = -\bar{X}\cos(\alpha) - \bar{Z}\sin(\alpha)$ is the drag, $\bar{L} = \bar{X}\sin(\alpha) - \bar{Z}\cos(\alpha)$ is the lift, $\bar{X}$, $\bar{Y}$ and $\bar{Z}$ are the aerodynamic parameters and $\bar{X} = \bar{q}S_rC_{X_{\bar{T}}}$, $\bar{Y} = \bar{q}S_rC_{Y_{\bar{T}}}$ and $\bar{Z} = \bar{q}S_rC_{Z_{\bar{T}}}$, $\bar{q} = \frac{1}{2}\bar{\rho}V^2$ is the aerodynamic pressure, S_r is the area of the wing, $\bar{\rho}$ is the airflow density of the UAV at the altitude V_e and $\bar{\rho} = \rho_0 e^{-\frac{g_0}{287.05\bar{\bar{T}}}V_e}$, $\rho_0 = 1.225$ is the density of sea level air, $\bar{\bar{T}}$ is the temperature of UAV at the altitude V_e, $C_{X_{\bar{T}}}$, $C_{Y_{\bar{T}}}$ and $C_{Z_{\bar{T}}}$ are axial force coefficient, lateral force coefficient and normal force coefficient respectively; $l_{\Delta 0} = \bar{l}_0 + \bar{l}_{\delta_a}\delta_a + \bar{l}_{\delta_r}\delta_r$, $\bar{l}_0 = \bar{q}S_r\bar{b}C_{l_{\bar{T}}}$, $\bar{l}_{\delta_a} = \bar{q}S_r\bar{b}C_{l\delta_a}$, $\bar{l}_{\delta_r} = \bar{q}S_r\bar{b}C_{l\delta_r}$, $m_{\Delta 0} = \bar{m}_0 + \bar{m}_{\delta_e}\delta_e - rh_e$, $\bar{m}_0 = \bar{q}S_r\bar{c}C_{m_{\bar{T}}}$, $\bar{m}_{\delta_e} = \bar{q}S_r\bar{c}C_{m\delta_e}$, $n_{\Delta 0} = \bar{n}_0 + \bar{n}_{\delta_a}\delta_a + \bar{n}_{\delta_r}\delta_r + qh_e$, $\bar{n}_0 = \bar{q}S_r\bar{b}C_{n_{\bar{T}}}$, $\bar{n}_{\delta_a} = \bar{q}S_r\bar{b}C_{n\delta_a}$, $\bar{n}_{\delta_r} = \bar{q}S_r\bar{b}C_{n\delta_r}$, where h_e is the angular momentum of the engine, $\bar{b}$ is the UAV wing length, $\bar{c}$ is the average aerodynamic string length; δ_e, δ_a and δ_r are elevator deflection angle, aileron deflection angle and rudder deflection Angle respectively; $C_{l_{\bar{T}}}$, $C_{m_{\bar{T}}}$ and $C_{n_{\bar{T}}}$ are total rolling moment coefficient, total pitching moment coefficient and total yaw moment coefficient respectively; $C_{l\delta_a}$, $C_{l\delta_r}$, $C_{m\delta_e}$, $C_{n\delta_a}$ and $C_{n\delta_r}$ are torque coefficients.

2.1.3 Zero Input Response Characteristics of UAV

In the case of zero input response, this section simulates the established UAV model (2.28)–(2.39) with external wind disturbance, the initial values of $\bar{x}_g$, $\bar{y}_g$, $\bar{z}_g$, $\bar{V}_A$, χ, γ, α, β, μ, p, q and r are chosen as 500 m, 500 m, 1000 m, 100 m/sec, 0 deg, 0 deg, 2 deg, 2 deg, 2 deg, 0 deg/sec, 0 deg/sec and 0 deg/sec, where deg denotes the degree, and sec denotes the second. In addition, for this simulation, only the influence of gust on the control of UAV system is considered. According to the result in [1, 2], the following gust pattern can be obtained:

$$V_{Wi} = \begin{cases} 0, & t < 0 \\ \frac{\bar{V}_{Wmi}}{2}\left[1 - \cos\left(\frac{\pi t}{t_m}\right)\right], & 0 \le t \le t_m \\ \frac{\bar{V}_{Wmi}}{2}, & t > t_m \end{cases} \tag{2.40}$$

where $i = 1, 2, 3$, $V_W = [V_{W1}, V_{W2}, V_{W3}]^{\mathrm{T}}$, $\bar{V}_{Wm} = [\bar{V}_{Wm1}, \bar{V}_{Wm2}, \bar{V}_{Wm3}]^{\mathrm{T}}$ is the maximum wind speed vector, and when $t > t_m$, no wind gradient appears. In addition, in this numerical simulation, the maximum external wind speed is assumed to be $V_{Wm} = [5, 4, 2]^{\mathrm{T}}$ m/sec, and $t_m = 3$ sec.

According to the simulation conditions, the simulation results of the UAV with zero input response are shown in Fig. 2.2. The Fig. 2.2 shows that the UAV in the absence of reasonable control signal input, the state of the UAV range is bigger,the flight of UAV is instability. In addition, it can be seen from the Fig. 2.2, the UAV in the complex features of zero input response is mainly manifested for the attitude control system. It can be seen that the UAV attitude angles β and α change slowly, and the roll angle μ is divergent. Therefore, it is necessary to design a reasonable controller to ensure the reliable flight of UAV. In addition, the simulation parameters of the UAV attitude dynamic mathematical model in this book are shown in Table 2.1 [3].

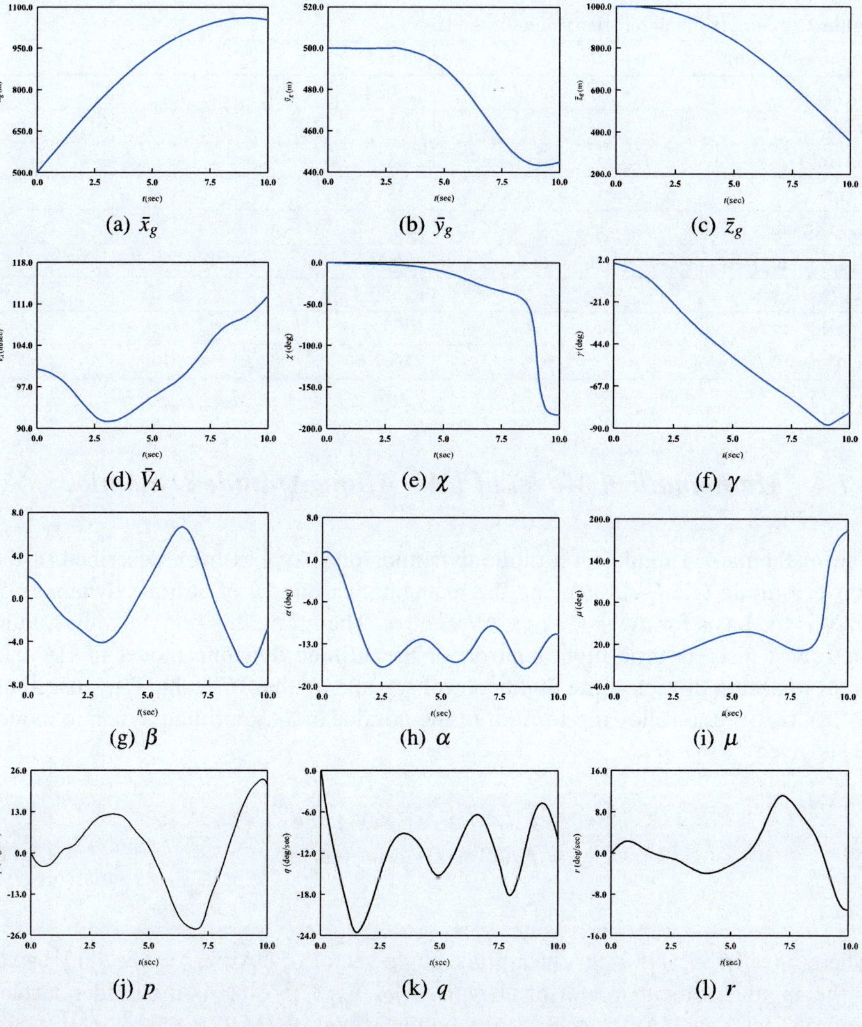

Fig. 2.2 The simulation results of UAV with zero input response

Table 2.1 Simulation parameter values of the UAV

Parameters (units)	Values
M (kg)	9295.44
$\bar{b}$ (m)	9.144
S_r (m^2)	27.87
$\bar{c}$ (m)	3.45
I_{xx} (kg $\cdot$ m^2)	12874.8
I_{yy} (kg $\cdot$ m^2)	75673.6
I_{zz} (kg $\cdot$ m^2)	85552.1
I_{xz} (kg $\cdot$ m^2)	1331.4
h_e (kg $\cdot$ m^2/s)	216.9

2.1.4 Mathematical Model of UAV Affine Attitude Dynamics

The mathematical model of attitude dynamics of UAV has been described in the above formula (2.28)–(2.39), and the mathematical model of attitude dynamics of UAV is the basis for the study of UAV control. Therefore, this book mainly studies the robust discrete-time flight control for the attitude dynamic model of UAV. In order to design discrete-time flight control scheme by the BC technology, based on (2.28)–(2.39), the following form of affine nonlinear UAV attitude dynamic model [3] is given:

$$\begin{cases} \dot{x}_1 = F_1(x_1) + G_1(x_1)x_2 + d_1^0 \\ \dot{x}_2 = F_2(x) + G_2(x)u + d_2^0 \\ y = x_1 \end{cases} \tag{2.41}$$

where $x_1 = [\beta, \alpha, \mu]^{\mathrm{T}} \in \Re^3$ the attitude angle vector of UAV; $x_2 = [p, q, r]^{\mathrm{T}} \in \Re^3$ is the angular velocity vector of UAV, $u = [\delta_e, \delta_a, \delta_r]^{\mathrm{T}} \in \Re^3$ is the rudder surface control vector of UAV; $y \in \Re^3$ is the output signal of UAV; $x = [x_1^{\mathrm{T}}, x_2^{\mathrm{T}}]^{\mathrm{T}} \in \Re^6$; $F_1(x_1) \in \Re^3$ and $F_2(x) \in \Re^3$ are known nonlinear function vectors; $G_1(x_1) \in \Re^{3\times 3}$ and $G_2(x) \in \Re^{3\times 3}$ are known control matrices; $d_1^0 = \left[d_1^1, d_1^2, d_1^3\right]^{\mathrm{T}} \in \Re^3$ and $d_2^0 = \left[d_2^1, d_2^2, d_2^3\right]^{\mathrm{T}} \in \Re^3$ are the influence of wind disturbance. In addition, the above mentioned nonlinear functions $F_1(x_1) = [F_{11}, F_{12}, F_{13}]^{\mathrm{T}}$ and $F_2(x) = [F_{21}, F_{22}, F_{23}]^{\mathrm{T}}$, control matrices $G_1(x_1)$ and $G_2(x)$ and external disturbances d_1^0 and d_2^0 can be written as follows:

$$F_{11} = \frac{1}{M\bar{V}_A}\left(\bar{Y} + Mg\cos\gamma\cos\mu - \bar{T}_{\bar{x}}\sin\beta\cos\alpha + \bar{T}_{\bar{y}}\cos\beta\right) \tag{2.42}$$

$$F_{12} = \frac{1}{M\bar{V}_A\cos\beta}\left(-\bar{L} + Mg\cos\gamma\cos\mu - \bar{T}_{\bar{x}}\sin\alpha + \bar{T}_{\bar{z}}\cos\alpha\right) \tag{2.43}$$

$$F_{13} = \frac{1}{M\bar{V}_A}\left[\bar{L}\left(\tan\gamma\sin\mu + \tan\beta\right) - Mg\cos\gamma\cos\mu\tan\beta + \left(\bar{Y} + \bar{T}_{\bar{y}}\right)\tan\gamma\cos\mu\right.$$
$$\left(\bar{T}_{\bar{x}}\sin\alpha - \bar{T}_{\bar{z}}\cos\alpha\right)\left(\tan\gamma\sin\mu + \tan\beta\right) - \bar{T}_{\bar{x}}\cos\alpha\tan\gamma\cos\mu\sin\beta$$
$$\left.-\bar{T}_{\bar{z}}\sin\alpha\tan\gamma\cos\mu\sin\beta\right] \tag{2.44}$$

$$F_{21} = \left(\bar{C}_1 r + \bar{C}_2 p\right) q + \bar{C}_3 \bar{l}_0 + \bar{C}_4\left(\bar{n}_0 + q h_e\right) \tag{2.45}$$

$$F_{22} = \bar{C}_5 pq - \bar{C}_6\left(p^2 - r^2\right) + \bar{C}_7\left(\bar{m}_0 - r h_e\right) \tag{2.46}$$

$$F_{23} = \left(\bar{C}_8 p - \bar{C}_2 r\right) q + \bar{C}_4 \bar{l}_0 + \bar{C}_9\left(\bar{n}_0 + q h_e\right) \tag{2.47}$$

$$G_1(x_1) = \begin{bmatrix} \sin\alpha & 0 & -\cos\alpha \\ -\tan\beta\cos\alpha & 1 & -\tan\beta\sin\alpha \\ \cos\alpha/\cos\beta & 0 & \sin\alpha/\cos\beta \end{bmatrix} \tag{2.48}$$

$$d_1^1 = -\frac{1}{\bar{V}_A}\left[\dot{u}_{Wg}\left(\sin\mu\sin\gamma\cos\chi - \cos\mu\,\sin\,\chi\right)\right.$$
$$\left. + \dot{v}_{Wg}\left(\sin\mu\sin\gamma\sin\chi + \cos\mu\cos\chi\right) + \dot{w}_{Wg}\sin\mu\cos\gamma\right] \tag{2.49}$$

$$d_1^2 = -\frac{1}{\bar{V}_A\cos\beta}\left[\dot{u}_{Wg}\left(\cos\mu\sin\gamma\cos\chi + \sin\mu\,\sin\,\chi\right) + \dot{w}_{Wg}\cos\mu\cos\gamma\right.$$
$$\left. + \dot{v}_{Wg}\left(\cos\mu\sin\gamma\sin\chi - \sin\mu\cos\chi\right)\right] \tag{2.50}$$

$$d_1^3 = \frac{1}{\bar{V}_A}\left[\dot{u}_{Wg}\tan\gamma\left(\sin\mu\sin\gamma\cos\chi - \cos\mu\,\sin\,\chi\right) + \dot{w}_{Wg}\sin\mu\sin\gamma\right.$$
$$\left. + \dot{v}_{Wg}\tan\beta\left(\sin\mu\sin\gamma\sin\chi + \cos\mu\cos\chi\right) + \dot{u}_{Wg}\tan\beta\cos\mu\sin\gamma\cos\chi\right]$$
$$+\frac{1}{\bar{V}_A}\left[\dot{u}_{Wg}\tan\beta\sin\mu\,\sin\,\chi + \dot{v}_{Wg}\tan\beta\left(\cos\mu\sin\gamma\sin\chi - \sin\mu\cos\chi\right)\right.$$
$$\left. + \dot{w}_{Wg}\tan\beta\cos\mu\sin\gamma\right] \tag{2.51}$$

$$G_2(x) = \begin{bmatrix} \bar{C}_3\bar{l}_{\delta_a} + \bar{C}_4\bar{n}_{\delta_a} & 0 & \bar{C}_3\bar{l}_{\delta_r} + \bar{C}_4\bar{n}_{\delta_r} \\ 0 & \bar{C}_7\bar{m}_{\delta_e} & 0 \\ \bar{C}_4\bar{l}_{\delta_a} + \bar{C}_9\bar{n}_{\delta_a} & 0 & \bar{C}_4\bar{l}_{\delta_r} + \bar{C}_9\bar{n}_{\delta_r} \end{bmatrix} \tag{2.52}$$

$$d_2^0 = \left[\frac{I_{zz}\Delta l_\Delta + I_{xz}\Delta n_\Delta}{I_{xx}I_{zz} - I_{xz}^2}, \frac{\Delta m_\Delta}{I_{yy}}, \frac{I_{xz}\Delta l_\Delta + I_{xx}\Delta n_\Delta}{I_{xx}I_{zz} - I_{xz}^2}\right]^{\mathrm{T}} \tag{2.53}$$

where $\bar{C}_1 = \frac{I_{zz}(I_{yy}-I_{zz})-I_{xz}^2}{\Gamma}$, $\bar{C}_2 = \frac{I_{xz}(I_{xx}-I_{yy}+I_{zz})}{\Gamma}$, $\bar{C}_3 = \frac{I_{zz}}{\Gamma}$, $\bar{C}_4 = \frac{I_{xz}}{\Gamma}$, $\bar{C}_5 = \frac{I_{zz}-I_{xx}}{I_{yy}}$, $\bar{C}_6 = \frac{I_{xz}}{I_{yy}}$, $\bar{C}_7 = \frac{1}{I_{yy}}$, $\bar{C}_8 = \frac{I_{xx}(I_{xx}-I_{yy})+I_{xz}^2}{\Gamma}$, $\bar{C}_9 = \frac{I_{xx}}{\Gamma}$ and $\Gamma = I_{xx}I_{zz} - I_{xz}^2$.

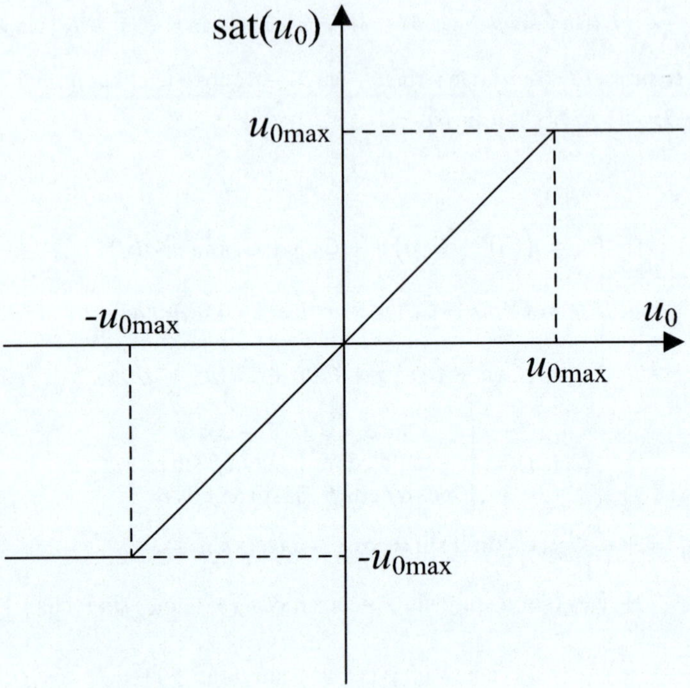

Fig. 2.3 Symmetric nonlinear input saturation model

2.2 Input Nonlinearities

The nonlinear characteristics of control inputs in a practical control system mainly refer to the nonlinear characteristics of the relationship between the control inputs and the actuator outputs. In the past studies of input nonlinearity mainly include the input saturation, the dead-zone input, the input nonlinearity combing the asymmetric saturation input and the dead-zone input and so on, and three input nonlinearities of the above are given as follows:

2.2.1 Input Saturation

According to Fig. 2.3, the symmetric nonlinear input saturation function sat(·) can be defined as [4]:

$$\text{sat}(u_0) = \text{sign}(u_0)\min\{u_{0\,\max}, |u_0|\} \tag{2.54}$$

where u_0 is the control input, $u_{0\,\max}$ is the saturation level of the input, sign(·) is the standard sign function and min{·} is the minimum value.

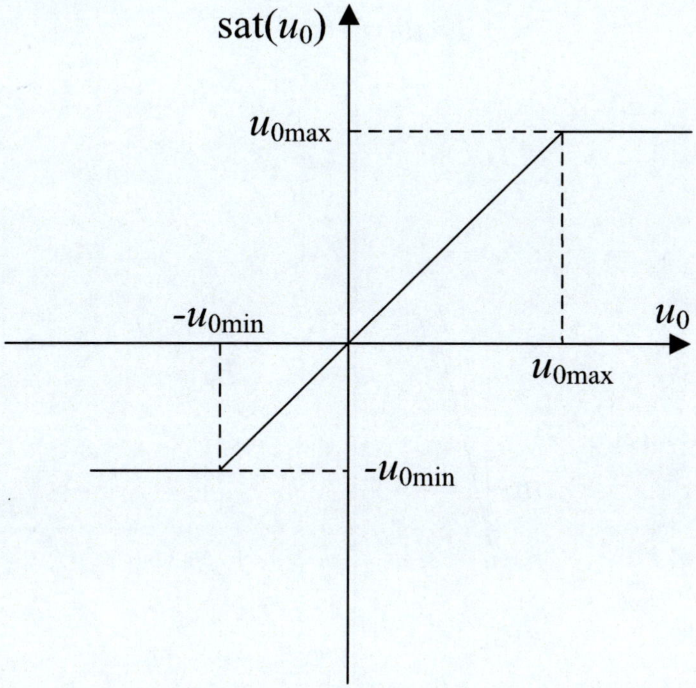

Fig. 2.4 Asymmetric nonlinear input saturation model

Furthermore, the input saturation function sat(·) can be described as [5]

$$\operatorname{sat}(u_0) = \begin{cases} \operatorname{sign}(u_0)u_{0\,\max}, & |u_0| \geq u_{0\,\max} \\ u_0, & |u_0| < u_{0\,\max} \end{cases} \tag{2.55}$$

Based on the symmetric nonlinear input saturation functions (2.54) and (2.55), the characteristic of the non-symmetric saturation function sat(·) is shown in Fig. 2.4, and the expression can be written as [6]

$$\operatorname{sat}(u_0) = \begin{cases} u_{0\,\max}, & u_0 > u_{0\,\max} \\ u_0, & -u_{0\,\min} \leq u_0 \leq u_{0\,\max} \\ -u_{0\,\min}, & u_0 < -u_{0\,\min} \end{cases} \tag{2.56}$$

where $u_{0\,\min}$ is the saturation level of the input saturation.

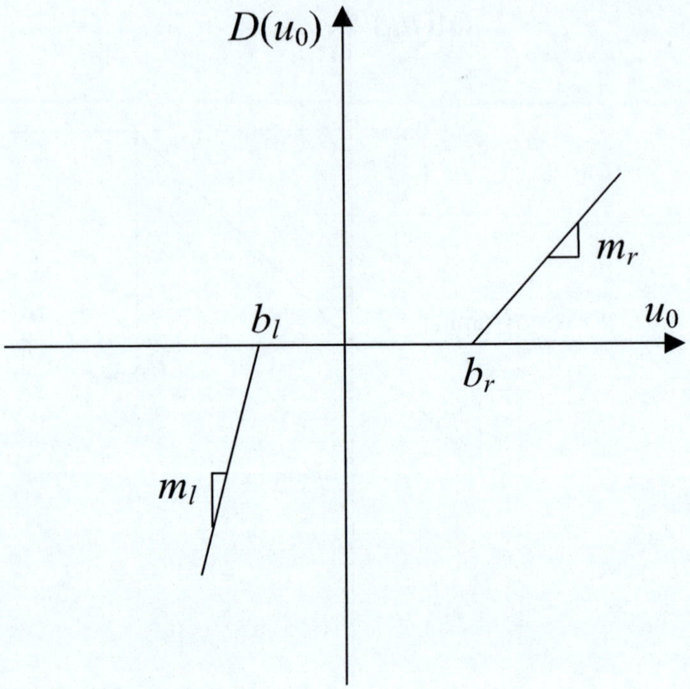

Fig. 2.5 Dead-zone nonlinear model

2.2.2 Dead-Zone Nonlinearity

From the structure of dead-zone in Fig. 2.5, the mathematical expression can be described as [7]

$$D_b(u_0) = \begin{cases} m_r(u_0 - b_r), & u_0 \geq b_r \\ 0, & b_l < u_0 < b_r \\ m_l(u_0 - b_l), & u_0 \leq b_l \end{cases} \quad (2.57)$$

where $D_b(\cdot)$ is the dead-zone nonlinear function, $b_l < 0$ is the left dead-zone boundary, $b_r > 0$ is the right dead-zone boundary, $m_l > 0$ is the left slope of the dead zone, and $m_r > 0$ is the left slope of the dead zone.

2.2.3 Asymmetric Saturation and Dead-Zone

This section describes the input nonlinear function satisfying the asymmetric saturation and the dead-zone, and the characteristic is shown in Fig. 2.6. In Fig. 2.6,

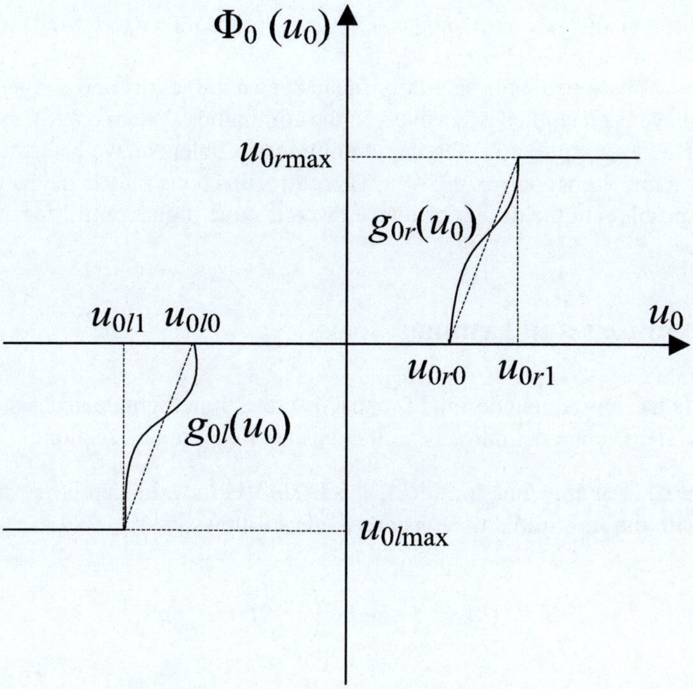

Fig. 2.6 Asymmetric saturated and dead-zone nonlinear model

$u_{0r\max} > 0$ and $u_{0l\max} < 0$ are saturation values, $u_{0r0} > 0$ and $u_{0l0} < 0$ are dead-zone boundaries, $g_{0r}(u_0)$ and $g_{0l}(u_0)$ are smooth nonlinear functions. According to Fig. 2.6, the input function $\Phi_0(u_0)$ can written as follows [8]:

$$\Phi_0(u_0) = \begin{cases} \bar{\chi}_0(\bar{u}_0)(u_0 - u_{0r0}), & u_0 \geq u_{0r0} \\ \bar{\chi}_0(\bar{u}_0)(u_0 - u_{0l0}), & u_0 \leq u_{0l0} \\ 0, & u_{0l0} < u_0 < u_{0r0} \end{cases} \tag{2.58}$$

According to the differential mean value theorem, we have that $\psi_{0r}(u_0) \in [u_{0r0}, u_{0r1}]$ and $\psi_{0l}(u_0) \in [u_{0l0}, u_{0l1}]$. Furthermore, the function $\bar{\chi}_0(\bar{u}_0)$ satisfies the following expression:

$$\bar{\chi}_0(\bar{u}_0) = \begin{cases} \frac{u_{0r\max}}{u_0 - u_{0r0}}, & u_0 > u_{0r1} \\ g'_{0r}\left[\psi_{0r}(u_0)\right], & u_{0r0} \leq u_0 \leq u_{0r1} \\ g'_{0l}\left[\psi_{0l}(u_0)\right], & u_{0l1} \leq u_0 \leq u_{0l0} \\ \frac{u_{0l\max}}{u_0 - u_{0l0}}, & u_0 < u_{0l1} \end{cases} \tag{2.59}$$

where $g'_{0r}\,[\psi_{0r}(u_0)] = dg_{0r}(z_0)/dz_0|_{z_0=\psi_{0r}(u_0)}$, $g'_{0l}\,[\psi_{0l}(u_0)] = dg_{0l}(z_0)/dz_0|_{z_0=\psi_{0l}(u_0)}$ and $\bar{u}_0 = [u_0, u_{0r\max}, u_{0l\max}, u_{0r0}, u_{0l0}]^{\mathrm{T}}$.

The above mentioned input saturation, dead zone nonlinearity and non-symmetric saturation have been studied extensively in the continuous system control system [6, 7, 9–30]. However, up to now, few input nonlinear problems have been studied for the discrete-time flight control of UAVs. Therefore, this book only consider the input saturation problem in the design of robust discrete-time flight control for the UAV.

2.3 Definitions and Lemmas

To facilitate the subsequent design of robust discrete-time flight control schemes for the UAV system, some definitions and lemmas are given in this section.

Definition 2.1 For the function $f(k)$, the DTFO Grünwald-Letnikow difference operator with the zero initial time is described as follows [31]:

$$\nabla^{\gamma_0} f(k) = \sum_{j=0}^{k} (-1)^j \binom{\gamma_0}{j} f(k-j) \tag{2.60}$$

where k is a positive constant, γ_0 is the fractional order, $\binom{\gamma_0}{j}$ is the binomial coefficient and satisfies the following expression [31]:

$$\binom{\gamma_0}{j} = \begin{cases} 1, & j = 0 \\ \frac{\gamma_0(\gamma_0-1)\cdots(\gamma_0-j+1)}{j!}, & j > 0 \end{cases} \tag{2.61}$$

Definition 2.2 The fractional order is γ_0 and the corresponding expression can be written as [31]

$$\nabla^{-\gamma_0} f(k) = \sum_{j=0}^{k} \binom{\gamma_0}{k-j} f(j) \tag{2.62}$$

Definition 2.3 If the smooth function $\varpi_1(t)$ satisfies that: (i) $\varpi_1(t)$ is positive and strictly decreasing, (ii) $\lim_{t\to\infty} \varpi_1(t) = \varpi_{1\infty} > 0$, then the function $\varpi_1(t)$ can be expressed as a performance function [32].

Lemma 2.1 *For two positive constants γ_{01} and γ_{02}, one has [31]*

$$\nabla^{\gamma_{01}} \nabla^{-\gamma_{02}} f(k) = \nabla^{\gamma_{01}-\gamma_{02}} f(k) \tag{2.63}$$

Based on (2.63), we have $\nabla^0 f(k) = f(k)$.

Lemma 2.2 *On the basis of the inherent ability of RBFNN to approximate nonlinear functions, it has been widely used in the modeling and control of nonlinear systems. Thus, the NN is employed to approximate the uncertain smooth nonlinear function* $\xi(\mathscr{X}(k)) : \Re^{\ell} \rightarrow \Re$ *as follows [33]:*

$$\begin{aligned}\xi(\mathscr{X}(k)) &= \zeta^T(k)\varphi(\mathscr{X}(k)), \zeta(k) \in R^{\bar{l}},\ \varphi(\mathscr{X}(k)) \in \Re^{\bar{l}}\\ \varphi(\mathscr{X}(k)) &= (\varphi_{01}(\mathscr{X}(k)), \ldots, \varphi_{0j}(\mathscr{X}(k)))^{\mathrm{T}}\end{aligned} \tag{2.64}$$

where $\mathscr{X}(k) = [\mathscr{X}_1(k), \mathscr{X}_2(k), \ldots, \mathscr{X}_{\bar{p}}(k)]^{\mathrm{T}} \in \Omega_{\mathscr{X}} \subset \Re^{\bar{p}}$, *the positive integer* $\bar{l}$ *is the node number of the NN,* $\zeta(k)$ *is the weight vector, and* $\varphi_{0j}(\mathscr{X}(k)) \in \Re$ *denotes the basis function, which is given by*

$$\varphi_{0j}(\mathscr{X}(k)) = \exp\left(\frac{-(\mathscr{X}(k) - \bar{\mu}_{0j})^T(\mathscr{X}(k) - \bar{\mu}_{0j})}{\varsigma_{0j}^2}\right) \tag{2.65}$$

where $\bar{\mu}_{0j} = [\bar{\mu}_{0j1}, \bar{\mu}_{0j2}, \ldots, \bar{\mu}_{0j\bar{p}}]^T$ *and* ς_{0j} *represent the width and the center of the Gaussian function, respectively.*

Using the approximation performance of the NN, a smooth nonlinear function $f(\mathscr{X}(k))$ *is written as*

$$f(\mathscr{X}(k)) = \zeta^{*T}(k)\varphi(\mathscr{X}(k)) + \varepsilon(k) \tag{2.66}$$

where $\zeta^*(k)$ *is the optimal weight vector with* $\zeta^*(k) = \zeta^*(k+1)$ *and* $\varepsilon(k)$ *is the approximation error. Furthermore, according to the definition of* $\varphi_{0j}(\mathscr{X}(k))$, *one has* $\varphi^T(\mathscr{X})\varphi(\mathscr{X}) \leq \bar{l}$ *with* $\bar{l} > 0$.

Lemma 2.3 *A discrete-time tracking differentiator can be written as follows [34]:*

$$\begin{aligned}\mathscr{B}_1(k+1) &= \mathscr{B}_1(k) + q_0\mathscr{B}_2(k)\\ \mathscr{B}_2(k+1) &= \mathscr{B}_2(k) - q_0 p_0 \mathscr{B}_3(\mathscr{B}_4(k), \delta_0)\end{aligned} \tag{2.67}$$

where $\mathscr{B}_1(k)$ *and* $\mathscr{B}_2(k)$ *are the variables of the discrete-time tracking differentiator (2.67),* $\bar{q}_0$ *and* $\bar{p}_0$ *denote the constants and* $\delta_0 = q_0 p_0$, $\mathscr{B}_3(\mathscr{B}_4(k)\delta_0)$ *and* $\mathscr{B}_4(k)$ *satisfy that*

$$\mathscr{B}_3(\mathscr{B}_4(k), \delta_0) = \begin{cases} \mathrm{sign}(\mathscr{B}_4(k)), & |\mathscr{B}_4(k)| > \delta_0\\ \frac{\mathscr{B}_4(k)}{\delta_0}, & |\mathscr{B}_4(k)| \leq \delta_0 \end{cases} \tag{2.68}$$

$$\mathscr{B}_4(k) = \begin{cases} \mathscr{B}_2(k) - \mathrm{sign}(\mathscr{B}_5(k))\frac{p_0\left(q_0 - \sqrt{\frac{8|\mathscr{B}_5(k)|}{p_0} + q_0^2}\right)}{2}, & |\mathscr{B}_5(k)| \geq \bar{\delta}_0\\ \mathscr{B}_2(k) + \frac{\mathscr{B}_5(k)}{q_0}, & |\mathscr{B}_5(k)| \leq \bar{\delta}_0 \end{cases} \tag{2.69}$$

where $\mathscr{B}_5(k) = \mathscr{B}_1(k) - \mathscr{B}_6(k) + q_0\mathscr{B}_2(k)$, $\mathscr{B}_6(k)$ *is the bounded input signal, and* $\bar{\delta}_0 = q_0\delta_0$.

From (2.67)–(2.69), it can be obtained that $\mathscr{B}_1(k)$ *can approximate* $\mathscr{B}_6(k)$ *well.*

Lemma 2.4 *The Lyapunov function* $\mathscr{V}(\mathscr{J}(k))$ *is chosen for a discrete-time system, which satisfies the following property [35]:*

$$\begin{aligned} \mathscr{T}_1(\|\mathscr{J}(k)\|) \leq \mathscr{V}(\mathscr{J}(k)) \leq \mathscr{T}_2(\|\mathscr{J}(k)\|) \\ \mathscr{V}(\mathscr{J}(k+1)) - V(\mathscr{J}(k)) = \Delta\mathscr{V}(\mathscr{J}(k)) \\ \leq -\mathscr{T}_3(\|\mathscr{J}(k)\|) + \mathscr{T}_3(\bar{\mathscr{J}}) \end{aligned} \tag{2.70}$$

where $\bar{\mathscr{J}}$ *is a positive constant,* $\mathscr{T}_1(\cdot)$ *and* $\mathscr{T}_2(\cdot)$ *is a strictly increasing function, and* $\mathscr{T}_3(\cdot)$ *is a non-decreasing function. If the following form is satisfied:*

$$\Delta\mathscr{V}(\mathscr{J}(k)) < 0, \quad \|\mathscr{J}(k)\| > \bar{\mathscr{J}} \tag{2.71}$$

then $\mathscr{J}(k)$ *is finally uniformly bounded and stable in a compact set, and there is a time* k_T *such that* $\|\mathscr{J}(k)\| < \bar{\mathscr{J}}, \forall k > k_T$.

2.4 Conclusions

In this chapter, a six-degree-of-freedom dynamic model of fixed-wing UAV with wind disturbance has been established, and the attitude dynamics nonlinear affine model of the UAV has been given. The derived model provides a model basis for the subsequent discrete-time flight control scheme design for the UAV system. Then the input saturation, the dead-zone input and the input nonlinearity combing the asymmetric saturation input and the dead-zone input have been introduced. The given input nonlinearities provide a basis for further research on the input nonlinearity control of UAV system. Finally, some lemmas and definitions have been presented, including the definition of discrete-time fractional-order Grünwald-Letnikow difference, the approximation theory of RBFNN, the definition of discrete-time tracking differentiator and the theory of discrete-time Lyapunov function. The above definitions and theorems provide a theoretical basis for the design of robust discrete-time flight control scheme for UAV nonlinear model under wind disturbances.

References

1. Xiao, Y., Jin, C.: Principles of Flight in Atmospheric Disturbances. National defense industry press, Beijing (1993)
2. Du, Y.: Study of Nonlinear Adaptive Attitude and Trajectory Control for Near Space Vehicles. Nanjing University of Aeronautics and Astronautics, Nanjing (2010)

3. Sonneveldt, L.: Nonlinear F-16 model description. Technical report, Delft University of Technology, Netherlands (2006)
4. Chen, M., Shao, S.-Y., Shi, P., et al.: Disturbance-observer-based robust synchronization control for a class of fractional-order chaotic systems. IEEE Trans. Circuits Syst. II: Express Br. **64**(4), 417–421 (2017)
5. Chen, M., Ge, S.S.: Direct adaptive neural control for a class of uncertain nonaffine nonlinear systems based on disturbance observer. IEEE Trans. Cybern. **43**(4), 1213–1225 (2013)
6. Wen, C., Zhou, J., Liu, Z., et al.: Robust adaptive control of uncertain nonlinear systems in the presence of input saturation and external disturbance. IEEE Trans. Autom. Control **56**(7), 1672–1678 (2011)
7. Wang, X.-S., Su, C.-Y., Hong, H.: Robust adaptive control of a class of nonlinear systems with unknown dead-zone. Automatica **40**(3), 407–413 (2004)
8. Hu, Q., Ma, G., Xie, L.: Robust and adaptive variable structure output feedback control of uncertain systems with input nonlinearity. Automatica **44**(2), 552–559 (2008)
9. Grimm, G., Hatfield, J., Postlethwaite, I., et al.: Antiwindup for stable linear systems with input saturation: an LMI-based synthesis. IEEE Trans. Autom. Control **48**(9), 1509–1525 (2003)
10. Chen, B.M., Lee, T.H., Peng, K., et al.: Composite nonlinear feedback control for linear systems with input saturation: theory and an application. IEEE Trans. Autom. Control **48**(3), 427–439 (2003)
11. Cao, Y.-Y., Lin, Z., Hu, T.: Stability analysis of linear time-delay systems subject to input saturation. IEEE Trans. Circuits Syst. I: Fundam. Theory Appl. **49**(2), 233–240 (2002)
12. He, W., Dong, Y., Sun, C.: Adaptive neural impedance control of a robotic manipulator with input saturation. IEEE Trans. Syst. Man Cybern. Syst. **46**(3), 334–344 (2016)
13. Hu, Q.: Adaptive output feedback sliding-mode manoeuvring and vibration control of flexible spacecraft with input saturation. IET Control Theory Appl. **2**(6), 467–478 (2008)
14. Zhu, Z., Xia, Y., Fu, M., et al.: Adaptive sliding mode control for attitude stabilization with actuator saturation. IEEE Trans. Ind. Electron. **58**(10), 4898–4907 (2011)
15. Bošković, J.D., Li, S.-M., Mehra, R.K.: Robust adaptive variable structure control of spacecraft under control input saturation. J. Guid. Control Dyn. **24**(1), 14–22 (2001)
16. Hu, Q., Xiao, B., Friswell, M.: Robust fault-tolerant control for spacecraft attitude stabilisation subject to input saturation. IET Control Theory Appl. **5**(2), 271–282 (2011)
17. Bustan, D., Pariz, N., Sani, S.K.H.: Robust fault-tolerant tracking control design for spacecraft under control input saturation. ISA Trans. **53**(4), 1073–1080 (2014)
18. Boskovic, J.D., Li, S.-M., Mehra, R.K.: Robust tracking control design for spacecraft under control input saturation. J. Guid. Control Dyn. **27**(4), 627–633 (2004)
19. Chen, M., Yu, J.: Adaptive dynamic surface control of NSVs with input saturation using a disturbance observer. Chin. J. Aeronaut. **28**(3), 853–864 (2015)
20. Zhou, J., Wen, C., Zhang, Y.: Adaptive output control of nonlinear systems with uncertain dead-zone nonlinearity. IEEE Trans. Autom. Control **51**(3), 504–511 (2006)
21. Zhang, T.-P., Ge, S.S.: Adaptive dynamic surface control of nonlinear systems with unknown dead zone in pure feedback form. Automatica **44**(7), 1895–1903 (2008)
22. Ibrir, S., Xie, W.F., Su, C.-Y.: Adaptive tracking of nonlinear systems with non-symmetric dead-zone input. Automatica **43**(3), 522–530 (2007)
23. Tong, S., Li, Y.: Adaptive fuzzy output feedback control of MIMO nonlinear systems with unknown dead-zone inputs. IEEE Trans. Fuzzy Syst. **21**(1), 134–146 (2013)
24. Hu, Q., Meng, Y., Wang, C., et al.: Adaptive backstepping control for air-breathing hypersonic vehicles with input nonlinearities. Aerosp. Sci. Technol. **73**, 289–299 (2018)
25. An, H., Xia, H., Wang, C.: Barrier lyapunov function-based adaptive control for hypersonic flight vehicles. Nonlinear Dyn. **88**(3), 1833–1853 (2017)
26. Wang, Y., Hu, J.: Improved prescribed performance control for air-breathing hypersonic vehicles with unknown deadzone input nonlinearity. ISA Trans. **79**, 95–107 (2018)
27. Xu, B.: Robust adaptive neural control of flexible hypersonic flight vehicle with dead-zone input nonlinearity. Nonlinear Dyn. **80**(3), 1509–1520 (2015)

28. Yang, Q., Chen, M.: Adaptive neural prescribed performance tracking control for near space vehicles with input nonlinearity. Neurocomputing **174**, 780–789 (2016)
29. Chen, M., Ge, S.S., Ren, B.: Adaptive tracking control of uncertain MIMO nonlinear systems with input constraints. Automatica **47**(3), 452–465 (2011)
30. Chen, M., Ge, S.S., How, B.V.E.: Robust adaptive neural network control for a class of uncertain MIMO nonlinear systems with input nonlinearities. IEEE Trans. Neural Netw. **21**(5), 796–812 (2010)
31. Cheng, J.: Fractional Difference Equation Theory. xiamen university press, Fujian (2011)
32. Chen, M., Shao, S.-Y., Jiang, B.: Adaptive neural control of uncertain nonlinear systems using disturbance observer. IEEE Trans. Cybern. **47**(10), 3110–3123 (2017)
33. Zhang, J., Ge, S.S., Lee, T.H.: Direct RBF neural network control of a class of discrete-time non-affine nonlinear systems. In: Proceedings of the American Control Conference, pp. 424–429 (2002)
34. Han, J., Yuan, L.: The discrete form of tracking-differentiator. J. Syst. Sci. Math. Sci. **19**(3), 268–273 (1999)
35. Ge, S.S., Zhang, J., Lee, T.H.: Adaptive neural network control for a class of MIMO nonlinear systems with disturbances in discrete-time. IEEE Trans. Syst. Man Cybern. Part B (Cybern.) **34**(4), 1630–1645 (2004)

Chapter 3
Discrete-Time BC-Based Methods for Fixed-Wing UAV System

3.1 Introduction

The BC is a control method based on the recursive Lyapunov function for a class of strict feedback systems. The BC method was studied for the design of adaptive controller in the early 1990s [1]. Furthermore, the global or local tuning and the tracking performance of the system can be guaranteed by using the BC method when the controlled object belongs to a class of systems that can be converted into strict feedback form. Therefore, the BC method can construct the feedback control law and the Lyapunov function systematically, and some conclusions on the flight control of the UAV systems were obtained based on the BC method [2–6]. In addition, the design idea of SMC was to use high-frequency switch control signal to drive the controlled signal to arrive and stay near SM surface in the limited time [7]. It is well known that the SMC method has been widely used in the design of flight control schemes and some meaningful conclusions have been obtained [8–16]. According to the above descriptions, the discrete-time flight control schemes are introduced for the attitude dynamics model of the UAV system in this chapter.

3.2 Discrete-Time BC for the Fixed-Wing UAV System Without Wind Disturbances

In this section, the discrete-time BC without considering wind disturbances and system uncertainties is studied. According to (2.41), the attitude dynamics model of UAV without wind disturbances and system uncertainties is described by

$$\begin{cases} \dot{x}_1 = F_1(x_1) + G_1(x_1)x_2 \\ \dot{x}_2 = F_2(x) + G_2(x)u \\ y = x_1 \end{cases} \tag{3.1}$$

To design a discrete-time BC method using a discrete-time model of Euler approximation [17], the continuous time model of the UAV (3.1) is converted to

S. Shao et al., *Robust Discrete-Time Flight Control of UAV with External Disturbances*, Studies in Systems, Decision and Control 317,
https://doi.org/10.1007/978-3-030-57957-9_3

an approximated discrete-time model. The discrete-time form is given by

$$\begin{cases} x_1(k+1) = x_1(k) + \Delta T(F_1(x_1(k)) + G_1(x_1(k))x_2(k)) \\ x_2(k+1) = x_2(k) + \Delta T(F_2(x(k)) + G_2(x(k))u(k)) \\ y(k) = x_1(k) \end{cases} \tag{3.2}$$

where ΔT is the sampling period, and k denotes the kth signal. Furthermore, for the convenience of control method analysis, the following forms are given as $\Delta T F_1(x_1(k)) + x_1(k) = F_1(k)$, $\Delta T G_1(x_1(k)) = G_1(k)$, $\Delta T F_2(x(k)) + x_2(k) = F_2(k)$ and $\Delta T G_2(x(k)) = G_2(k)$ are defined.

In this section, the discrete-time BC is analyzed for the attitude dynamics model of UAV without wind disturbances and system uncertainties (3.1). The aim is to design a controller such that: (1) the bounded and known signal $x_d(k) = [\beta_d, \alpha_d, \mu_d]^{\mathrm{T}}$ can be tracked by the output signal $y(k)$ asymptotically; (2) the asymptotic convergence of all the signals is ensured in the closed-loop system. To facilitate the design of controller, the following assumption is introduced:

Assumption 3.1 According to the physical properties of the UAV system (3.1), the control coefficient matrix $\bar{G}_1(k)$ is invertible, and $\bar{G}_1(k)$ satisfies that $\underline{g}_1 \leq \|\bar{G}_1(k)\| \leq \bar{g}_1$, where $\underline{g}_1 > 0$ and $\bar{g}_1 > 0$ are positive constants.

3.2.1 Discrete-Time Controller Design

A discrete-time controller is designed by using the BC technology in this subsection, and the designed discrete-time controller can ensure that the tracking error signals converge asymptotically. The following is the detailed process of designing discrete-time controller by using the BC technology.

Step 1: Defining error variables as $e_1(k) = x_1(k) - x_d(k)$ and $e_2(k) = x_2(k) - x_{vd}(k)$, where $x_{vd}(k)$ denotes the virtual control law, and based on (3.2), one has

$$\begin{aligned} e_1(k+1) &= x_1(k+1) - x_d(k+1) \\ &= F_1(k) + G_1(k)(e_2(k) + x_{vd}(k)) - x_d(k+1) \end{aligned} \tag{3.3}$$

The virtual controller $x_{vd}(k)$ is chosen as

$$x_{vd}(k) = -K_0 {G_1}^{-1}(k)e_1(k) - {G_1}^{-1}(k)F_1(k) + {G_1}^{-1}(k)x_d(k+1) \tag{3.4}$$

where K_0 is the designed constant.

Combining (3.3) and (3.4), one can obtain

$$e_1(k+1) = G_1(k)e_2(k) - K_0 e_1(k) \tag{3.5}$$

To prove the effectiveness of the virtual law $x_{vd}(k)$ (3.4), the following Lyapunov function is chosen as

$$V_1(k) = \frac{1}{K_1} e_1^{\mathrm{T}}(k) e_1(k) \tag{3.6}$$

where K_1 is a positive constant.

On the basis of (3.6), the first difference of the Lyapunov function $V_1(k)$ can be written as

$$\begin{aligned} \varDelta V_1(k) &= V_1(k+1) - V_1(k) \\ &= \frac{1}{K_1} e_1^{\mathrm{T}}(k+1) e_1(k+1) - \frac{1}{K_1} e_1^{\mathrm{T}}(k) e_1(k) \end{aligned} \tag{3.7}$$

According to (3.5) and (3.7), one has

$$\begin{aligned} K_1 \varDelta V_1(k) = {} & e_2^{\mathrm{T}}(k) G_1^{\mathrm{T}}(k) G_1(k) e_2(k) - 2 e_2^{\mathrm{T}}(k) G_1^{\mathrm{T}}(k) K_0 e_1(k) \\ & + K_0 e_1^{\mathrm{T}}(k) K_0 e_1(k) - e_1^{\mathrm{T}}(k) e_1(k) \end{aligned} \tag{3.8}$$

Step 2: According to (3.2) and $e_2(k) = x_2(k) - x_{vd}(k)$, one has

$$e_2(k+1) = F_2(k) + G_2(k) u(k) - x_{vd}(k+1) \tag{3.9}$$

From (3.2) and (3.4), it yields

$$\begin{aligned} x_{vd}(k+1) = {} & -\frac{1}{\varDelta T} {G_1}^{-1}(x_1(k+1))(\varDelta T F_1(x_1(k+1)) + x_1(k+1)) \\ & + \frac{1}{\varDelta T} {G_1}^{-1}(x_1(k+1)) x_d(k+2) \end{aligned} \tag{3.10}$$

On the basis of (3.1) and (3.2), the state variable $x_1(k+1)$, the control gain $G_1(x_1(k+1))$ and the nonlinear function $F_1(x_1(k+1))$ are known. Thus, $x_{vd}(k+1)$ can be used to the design of the discrete-time controller.

Then, the discrete-time controller is designed by

$$u(k) = -{G_2}^{-1}(k) F_2(k) + {G_2}^{-1}(k) x_{vd}(k+1) - K_2 {G_2}^{-1}(k) e_2(k) \tag{3.11}$$

Substituting (3.11) into (3.9), one has

$$e_2(k+1) = -K_2 e_2(k) \tag{3.12}$$

To prove the effectiveness of the controller $u(k)$ (3.11), the following Lyapunov function is chosen as

$$V_2(k) = \frac{1}{K_3} e_2^{\mathrm{T}}(k) e_2(k) \tag{3.13}$$

where K_3 is a positive constant.

Based on (3.13), the first difference of the Lyapunov function $V_2(k)$ can be written as

$$\begin{aligned} \Delta V_2(k) &= V_2(k+1) - V_2(k) \\ &= \frac{1}{K_3} e_2^{\mathrm{T}}(k+1) e_2(k+1) - \frac{1}{K_3} e_2^{\mathrm{T}}(k) e_2(k) \end{aligned} \tag{3.14}$$

According to (3.12) and (3.14), one has

$$\Delta V_2(k) = \frac{K_2^2}{K_3} e_2^{\mathrm{T}}(k) e_2(k) - \frac{1}{K_3} e_2^{\mathrm{T}}(k) e_2(k) \tag{3.15}$$

For the attitude dynamics model of UAV without wind disturbances and system uncertainties (3.1), the above designed discrete-time control scheme based on the BC technology can be summarized as follows:

Theorem 3.1 *For the discrete-time attitude dynamic model of UAV (3.2), the virtual control law is chosen as (3.4) and the discrete-time controller is deigned as (3.11), the tracking error $e_1(k)$ can be guaranteed to converge asymptotically, and all signals in the closed-loop system can also be ensured to converge asymptotically.*

Proof For the entire closed-loop system, the following Lyapunov function is selected as

$$V(k) = \frac{1}{K_1} e_1^{\mathrm{T}}(k) e_1(k) + \frac{1}{K_3} e_s^{\mathrm{T}}(k) e_2(k) \tag{3.16}$$

According to (3.8) and (3.15), one has

$$\begin{aligned} \Delta V(k) &= \frac{1}{K_1} e_1^{\mathrm{T}}(k+1) e_1(k+1) + \frac{1}{K_3} e_2^{\mathrm{T}}(k+1) e_2(k+1) \\ &\quad - \frac{1}{K_1} e_1^{\mathrm{T}}(k) e_1(k) - \frac{1}{K_3} e_2^{\mathrm{T}}(k) e_2(k) \\ &\leq 2 \frac{\bar{g}_1^2}{K_1} e_2^{\mathrm{T}}(k) e_2(k) + 2 \frac{K_0^2}{K_1} e_1^{\mathrm{T}}(k) e_1(k) \\ &\quad - \frac{1}{K_1} e_1^{\mathrm{T}}(k) e_1(k) + \frac{K_2^2}{K_3} e_2^{\mathrm{T}}(k) e_2(k) - \frac{1}{K_3} e_2^{\mathrm{T}}(k) e_2(k) \\ &= -\bar{K}_1 e_1^{\mathrm{T}}(k) e_1(k) - \bar{K}_2 e_2^{\mathrm{T}}(k) e_2(k) \end{aligned} \tag{3.17}$$

where $\|\cdot\|$ denotes the 2-norm, $\bar{K}_1 = \frac{1}{K_1} - 2\frac{K_0^2}{K_1}$ and $\bar{K}_2 = \frac{1}{K_3} - 2\frac{\bar{g}_1^2}{K_1} - \frac{K_2^2}{K_3}$. On the basis of (3.17), if the appropriate control parameters can be selected to ensure that $\bar{K}_1 > 0$ and $\bar{K}_2 > 0$, the tracking error $e_1(k)$ is asymptotically convergent. ◇

3.2.2 *Simulation Study of the BC Method*

In this section, to illustrate the performance of the designed BC-based discrete-time flight control scheme, the simulations of the UAV system without wind disturbances (3.1) are investigated.

In the simulation of the UAV attitude dynamics system (3.1), the discrete-time sampling period ΔT is chosen as $\Delta T = 0.01$. The initial attitude values of the UAV system are set as $\beta_0 = 0.2\text{deg}, \alpha_0 = 0\text{deg}, \mu_0 = 0\text{deg}$ and $p_0 = q_0 = r_0 = 0\text{deg/sec}$. The desired attitude angle signal x_d of the UAV system is assumed as $\beta_d = 0\text{deg}$, $\alpha_d = 2\text{erf}(0.4t)\text{deg}$ and $\mu_d = 3\text{erf}(0.4t)\text{deg}$, where $\text{erf}(\cdot)\text{deg}$ denotes the Gaussian error function. The control parameters are set as $K_0 = -0.6$ and $K_2 = 0.2$, and the tracking errors are defined as $e_\beta(k) = \beta(k) - \beta_d(k)$, $e_\alpha(k) = \alpha(k) - \alpha_d(k)$ and $e_\mu(k) = \mu(k) - \mu_d(k)$.

According to the above designed control parameters, the numerical simulation results are shown in the Figs. 3.1, 3.2 and 3.3. The tracking control results are presented in Fig. 3.1. As can be seen from Fig. 3.1, the output signal $y(k)$ can quickly track the reference signal $x_d(k)$. The tracking errors $e_\beta(k)$, $e_\alpha(k)$ and $e_\mu(k)$ are shown in Fig. 3.2, and the tracking errors can quickly converge. Moreover, the attitude angular velocities $p(k)$, $q(k)$ and $r(k)$ are convergent based on the simulation results in Fig. 3.3. Therefore, according to the above numerical simulation results, the discrete-time BC-based control method can obtain the effective control performance of the UAV attitude dynamic system.

3.3 Discrete-Time BC-Based SMC for the Fixed-Wing UAV System with Time-Varying Disturbance

In this section, the discrete-time BC with considering the time-varying disturbance is studied. According to (2.41), the attitude dynamics model of UAV with time-varying disturbance is described by

$$\begin{cases} \dot{x}_1 = F_1(x_1) + G_1(x_1)x_2 \\ \dot{x}_2 = F_2(x) + G_2(x)u + d \\ y = x_1 \end{cases} \tag{3.18}$$

where d is the time-varying disturbance.

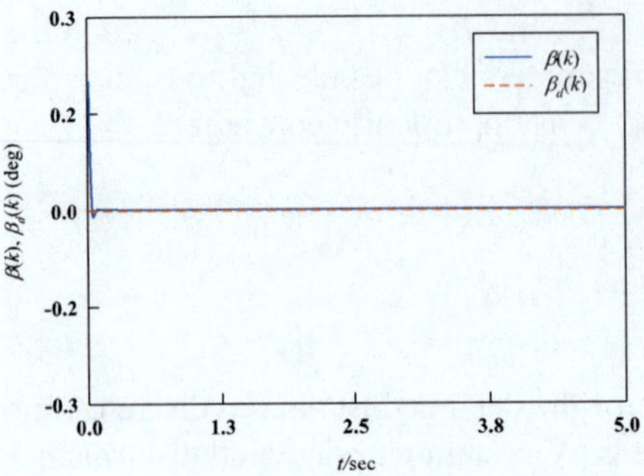

(a) $\beta(k)$ and $\beta_d(k)$

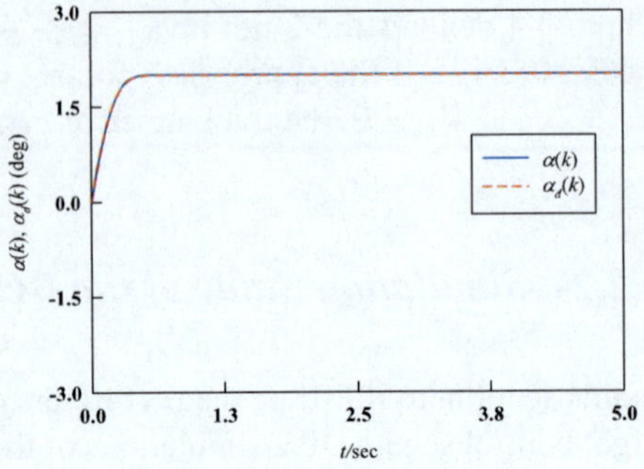

(b) $\alpha(k)$ and $\alpha_d(k)$

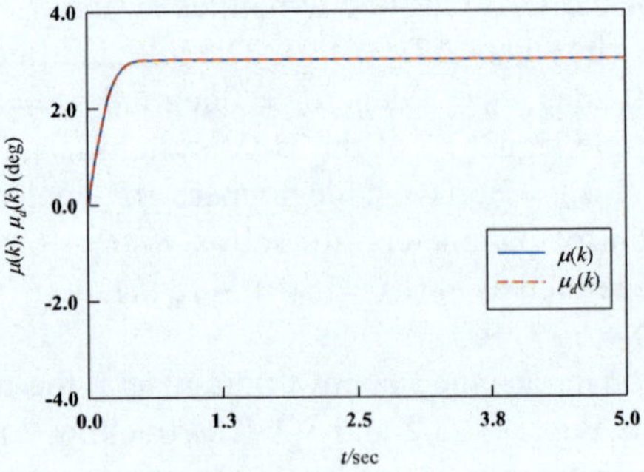

(c) $\mu(k)$ and $\mu_d(k)$

Fig. 3.1 Attitude angle tracking response

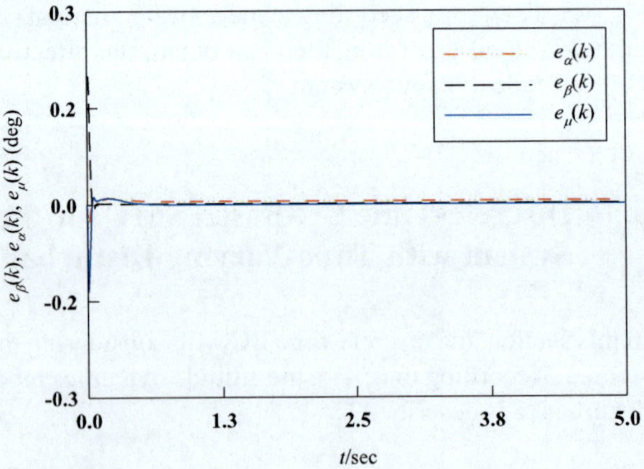

Fig. 3.2 Attitude angle tracking errors $e_\beta(k)$, $e_\alpha(k)$ and $e_\mu(k)$

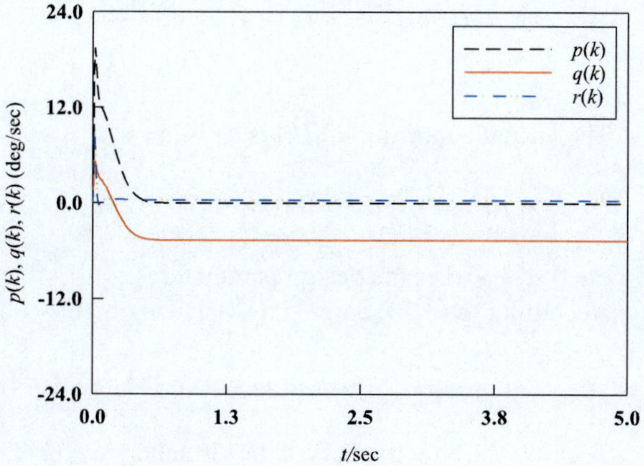

Fig. 3.3 Response of attitude angular velocities $p(k)$, $q(k)$ and $r(k)$

To design a discrete-time backstepping control method using a discrete-time model of Euler approximation [17], the continuous time model of the UAV (3.18) is converted to an approximated discrete-time model. The discrete-time form is described as

$$\begin{cases} x_1(k+1) = x_1(k) + \Delta T(F_1(x_1(k)) + G_1(x_1(k))x_2(k)) \\ x_2(k+1) = x_2(k) + \Delta T(F_2(x(k)) + G_2(x(k))u(k)) + d(k) \\ y(k) = x_1(k) \end{cases} \tag{3.19}$$

where $d(k) = \Delta T d$.

In this section, the discrete-time BC-based SMC is analyzed for the attitude dynamics model of UAV system with time-varying disturbance (3.18). The aim is to design a discrete-time SM controller such that: (1) the tracking error between the bounded and known signal $x_d(k) = [\beta_d, \alpha_d, \mu_d]^{\mathrm{T}}$ and the output signal $y(k)$ are bounded; (2) all the signals in the closed-loop system are also bounded. To facilitate the design of controller, the following assumption is introduced:

Assumption 3.2 Assume that the time-varying disturbance is bounded, that $d(k)$ and $d(k+1)$ are bounded disturbances.

3.3.1 Discrete-Time SM Controller Design

A discrete-time SM controller is designed by using the BC technology in this subsection, and the designed SM discrete-time controller can ensure that the tracking error signals are bounded. The following is the detailed process of designing discrete-time SM controller by using the BC technology.

Step 1: Defining error variables as $e_1(k) = x_1(k) - x_d(k)$ and $e_2(k) = x_2(k) - x_{vd}(k)$, where $x_{vd}(k)$ denotes the virtual control law, and based on (3.19), one has

$$\begin{aligned} e_1(k+1) &= x_1(k+1) - x_d(k+1) \\ &= F_1(k) + G_1(k)(e_2(k) + x_{vd}(k)) - x_d(k+1) \end{aligned} \tag{3.20}$$

The virtual controller $x_{vd}(k)$ is chosen as

$$x_{vd}(k) = qG_1{}^{-1}(k)e_1(k) - G_1{}^{-1}(k)F_1(k) + G_1{}^{-1}(k)x_d(k+1) \tag{3.21}$$

where $0 < q < 1$ is the design parameter.

According to (3.20) and (3.21), one can obtain

$$e_1(k+1) = G_1(k)e_2(k) + qe_1(k) \tag{3.22}$$

To prove the effectiveness of the virtual law $x_{vd}(k)$ (3.21), the following Lyapunov function is chosen as

$$V_1(k) = \frac{1}{q_1}e_1^{\mathrm{T}}(k)e_1(k) \tag{3.23}$$

where q_1 is a positive constant.

On the basis of (3.23), the first difference of the Lyapunov function $V_1(k)$ can be written as

$$\begin{aligned} \Delta V_1(k) &= V_1(k+1) - V_1(k) \\ &= \frac{1}{q_1}e_1^{\mathrm{T}}(k+1)e_1(k+1) - \frac{1}{q_1}e_1^{\mathrm{T}}(k)e_1(k) \end{aligned} \tag{3.24}$$

According to (3.22) and (3.24), one has

$$\Delta V_1(k) \le \frac{2\bar{g}_1^2}{q_1}e_2^{\mathrm{T}}(k)e_2(k) + \frac{2q^2}{q_1}e_1^{\mathrm{T}}(k)e_1(k) - \frac{1}{q_1}e_1^{\mathrm{T}}(k)e_1(k) \tag{3.25}$$

Step 2: According to (3.19) and $e_2(k) = x_2(k) - x_{vd}(k)$, one has

$$e_2(k+1) = F_2(k) + G_2(k)u(k) + d(k) - x_{vd}(k+1) \tag{3.26}$$

From (3.19) and (3.21), it yields

$$\begin{aligned} x_{vd}(k+1) = &-\frac{1}{\Delta T}G_1{}^{-1}(x_1(k+1))(\Delta T F_1(x_1(k+1)) + x_1(k+1)) \\ &+\frac{1}{\Delta T}G_1{}^{-1}(x_1(k+1))x_d(k+2) \end{aligned} \tag{3.27}$$

On the basis of (3.18) and (3.19), the state variable $x_1(k+1)$, the control gain $G_1(x_1(k+1))$ and the nonlinear function $F_1(x_1(k+1))$ are known. Thus, $x_{vd}(k+1)$ can be used to the design of the discrete-time controller.

Then, the following SM surface $s_2(k)$ is defined as

$$s_2(k) = ce_1(k) + e_2(k) \tag{3.28}$$

where $c > 0$ is a design constant.

On the basis of (3.28), if the SM surface $s_2(k)$ is also bounded then the error signal $e_2(k)$ is also bounded. On the basis of (3.20), (3.26) and (3.28), one has

$$\begin{aligned} s_2(k+1) &= ce_1(k+1) + e_2(k+1) \\ &= c(F_1(k) + G_1(k)(e_2(k) + x_{vd}(k)) - x_d(k+1)) \\ &\quad + F_2(k) + G_2(k)u(k) + d(k) - x_{vd}(k+1) \end{aligned} \tag{3.29}$$

Then, the discrete-time controller is designed by

$$\begin{aligned} u(k) = &-G_2{}^{-1}(c(F_1(k) + G_1(k)(e_2(k) + x_{vd}(k)) - x_d(k+1))) \\ &-G_2{}^{-1}(k)F_2(k) + G_2{}^{-1}(k)x_{vd}(k+1) \\ &+q_2 G_2{}^{-1}(k)s_2(k) - \eta_1 G_2{}^{-1}(k)\text{sign}(s_2(k)) \end{aligned} \tag{3.30}$$

where $0 < q_2 < 1$ and $\eta_1 > 0$ is the design parameters.

According to (3.29) and (3.30), one has

$$s_2(k+1) = q_2 s_2(k) - \eta_1 \text{sign}(s_2(k)) + d(k) \tag{3.31}$$

To prove the effectiveness of the controller $u(k)$ (3.30), the following Lyapunov function is chosen as

$$V_2(k) = \frac{1}{q_3} s_2^{\mathrm{T}}(k)s_2(k) \tag{3.32}$$

where q_3 is a positive constant.

Based on (3.32), the first difference of the Lyapunov function $V_2(k)$ can be written as

$$\begin{aligned} \Delta V_2(k) &= V_2(k+1) - V_2(k) \\ &= \frac{1}{q_3} s_2^{\mathrm{T}}(k+1)s_2(k+1) - \frac{1}{q_3} s_2^{\mathrm{T}}(k)s_2(k) \end{aligned} \tag{3.33}$$

According to (3.31) and (3.33), one has

$$\Delta V_2(k) = \frac{3q_2^2}{q_3} s_2^{\mathrm{T}}(k)s_2(k) + \frac{3\eta_1^2}{q_3} + \frac{3\bar{d}^2}{q_3} - \frac{1}{q_3} s_2^{\mathrm{T}}(k)s_2(k) \tag{3.34}$$

where $\|d(k)\| \leq \bar{d}$, and $\bar{d}$ is a positive constant.

On the basis of the attitude dynamics model of UAV with wind disturbances (3.18), the above designed discrete-time SMC scheme based on the BC technology can be summarized as follows:

Theorem 3.2 *For the discrete-time attitude dynamic model of UAV (3.19), the SM surface $s_2(k)$ is chosen as (3.28), the virtual control law is designed as (3.27) and the discrete-time SM controller is chosen as (3.30), the tracking error $e_1(k)$ and the SM surfaces $s_1(k)$ are bounded, and all signals in the closed-loop system are also bounded.*

Proof For the entire closed-loop system, the following Lyapunov function is selected as

$$V(k) = \frac{1}{q_1} e_1^{\mathrm{T}}(k) e_1(k) + \frac{1}{q_3} s_2^{\mathrm{T}}(k) s_2(k) \tag{3.35}$$

According to (3.25), (3.34) and (3.35), one has

$$\begin{aligned}
\Delta V(k) &= \frac{1}{q_1} e_1^{\mathrm{T}}(k+1) e_1(k+1) + \frac{1}{q_3} s_2^{\mathrm{T}}(k+1) s_2(k+1) \\
&\quad - \frac{1}{q_1} e_1^{\mathrm{T}}(k) e_1(k) - \frac{1}{q_3} s_2^{\mathrm{T}}(k) s_2(k) \\
&\leq \frac{2\bar{g}_1^2}{q_1} e_2^{\mathrm{T}}(k) e_2(k) + \frac{2q^2}{q_1} e_1^{\mathrm{T}}(k) e_1(k) - \frac{1}{q_1} e_1^{\mathrm{T}}(k) e_1(k) \\
&\quad + \frac{3q_2^2}{q_3} s_2^{\mathrm{T}}(k) s_2(k) + \frac{3\eta_1^2}{q_3} - \frac{1}{q_3} s_2^{\mathrm{T}}(k) s_2(k) + \frac{3\bar{d}^2}{q_3} \\
&= -\bar{q}_1 s_1^{\mathrm{T}}(k) s_1(k) - \bar{q}_2 s_2^{\mathrm{T}}(k) s_2(k) + \bar{q}_3
\end{aligned} \tag{3.36}$$

where $\bar{q}_1 = \frac{1-2q^2-4c^2\bar{g}_1^2}{q_1}$, $\bar{q}_2 = \frac{1-3q_2^2}{q_3} - \frac{4\bar{g}_1^2}{q_1}$ and $\bar{q}_3 = \frac{2\eta_1^2}{q_3} + \frac{3\bar{d}^2}{q_3}$. On the basis of (3.36), if the appropriate control parameters can be selected to ensure that $\bar{q}_1 > 0$ and $\bar{q}_2 > 0$, the tracking error $e_1(k)$ and the SM surface $s_2(k)$ are bounded. ◇

3.3.2 Simulation Study of the Discrete-Time SMC Method

In this section, to illustrate the performance of the designed SMC-based discrete-time flight control scheme, the simulations of the UAV system with wind disturbances (3.18) are investigated.

In the simulation of the UAV attitude dynamics system (3.19), the discrete-time sampling period ΔT is chosen as $\Delta T = 0.01$ and the disturbance $d = [0.3\sin(0.2t) - 0.4\cos(0.3t), 0.4\sin(0.1t) - 0.4\cos(0.4t), 0.4\sin(0.5t) - 0.6\cos(0.2t)]^{\mathrm{T}}$. The initial attitude values of the UAV system are set as β_0=0.2deg, $\alpha_0 =$ 0deg, $\mu_0 =$ 0deg and $p_0 = q_0 = r_0 =$ 0deg/sec. The desired attitude angle signal x_d of the

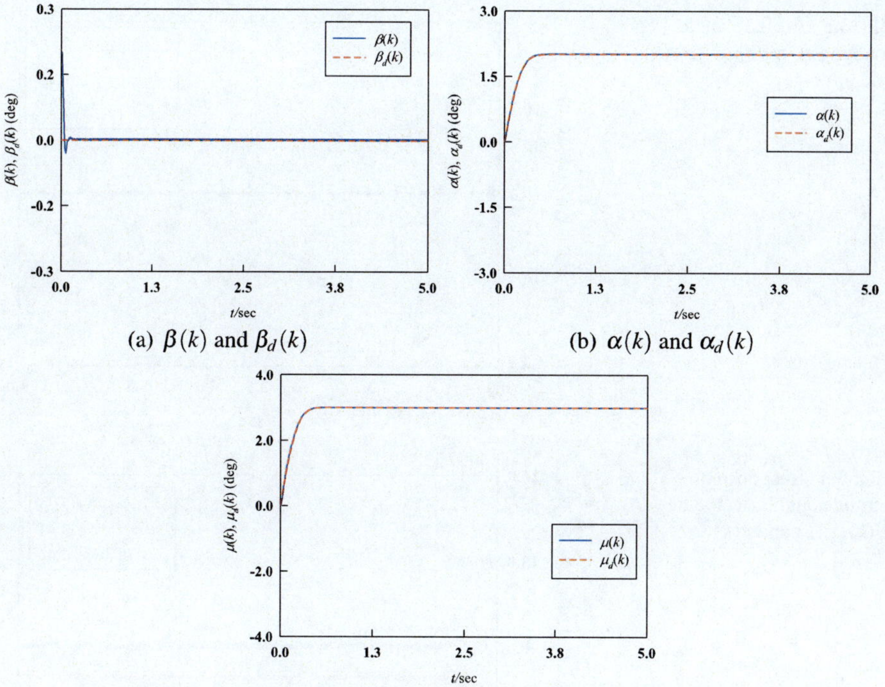

(a) $\beta(k)$ and $\beta_d(k)$

(b) $\alpha(k)$ and $\alpha_d(k)$

(c) $\mu(k)$ and $\mu_d(k)$

Fig. 3.4 Attitude angle tracking response

UAV system is assumed as $\beta_d = 0$deg, $\alpha_d = 2\text{erf}(0.4t)$deg and $\mu_d = 3\text{erf}(0.4t)$deg. The control parameters are set as $q = 0.5$, $c = 0.1$, $q_2 = 0.4$ and $\eta_1 = 0.01$, and the tracking errors are defined as $e_\beta(k) = \beta(k) - \beta_d(k)$, $e_\alpha(k) = \alpha(k) - \alpha_d(k)$ and $e_\mu(k) = \mu(k) - \mu_d(k)$.

According to the above designed control parameters, the numerical simulation results are shown in the Figs. 3.4 and 3.6. The tracking control results are presented in Fig. 3.4. On the simulation results in Fig. 3.4, although the output signal $y(k)$ can track the reference signal $x_d(k)$, the output signal $y(k)$ is oscillating due to the SM term. The tracking errors $e_\beta(k)$, $e_\alpha(k)$ and $e_\mu(k)$ are shown in Fig. 3.5, and the tracking errors are also oscillating. Furthermore, the attitude angular velocities $p(k)$, $q(k)$ and $r(k)$ converge in a bounded range based on the simulation results in Fig. 3.6. Therefore, according to the above numerical simulation results, the discrete-time BC-based control method can ensure that the tracking errors are ultimately bounded, but the output signals are oscillatory and need to be addressed further.

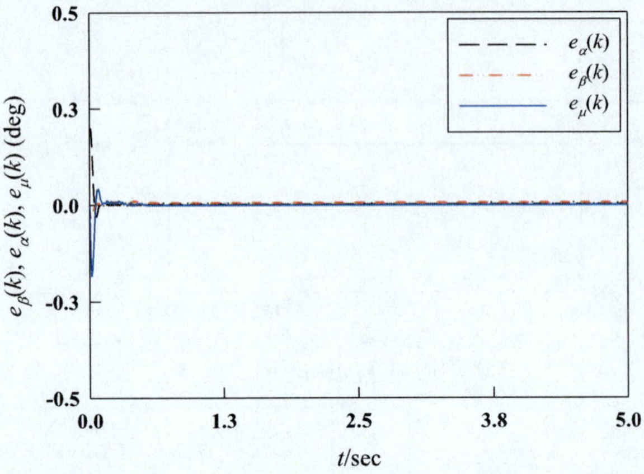

Fig. 3.5 Attitude angle tracking errors $e_\beta(k)$, $e_\alpha(k)$ and $e_\mu(k)$

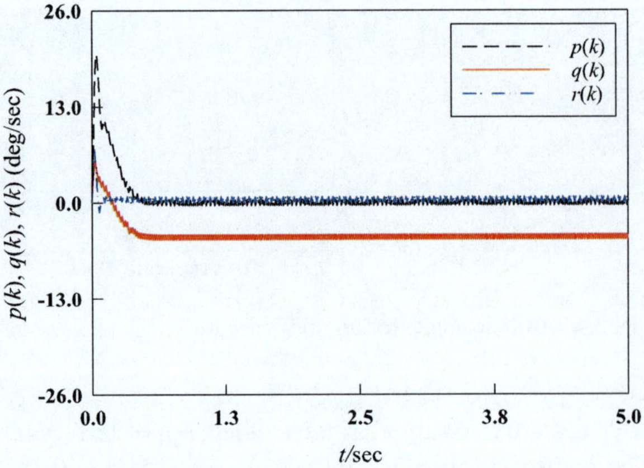

Fig. 3.6 Response of attitude angular velocities $p(k)$, $q(k)$ and $r(k)$

3.4 Conclusions

The discrete-time BC and the discrete-time SMC have been introduced for the fixed-wing UAV system. According to the BC technology, the discrete-time controllers have been designed, and the stability theory of Lyapunov in the form of discrete-time has been used to prove that the discrete-time controller designed can ensure the boundedness of the closed-loop system signal. Finally, the effectiveness of the discrete-time control schemes has been illustrated by using the simulation results.

References

1. Zhou, J., Wen, C.: Adaptive Backstepping Control of Uncertain Systems: Nonsmooth Nonlinearities, Interactions or Time-variations. Springer, Berlin (2008)
2. Choi, I.-H., Bang, H.-C.: Adaptive command filtered backstepping tracking controller design for quadrotor unmanned aerial vehicle. Proc. Inst. Mech. Eng. Part G: J. Aerosp. Eng. **226**(5), 483–497 (2012)
3. Lungu, M., Lungu, R.: Adaptive backstepping flight control for a mini-UAV. Int. J. Adapt. Control Signal Process. **27**(8), 635–650 (2013)
4. Gong, X., Hou, Z.-C., Zhao, C.-J., et al.: Adaptive backstepping sliding mode trajectory tracking control for a quad-rotor. Int. J. Autom. Comput. **9**(5), 555–560 (2012)
5. Xian, B., Guo, J., Zhang, Y.: Adaptive backstepping tracking control of a 6-DOF unmanned helicopter. IEEE/CAA J. Autom. Sin. **2**(1), 19–24 (2015)
6. Choi, Y.H., Yoo, S.J.: A simple fuzzy-approximation-based adaptive control of uncertain unmanned helicopters. Int. J. Control, Autom. Syst. **14**(1), 340–349 (2016)
7. Chen, F., Jiang, R., Zhang, K., et al.: Robust backstepping sliding-mode control and observer-based fault estimation for a quadrotor UAV. IEEE Trans. Ind. Electron. **63**(8), 5044–5056 (2016)
8. Ma, D., Xia, Y., Shen, G., et al.: Flatness-based adaptive sliding mode tracking control for a quadrotor with disturbances. J. Frankl. Inst. **355**(14), 6300–6322 (2018)
9. Zheng, Z., Jin, Z., Sun, L., et al.: Adaptive sliding mode relative motion control for autonomous carrier landing of fixed-wing unmanned aerial vehicles. IEEE Access **5**, 5556–5565 (2017)
10. Espinoza, T., Dzul, A., Lozano, R., et al.: Backstepping-sliding mode controllers applied to a fixed-wing UAV. J. Intell. Robot. Syst. **73**(1–4), 67–79 (2014)
11. Oh, H., Kim, S., Tsourdos, A., et al.: Decentralised standoff tracking of moving targets using adaptive sliding mode control for UAVs. J. Intell. Robot. Syst. **76**(1), 169–183 (2014)
12. Mu, B., Zhang, K., Shi, Y.: Integral sliding mode flight controller design for a quadrotor and the application in a heterogeneous multi-agent system. IEEE Trans. Ind. Electron. **64**(12), 9389–9398 (2017)
13. Castañeda, H., Salas-Peña, O.S., de León-Morales, J.: Extended observer based on adaptive second order sliding mode control for a fixed wing UAV. ISA Trans. **66**, 226–232 (2017)
14. Xiong, J.-J., Zhang, G.-B.: Global fast dynamic terminal sliding mode control for a quadrotor UAV. ISA Trans. **66**, 233–240 (2017)
15. Wu, K., Cai, Z., Zhao, J., et al.: Target tracking based on a nonsingular fast terminal sliding mode guidance law by fixed-wing UAV. Appl. Sci. **7**(4), 333–1–18 (2017)
16. Hua, C.-C., Wang, K., Chen, J.-N., et al.: Tracking differentiator and extended state observer-based nonsingular fast terminal sliding mode attitude control for a quadrotor. Nonlinear Dyn. **94**(1), 343–354 (2018)
17. Mareels, I.M., Penfold, H., Evans, R.J.: Controlling nonlinear time-varying systems via euler approximations. Automatica **28**(4), 681–696 (1992)

Chapter 4
Discrete-Time Adaptive NN Tracking Control of an Uncertain UAV System Based on DTDO

4.1 Introduction

In the process of modeling the fixed wing UAV system, it is often difficult to obtain accurate aerodynamic parameters and body structure model, so there will be modeling errors between the accurate model and the constructed model, resulting in system uncertainties in the UAV system model. In addition, due to the variability of the flight environment of the fixed wing UAVs during the flight, the UAVs are often affected by unknown external disturbances (e.g. wind). Consequently, system uncertainties and external disturbances will not only reduce the control performance of the flight control systems, but also lead to the instability of the UAV systems and make the flight of UAVs unsafe. In recent years, some flight control methods based on the NN were proposed for the study of flight control under system uncertainties [1–7]. According to the above research results, some continuous time control schemes based on the NN were studied for the flight control, and the negative influence of system uncertainties on flight control performance were effectively offset. In addition, since DOs can enhance the disturbance suppression ability and robustness of the control systems, the design of DOs and flight control methods have been widely studied [8–22]. The above research works focus on the design of the continuous time DOs and the design of the flight control based on continuous time DOs. In addition, with the extensive use of digital computers and microprocessors in control applications, the study of discrete-time control methods has attracted more and more attention, although some research results on the design of discrete time flight control methods have been reported in recent years [23–29]. However, in previous studies, discrete-time flight control schemes based on DTDOs and NN considering the influence of system uncertainties and external disturbances are still few. Therefore, the discrete-time flight control methods based on DTDOs and NN for uncertain fixed-wing UAV systems with external disturbance need to be further studied. According to the above analysis, a discrete-time flight control scheme is proposed for the attitude dynamics model of UAV with wind disturbances and system uncertainties based on a designed DTDO and the NN in this chapter.

S. Shao et al., *Robust Discrete-Time Flight Control of UAV with External Disturbances*, Studies in Systems, Decision and Control 317,
https://doi.org/10.1007/978-3-030-57957-9_4

4.2 Problem Formulation

According to (2.41), the uncertain attitude dynamics model of UAV with wind disturbances and system uncertainties is described by

$$\begin{cases} \dot{x}_1 = F_1(x_1) + \Delta F_1(x_1) + G_1(x_1)x_2 + d_1 \\ \dot{x}_2 = F_2(x) + \Delta F_2(x) + G_2(x)u + d_2 \\ y = x_1 \end{cases} \tag{4.1}$$

where $F_1(x_1) \in \Re^3$ and $F_2(x) \in \Re^3$ are known nonlinear functions, $\Delta F_1(x_1) \in \Re^3$ are $\Delta F_2(x) \in \Re^3$ are unknown nonlinear functions, $\Delta F_1(x_1)$ is system uncertainty or modeling error, $\Delta F_2(x)$ is the system uncertainty caused by the wind gradient, d_1 is the disturbance caused by wind, and d_2 is unknown external disturbance.

To design a discrete-time control scheme using a discrete-time model of Euler approximation [30], the continuous time model of the UAV (4.1) with wind disturbance and system uncertainties is converted to an approximated discrete-time model subject to system uncertainties and ISED. The discrete-time nonlinear dynamics form is described as

$$\begin{cases} x_1(k+1) = \bar{F}_1(k) + \Delta\bar{F}_1(k) + \bar{G}_1(k)x_2(k) + \bar{d}_1(k) \\ x_2(k+1) = \bar{F}_2(k) + \Delta\bar{F}_2(k) + \bar{G}_2(k)u(k) + \bar{d}_2(k) \\ y(k) = x_1(k) \end{cases} \tag{4.2}$$

where $\bar{F}_1(k) = \Delta T F_1(x_1(k)) + x_1(k)$, $\bar{G}_1(k) = \Delta T G_1(x_1(k))$, $\bar{F}_2(k) = \Delta T F_2 (x(k)) + x_2(k)$, $\bar{G}_2(k) = \Delta T G_2(x(k))$, ΔT is a sampling period, $\Delta\bar{F}_1(k) = \Delta T \Delta F_1 (x_1)$ and $\Delta\bar{F}_2(k) = \Delta T \Delta F_2(x)$. Furthermore, define $\bar{d}_2(k) = [\bar{d}_{21}, \bar{d}_{22}, \bar{d}_{23}]^{\mathrm{T}}$, $\bar{u}_{Wg} (k) = u_{Wg}(k+1) - u_{Wg}(k)$, $\bar{d}_1(k) = [\bar{d}_{11}, \bar{d}_{12}, \bar{d}_{13}]^{\mathrm{T}}$, $\bar{v}_{Wg}(k) = v_{Wg}(k+1) - v_{Wg}(k)$ and $\bar{w}_{Wg}(k) = w_{Wg}(k+1) - w_{Wg}(k)$, and the following expressions are given as:

$$\begin{aligned} \bar{d}_{11} = &-\frac{1}{\bar{V}_A}[\bar{u}_{Wg}(k)(\sin\mu(k)\sin\gamma(k)\cos\chi(k) - \cos\mu(k)\sin\chi(k)) \\ &+\bar{v}_{Wg}(k)(\sin\mu(k)\sin\gamma(k)\sin\chi(k) + \cos\mu(k)\cos\chi(k)) \\ &+\bar{w}_{Wg}(k)\sin\mu(k)\cos\gamma(k)] \end{aligned}$$

$$\begin{aligned} \bar{d}_{12} = &-\frac{1}{\bar{V}_A\cos\beta(k)}[\bar{u}_{Wg}(k)(\cos\mu(k)\sin\gamma(k)\cos\chi(k) + \sin\mu(k)\sin\chi(k)) \\ &+\bar{v}_{Wg}(k)(\cos\mu(k)\sin\gamma(k)\sin\chi(k) - \sin\mu(k)\cos\chi(k)) \\ &+\bar{w}_{Wg}(k)\cos\mu(k)\cos\gamma(k)] \end{aligned}$$

$$\bar{d}_{13} = \frac{1}{\bar{V}_A}[\bar{u}_{Wg}(k)\tan\gamma(k)(\sin\mu(k)\sin\gamma(k)\cos\chi(k) - \cos\mu(k)\sin\chi(k))$$
$$+\bar{w}_{Wg}(k)\sin\mu(k)\sin\gamma(k) + \bar{v}_{Wg}(k)\tan\beta(k)(\sin\mu(k)\sin\gamma(k)\sin\chi(k)$$
$$+\cos\mu(k)\cos\chi(k)) + \bar{u}_{Wg}(k)\tan\beta(k)(\cos\mu(k)\sin\gamma(k)\cos\chi(k)$$
$$+\sin\mu(k)\sin\chi(k)) + \bar{v}_{Wg}(k)\tan\beta(k)(\cos\mu(k)\sin\gamma(k)\sin\chi(k)$$
$$-\sin\mu(k)\cos\chi(k)) + \bar{u}_{Wg}(k)\tan\beta(k)\cos\mu(k)\sin\gamma(k)]$$

In this chapter, an adaptive control method is proposed for the uncertain attitude dynamics model of UAV with wind disturbances and system uncertainties (4.2) by utilizing a designed DTDO and the tracking differentiator with discrete-time form. The control purpose is to develop a DTDO-based neural controller such that: (1) the output signal $y(k)$ tracks a known and bounded signal $x_d(k) = [\beta_d, \alpha_d, \mu_d]^{\mathrm{T}}$ to a bounded compact set; (2) the boundedness of all the signals is guaranteed in the closed-loop system. In order to facilitate the design of discrete-time controller, the following assumptions are given:

Assumption 4.1 Based on the physical properties of the UAV system (4.1), the control coefficient matrix $\bar{G}_1(k)$ is invertible, and $\bar{G}_1(k)$ satisfies that $\underline{g}_1 \leq \|\bar{G}_1(k)\| \leq \bar{g}_1$, where $\underline{g}_1 > 0$ and $\bar{g}_1 > 0$ are positive constants.

Assumption 4.2 Assume that the external wind speed and wind acceleration are bounded, that $\bar{d}_1(k)$, $\bar{d}_2(k)$, $\bar{d}_1(k+1)$ and $\bar{d}_2(k+1)$ are bounded disturbances.

Remark 4.1 According to the physical properties and controllability of the UAVs, it can be assumed that $\bar{d}_1(k)$ and $\bar{d}_2(k)$ in the uncertain UAV attitude dynamic model are bounded disturbances.

4.3 Discrete Tracking Control Based on NN and Nonlinear DTDO

Firstly, the NN-based nonlinear DTDO is designed for the attitude dynamics model of UAV with wind disturbances and system uncertainties (4.2) in this section. Secondly, a discrete-time adaptive NN controller based on the nonlinear DTDO is proposed, and the validity of the controller is proved by the discrete-time stability theory of Lyapunov.

4.3.1 Design of Nonlinear DTDO

Without loss of generality, (4.2) can be written as

$$\bar{x}(k+1) = \bar{F}(k) + \Delta\bar{F}(k) + \bar{v}(k) + \bar{d}(k) \tag{4.3}$$

where $\bar{x}(k+1) = [[x_1(k+1)]^{\mathrm{T}}, [x_2(k+1)]^{\mathrm{T}}]^{\mathrm{T}}$ is the state variable vector, $\bar{F}(k) = [[\bar{F}_1(k)]^{\mathrm{T}}, [\bar{F}_2(k)]^{\mathrm{T}}]^{\mathrm{T}}$ is the known nonlinear function vector, the unknown nonlinear function vector $\Delta\bar{F}(k) = [[\Delta\bar{F}_1(k)]^{\mathrm{T}}, [\Delta\bar{F}_2(k)]^{\mathrm{T}}]^{\mathrm{T}}$, $\bar{d}(k) = [[\bar{d}_1(k)]^{\mathrm{T}}, [\bar{d}_2(k)]^{\mathrm{T}}]^{\mathrm{T}}$ is the bounded disturbance vector, and $\bar{v}(k) = [[\bar{G}_1(k)x_2(k)]^{\mathrm{T}}, [\bar{G}_2(k)u(k)]^{\mathrm{T}}]^{\mathrm{T}}$. Furthermore, define $\bar{x}(k+1) = [\bar{x}_1(k+1), \ldots, \bar{x}_6(k+1)]^{\mathrm{T}} \in \Re^6$, $\bar{F}(k) = [\bar{f}_1(k), \ldots, \bar{f}_6(k)]^{\mathrm{T}} \in \Re^6$, $\Delta\bar{F}(k) = [\Delta\bar{f}_1(k), \ldots, \Delta\bar{f}_6(k)]^{\mathrm{T}} \in \Re^6$, $\bar{v}(k) = [\bar{v}_1(k), \ldots, \bar{v}_6(k)]^{\mathrm{T}} \in \Re^6$, $\bar{d}(k) = [\underline{d}_1(k), \ldots, \underline{d}_6(k)]^{\mathrm{T}} \in \Re^6$.

The ith variable in (4.3) can be described as

$$\bar{x}_j(k+1) = \bar{f}_j(k) + \Delta\bar{f}_j(k) + \bar{v}_j(k) + \underline{d}_j(k) \tag{4.4}$$

where $j = 1, 2, \ldots, 6$.

On the basis of Lemma 2.2 and using NN to approximate $\rho_j \Delta f_j(k)$, (4.4) can be written as

$$\bar{x}_j(k+1) = \bar{f}_j(k) + \frac{1}{\rho_j}\theta_j^{*\mathrm{T}}(k)\phi_j(Z_0(k)) + \frac{1}{\rho_j}\varepsilon_j(k) + \bar{v}_j(k) + \underline{d}_j(k) \tag{4.5}$$

where ρ_j is the designed constant, $\theta_j^*(k)$ is the optimal weight vector of the NN, $\varepsilon_j(k)$ is the minimum approximation error, $\phi_j(\cdot)$ is the basis function vector, and $Z_0(k) = [\bar{x}_1(k), \ldots, \bar{x}_j(k)]^{\mathrm{T}}$. To facilitate the design of DTDO, define $\phi_j(Z_0(k)) = \phi_j(k)$.

In addition, (4.5) can be rewritten as

$$\bar{x}_j(k+1) = \bar{f}_j(k) + \frac{1}{\rho_j}\theta_j^{*T}(k)\phi_j(k) + \bar{v}_j(k) + D_j(k) \tag{4.6}$$

where $D_j(k) = \frac{1}{\rho_j}\varepsilon_j(k) + \underline{d}_j(k)$ is the complex disturbance.

The $D_j(k)$ is unknown in (4.5), thus $D_j(k)$ cannot be directly applied to the design of discrete-time controller. In order to solve the above mentioned problem, a nonlinear DTDO based on the NN is designed to suppress the adverse effects of external disturbance.

For (4.6), the NN-based nonlinear DTDO can be designed by

$$\begin{cases} \eta_j(k+1) = \bar{f}_j(k) + \bar{v}_j(k) \\ \mathscr{H}_j(k+1) = -k_j\hat{\theta}_j^{\mathrm{T}}(k)\phi_j(k) \\ \hat{D}_j(k) = \bar{x}_j(k) - \eta_j(k) + \frac{1}{k_j\rho_j}\mathscr{H}_j(k) \end{cases} \tag{4.7}$$

where η_j and $\mathscr{H}_j$ are the state variables of nonlinear DTDO, k_j is a designed constant, $\hat{D}_j(k)$ is the output of DO, $\hat{\theta}_j(k)$ is the estimation of $\theta_j^*(k)$, and the adaptive law $\hat{\theta}_j$ will be designed in Sect. 4.3.2.

According to (4.6) and (4.7), one has

$$\bar{x}_j(k+1) - \eta_j(k+1) = \frac{1}{\rho_j}\theta_j^{*T}(k)\phi_j(k) + D_j(k) \tag{4.8}$$

Defining $\tilde{\theta}_j(k) = \hat{\theta}_j(k) - \theta_j^*(k)$ and $\tilde{D}_j(k) = D_j(k) - \hat{D}_j(k)$, $\tilde{D}_j(k+1)$ can be described as

$$\begin{aligned}\tilde{D}_j(k+1) &= D_j(k+1) - \hat{D}_j(k+1)\\ &= D_j(k+1) - (x_j(k+1) - \eta_j(k+1)) - \frac{1}{k_j\rho_j}H_j(k+1)\\ &= D_j(k+1) - \frac{1}{\rho_j}\theta_j^{*T}(k)\phi_j(k) - D_j(k) + \frac{1}{\rho_j}\hat{\theta}_j^{\mathrm{T}}(k)\phi_j(k)\\ &= \Delta D_j(k) + \frac{1}{\rho_j}\tilde{\theta}_j^{\mathrm{T}}(k)\phi_j(k)\end{aligned} \tag{4.9}$$

where $\Delta D_j(k) = D_j(k+1) - D_j(k)$. Furthermore, according to Assumption 4.2, $\Delta D_j(k)$ is bounded. $|\Delta D_j(k)| \le \bar{D}_j$, and $\bar{D}_j$ is a positive constant.

Remark 4.2 Based on the known information of the controlled object (such as system states and control inputs), the adverse effects of external interference are suppressed by designing DTDO. On the basis of the above design of nonlinear DTDO (4.7), the state variables of DO $\eta_j(k)(j = 1, \ldots, 6)$ and $\mathscr{H}_j(k)$ can be obtained by designing appropriate adaptive law $\hat{\theta}_j(k)$ and control input $\bar{v}_j(k)$. Thus, according to the known state variables $\bar{x}_j(k)$, $\eta_j(k)$ and $\mathscr{H}_j(k)$, the output signal $\hat{D}_j(k)$ of the DTDO can be obtained. Moreover, by selecting the appropriate parameters ρ_j and k_j, the boundedness of disturbance estimation errors is guaranteed.

4.3.2 *Design of NN-Based Controller and Stability Analysis*

A discrete-time controller based on the NN and the nonlinear DTDO is designed in this section, and the designed controller can ensure that all closed-loop system signals are bounded, that is, the error between the output signal $y(k)$ and the reference signal $x_d(k)$, and the disturbance estimation error $\tilde{D}_j(k)$ are bounded. The following is the detailed process of designing discrete-time controller based on the NN and the nonlinear DTDO by using the BC technology.

Step 1: Defining error variables as $s_1(k) = x_1(k) - x_d(k)$ and $s_2(k) = x_2(k) - \mathrm{sat}(x_{vd}(k))$, where sat$(\cdot)$ denotes the saturation function, $x_{vd}(k)$ denotes the designed signal, and based on (4.2), one has

$$\begin{aligned}s_1(k+1) &= x_1(k+1) - x_d(k+1)\\ &= \bar{F}_1(k) + \Delta\bar{F}_1(k) + \bar{G}_1(k)(s_2(k) + \mathrm{sat}(x_{vd}(k)))\\ &\quad + \bar{d}_1(k) - x_d(k+1)\end{aligned} \tag{4.10}$$

where the following input saturation nonlinearity $v_{di}(k) = \text{sat}(x_{vdi}(k))(i = 1, 2, 3)$ is given based on (2.55):

$$\text{sat}(x_{vdi}(k)) = \begin{cases} v_{di\max}(k), & v_i(\kappa) \geq v_{di\max}(k) \\ x_{vdi}(k), & v_{di\min}(k) < x_{vdi}(k) < v_{di\max}(k) \\ v_{di\min}(k), & x_{vdi}(k) \leq v_{di\min}(k) \end{cases} \tag{4.11}$$

where $v_d(k) = [v_{d1}(k), v_{d2}(k), v_{d3}(k)]^{\text{T}} \in \Re^3$, $x_{vdi}(k)$ is the ith element of $x_{vd}(k)$, $v_{di\max}(k) > 0$ and $v_{di\min}(k) < 0$ are upper and lower bounds of input saturation respectively, $\text{sat}(x_{vd}(k)) = [\text{sat}(x_{vd1}(k)), \text{sat}(x_{vd2}(k)), \text{sat}(x_{vd3}(k))]^{\text{T}} \in \Re^3$, and the variant Sigmoid function $\mathscr{S}(x_{vdi}(k))$ is described as follows [31]:

$$\mathscr{S}(x_{vdi}(k)) = \frac{\ell_1}{1 + e^{-\ell_2 x_{vdi}(k)}} - \ell_3 \tag{4.12}$$

where ℓ_1, ℓ_2 and ℓ_3 are designed positive constants. In addition, the input saturation can be approximated by using $\overline{\lambda}(x_{vdi}(k)) = v_{di\max}(k)(2\mathscr{S}(\ell_0 x_{vdi}(k)/v_{di\max}(k)) - 1)$ with $\ell_1 = 1$, $\ell_2 = 1$ and $\ell_3 = 0$, and ℓ_0 is a designed positive constant. Then, we define that $v_{di}(k) = \text{sat}(x_{vdi}(k)) = \mathscr{Y}(x_{vdi}(k)) + \overline{\lambda}(x_{vdi}(k))$ with $\mathscr{Y}(x_{vdi}(k)) = \text{sat}(x_{vdi}(k)) - \overline{\lambda}(x_{vdi}(k))$. According to (4.11) and (4.12), it yields that $|\mathscr{Y}(x_{vdi}(k))| = \left|\text{sat}(x_{vdi}(k)) - \overline{\lambda}(x_{vdi}(k))\right| \leq 4v_{di\max}(k) = \bar{\mathscr{Y}}_i$, and $\bar{\mathscr{Y}}_i$ is a positive constant.

Using the mean-value theorem, one has

$$\overline{\lambda}(x_{vdi}(k)) = \overline{\lambda}_{x_{vdi\eta}(k)}(k)(x_{vdi}(k) - x_{vdi0}(k)) + \overline{\lambda}(x_{vdi0}(k)) \tag{4.13}$$

where $x_{vdi0}(k)$ is the chosen parameter, $\overline{\lambda}_{x_{vdi\eta}(k)}(k)$ is evaluated at $x_{vdi}(k) = x_{vdi\eta}(k)$ with $x_{vdi\eta}(k) = \eta x_{vdi}(k) + (1 - \eta)x_{vdi0}(k)$, $0 < \eta < 1$ is a constant. Moreover, $\overline{\lambda}_{x_{vdi\eta}(k)}(k)$ satisfies that $\overline{\lambda}_{x_{vdi\eta}(k)}(k) \leq \ell_0/2$. Thus, $\overline{\lambda}_{x_{vdi\eta}(k)}(k)$ can be appropriately selected based on the system performance [31].

Define $x_{vdi0}(k) = 0$. From (4.12) and (4.13), $v_{di}(k)$ can be written as

$$v_{di}(k) = \mathscr{Y}(x_{vdi}(k)) + \overline{\lambda}_{x_{vdi\eta}(k)}(k)x_{vdi}(k) \tag{4.14}$$

From (4.14), one has

$$v_d(k) = \mathscr{Y}(k) + \overline{\lambda}(k)x_{vd}(k) \tag{4.15}$$

where $\overline{\lambda}(k) = \text{diag}[\overline{\lambda}_{x_{vd1\eta}(k)}(k), \overline{\lambda}_{x_{vd2\eta}(k)}(k), \overline{\lambda}_{x_{vd3\eta}(k)}(k)]$ with $\text{diag}[\cdot]$ being the diagonal matrix, $\mathscr{Y}(k) = [\mathscr{Y}(x_{vd1}(k)), \mathscr{Y}(x_{vd2}(k)), \mathscr{Y}(x_{vd3}(k))]^{\text{T}}$, $\left\|\overline{\lambda}(k)\right\| \leq \bar{\ell}$, and $\bar{\ell}$ is a positive constant.

Based on (4.15), (4.10) can be written as

$$\begin{aligned} s_1(k+1) = {} & \bar{F}_1(k) + \Delta\bar{F}_1(k) + \bar{G}_1(k)(s_2(k) + \mathscr{Y}(k) \\ & + \overline{\lambda}(k)x_{vd}(k)) + \bar{d}_1(k) - x_d(k+1) \end{aligned} \tag{4.16}$$

According to Lemma 2.2 and (4.15), the NN is used to approximate $\rho_{1i}\Delta\bar{F}_{1i}(k)$ $(i = 1, 2, 3)$ with $\Delta\bar{F}_{1i}(k)$ being the ith variable of $\Delta\bar{F}_1(k)$, it yields

$$s_1(k+1) = \bar{F}_1(k) + \Theta_1(k) + \bar{G}_1(k)(s_2(k) + \mathscr{Y}(k) + \bar{\lambda}(k)x_{vd}(k)) + \bar{d}_1(k) - x_d(k+1) + \mathrm{E}_1(k) \quad (4.17)$$

where $\Theta_1(k) = [\frac{1}{\rho_{11}}\theta_{11}^{*\mathrm{T}}(k)\phi_{11}(Z_1(k)), \frac{1}{\rho_{12}}\theta_{12}^{*\mathrm{T}}(k)\phi_{12}(Z_1(k)), \frac{1}{\rho_{13}}\theta_{13}^{*\mathrm{T}}(k)\phi_{13}(Z_1(k))]^{\mathrm{T}}$, $Z_1(k) = x_1(k)$, $\mathrm{E}_1(k) = [\frac{1}{\rho_{11}}\varepsilon_{11}(k), \frac{1}{\rho_{12}}\varepsilon_{12}(k), \frac{1}{\rho_{13}}\varepsilon_{13}(k)]^{\mathrm{T}}$, ρ_{1i} is the designed constant, $\theta_{1i}^{*}(k)$ is the optimal weight vector of the NN, $\varepsilon_{1i}(k)$ is the minimum approximation error, $\phi_{1i}(\cdot)$ is the basis function vector. To facilitate the design of control scheme, define $\phi_{1i}(Z_1(k)) = \phi_{1i}(k)$.

Furthermore, (4.17) can be written as

$$s_1(k+1) = \bar{F}_1(k) + \Theta_1(k) + \bar{G}_1(k)(s_2(k) + \mathscr{Y}(k) + \bar{\lambda}(k)x_{vd}(k)) + \mathscr{D}_1(k) - x_d(k+1) \quad (4.18)$$

where $\mathscr{D}_1(k) = \mathrm{E}_1(k) + \bar{d}_1(k)$ is a complex disturbance.

In order to deal with the complex disturbance $\mathscr{D}_1(k)$ in (4.18), the NN-based nonlinear DTDO is designed according to (4.7), which is given by

$$\begin{cases} \eta_{1i}(k+1) = \bar{F}_{1i}(k) + \mathscr{V}_{1i}(k) \\ \mathscr{H}_{1i}(k+1) = -k_{1i}\hat{\theta}_{1i}^{\mathrm{T}}(k)\phi_{1i}(k) \\ \hat{\mathscr{D}}_{1i}(k) = x_{1i}(k) - \eta_{1i}(k) + \frac{1}{k_{1i}\rho_{1i}}\mathscr{H}_{1i}(k) \end{cases} \quad (4.19)$$

where $i = 1, 2, 3$, $x_{1i}(k)$ is the ith variable of $x_1(k)$, $\bar{F}_{1i}(k)$ is the ith variable of $\bar{F}_1(k)$, $\hat{\mathscr{D}}_{1i}(k)$ is the output of DO, and $\hat{\mathscr{D}}_1(k) = [\hat{\mathscr{D}}_{11}(k), \hat{\mathscr{D}}_{12}(k), \hat{\mathscr{D}}_{13}(k)]^{\mathrm{T}}$. The disturbance estimation error $\tilde{\mathscr{D}}_{1i}(k) = \mathscr{D}_{1i}(k) - \hat{\mathscr{D}}_{1i}(k)$. $\mathscr{V}_{1i}(k)$ is the ith variable of $\mathscr{V}_1(k)$, and $\mathscr{V}_1(k) = \bar{G}_1(k)x_2(k)$. $\eta_{1i}(k)$ and $\mathscr{H}_{1i}(k)$ are the state variables of nonlinear DTDO, k_{1i} is a designed constant, $\hat{\theta}_{1i}(k)$ is the estimation of $\theta_{1i}^{*}(k)$.

Furthermore, (4.18) can be described as

$$\begin{aligned} s_1(k+1) = {} & \bar{F}_1(k) + \Theta_1(k) + \bar{G}_1(k)\bar{\lambda}(k)(\bar{F}_1(k) + \mathscr{B}s_1(k) + \hat{\mathscr{D}}(k) + \hat{\Theta}_1(k) \\ & -x_d(k+1)) + \bar{G}_1(k)[s_2(k) + \mathscr{Y}(k) - \bar{\lambda}(k)(\bar{F}_1(k) + \mathscr{B}s_1(k) \\ & +\hat{\mathscr{D}}(k) + \hat{\Theta}_1(k) + x_d(k+1)) - \bar{\lambda}(k)\bar{F}_1(k) + \bar{\lambda}(k)x_{vd}(k)] \\ & +\mathscr{D}_1(k) - x_d(k+1) \end{aligned} \quad (4.20)$$

where $\hat{\Theta}_1(k) = [\frac{1}{\rho_{11}}\hat{\theta}_{11}^{\mathrm{T}}(k)\phi_{11}(k), \frac{1}{\rho_{12}}\hat{\theta}_{12}^{\mathrm{T}}(k)\phi_{12}(k), \frac{1}{\rho_{13}}\hat{\theta}_{13}^{\mathrm{T}}(k)\phi_{13}(k)]^{\mathrm{T}}$, $\hat{\mathscr{D}}_1(k)$ is the estimation of $\mathscr{D}_1(k)$ and $\mathscr{B}$ is a designed diagonal matrix.

According to (4.20), the desired control signal $x_{vd}(k)$ is designed as follows:

$$x_{vd}(k) = -(\bar{\lambda}(k)\bar{G}_1(k))^{-1}(\bar{F}_1(k) + \mathscr{B}s_1(k) + \hat{\mathscr{D}}(k) + \hat{\Theta}_1(k) - x_d(k+1)) \quad (4.21)$$

Furthermore, (4.20) can be written as

$$\begin{aligned} x_{vd}(k) &= -\bar{G}_1^{-1}(k)(\bar{F}_1(k) + \mathscr{B}s_1(k) + \hat{\mathscr{D}}(k) + \hat{\Theta}_1(k) - x_d(k+1)) \\ &\quad + \bar{G}_1^{-1}(k)(\bar{F}_1(k) + \mathscr{B}s_1(k) + \hat{\mathscr{D}}(k) + \hat{\Theta}_1(k) - x_d(k+1)) \\ &\quad -(\bar{\lambda}(k)\bar{G}_1(k))^{-1}(\bar{F}_1(k) + \mathscr{B}s_1(k) + \hat{\mathscr{D}}(k) + \hat{\Theta}_1(k) - x_d(k+1)) \end{aligned} \tag{4.22}$$

Based on (4.22), $x_{vd}(k)$ cannot be applied directly to the system because the matrix $\bar{\lambda}(k)$ is unknown. Therefore, on the basis of Lemma 2.2 and (4.18), the NN is used to approximate $\rho_{0i}\Delta\bar{\mathscr{F}}_{1i}(k)(i = 1, 2, 3)$ with $\Delta\bar{\mathscr{F}}_{1i}(k)$ being the ith variable of $\Delta\bar{\mathscr{F}}_1(k) = -(\bar{\lambda}(k)\bar{G}_1(k))^{-1}(\bar{F}_1(k) + \mathscr{B}s_1(k) + \hat{\mathscr{D}}(k) + \hat{\Theta}_1(k) - x_d(k+1)) + \bar{G}_1^{-1}(k)(\bar{F}_1(k) + \mathscr{B}s_1(k) + \hat{\mathscr{D}}(k) + \hat{\Theta}_1(k) - x_d(k+1))$, it yields

$$\begin{aligned} x_{vd}(k) &= -\bar{G}_1^{-1}(k)(\bar{F}_1(k) + \mathscr{B}s_1(k) + \hat{\mathscr{D}}(k) + \hat{\Theta}_1(k) - x_d(k+1)) \\ &\quad + \Theta_0(k) + \mathrm{E}_0(k) \end{aligned} \tag{4.23}$$

where $\Theta_0(k) = [\frac{1}{\rho_{01}}\theta_{01}^{*\mathrm{T}}(k)\phi_{01}(Z_0(k)), \frac{1}{\rho_{02}}\theta_{02}^{*\mathrm{T}}(k)\phi_{02}(Z_0(k)), \frac{1}{\rho_{03}}\theta_{03}^{*\mathrm{T}}(k)\phi_{03}(Z_0(k))]^{\mathrm{T}}$, $Z_0(k) = [x_1^{\mathrm{T}}(k), s_1^{\mathrm{T}}(k), \hat{\Theta}_1^{\mathrm{T}}(k), \hat{\mathscr{D}}_1^{\mathrm{T}}(k)]^{\mathrm{T}}$, $\hat{\Theta}_1(k)$ is the estimation of $\hat{\Theta}(k)$, $\mathrm{E}_0(k) = [\frac{1}{\rho_{01}}\varepsilon_{01}(k), \frac{1}{\rho_{02}}\varepsilon_{02}(k), \frac{1}{\rho_{03}}\varepsilon_{03}(k)]^{\mathrm{T}}$, ρ_{0i} is the designed constant, $\theta_{0i}^*(k)$ is the optimal weight vector of the NN, $\varepsilon_{0i}(k)$ is the minimum approximation error, $\phi_{0i}(\cdot)$ is the basis function vector. To facilitate the design of control scheme, define $\phi_{0i}(Z_0(k)) = \phi_{0i}(k)$.

From (4.23), the control signal $x_{vd}(k)$ is designed as

$$x_{vd}(k) = -\bar{G}_1^{-1}(k)(\bar{F}_1(k) + \mathscr{B}s_1(k) + \hat{\mathscr{D}}(k) + \hat{\Theta}_1(k) - x_d(k+1)) + \hat{\Theta}_0(k) \tag{4.24}$$

where $\hat{\Theta}_0(k) = [\frac{1}{\rho_{01}}\hat{\theta}_{01}^{\mathrm{T}}(k)\phi_{01}(k), \frac{1}{\rho_{02}}\hat{\theta}_{02}^{\mathrm{T}}(k)\phi_{02}(k), \frac{1}{\rho_{03}}\hat{\theta}_{03}^{\mathrm{T}}(k)\phi_{03}(k)]^{\mathrm{T}}$, and $\hat{\Theta}_0(k)$ is the estimation of $\Theta_0(k)$.

According to the analysis above, one has

$$\begin{aligned} -\bar{\lambda}(k)(\bar{F}_1(k) + \mathscr{B}s_1(k) + \hat{\mathscr{D}}(k) + \hat{\Theta}_1(k) - x_d(k+1)) &= -\bar{\lambda}(k)\bar{G}_1(\Theta_0(k) \\ + \mathrm{E}_0(k)) - (\bar{F}_1(k) + \mathscr{B}s_1(k) + \hat{\mathscr{D}}(k) + \hat{\Theta}_1(k) - x_d(k+1)) \end{aligned} \tag{4.25}$$

Substituting (4.24) into (4.18), one can obtain

$$\begin{aligned} s_1(k+1) &= -\tilde{\Theta}_1(k) + \bar{\lambda}(k)\bar{G}_1(k)\tilde{\Theta}_0(k) - \mathscr{B}s_1(k) + \bar{G}_1(k)(s_2(k) \\ &\quad + \mathscr{Y}(k)) + \tilde{\mathscr{D}}_1(k) - \bar{\lambda}(k)\bar{G}_1(k)\mathrm{E}_0(k) \end{aligned} \tag{4.26}$$

where $\tilde{\Theta}_1(k) = [\frac{1}{\rho_{11}}\tilde{\theta}_{11}^{\mathrm{T}}(k)\phi_{11}(k), \frac{1}{\rho_{12}}\tilde{\theta}_{12}^{\mathrm{T}}(k)\phi_{12}(k), \frac{1}{\rho_{13}}\tilde{\theta}_{13}^{\mathrm{T}}(k)\phi_{13}(k)]^{\mathrm{T}}$, $\tilde{\Theta}_1(k) = \hat{\Theta}_1(k) - \Theta_1(k)$, $\tilde{\Theta}_0(k) = [\frac{1}{\rho_{01}}\tilde{\theta}_{01}^{\mathrm{T}}(k)\phi_{01}(k), \frac{1}{\rho_{02}}\tilde{\theta}_{02}^{\mathrm{T}}(k)\phi_{02}(k), \frac{1}{\rho_{03}}\tilde{\theta}_{03}^{\mathrm{T}}(k)\phi_{03}(k)]^{\mathrm{T}}$, and $\tilde{\Theta}_0(k) = \hat{\Theta}_0(k) - \Theta_0(k)$.

Furthermore, the adaptive law $\hat{\theta}_{1i}(k)$ is designed by

$$\hat{\theta}_{1i}(k+1) = \hat{\theta}_{1i}(k) + \mathscr{P}_{1i}\mathscr{Q}_{1i}\phi_{1i}(k)s_{1i}(k+1) - \mathscr{P}_{1i}\mathscr{R}_{1i}\hat{\theta}_{1i}(k) \tag{4.27}$$

where $s_{1i}(k+1)$ denotes the ith variable of $s_1(k+1)$, and $\mathscr{P}_{1i}$, $\mathscr{Q}_{1i}$ and $\mathscr{R}_{1i}$ are designed positive constants.

According to (4.27), one has

$$\tilde{\theta}_{1i}(k+1) = \tilde{\theta}_{1i}(k) + \mathscr{P}_{1i}\mathscr{Q}_{1i}\phi_{1i}(k)s_{1i}(k+1) - \mathscr{P}_{1i}\mathscr{R}_{1i}\hat{\theta}_{1i}(k) \tag{4.28}$$

where $\tilde{\theta}_{1i}(k+1) = \hat{\theta}_{1i}(k+1) - \theta^*_{1i}(k+1)$.

Moreover, the adaptive law $\hat{\theta}_{0i}(k)$ is designed by

$$\hat{\theta}_{0i}(k+1) = \hat{\theta}_{0i}(k) + \mathscr{P}_{0i}\mathscr{Q}_{1i}\phi_{0i}(k)s_{1i}(k+1) - \mathscr{P}_{0i}\mathscr{R}_{0i}\hat{\theta}_{0i}(k) \tag{4.29}$$

where $\mathscr{P}_{0i}$ and $\mathscr{R}_{0i}$ are designed positive constants.

From (4.29), one has

$$\tilde{\theta}_{0i}(k+1) = \tilde{\theta}_{0i}(k) + \mathscr{P}_{0i}\mathscr{Q}_{1i}\phi_{0i}(k)s_{1i}(k+1) - \mathscr{P}_{0i}\mathscr{R}_{0i}\hat{\theta}_{0i}(k) \tag{4.30}$$

where $\tilde{\theta}_{0i}(k+1) = \hat{\theta}_{0i}(k+1) - \theta^*_{0i}(k+1)$.

To prove the effectiveness of the nonlinear DTDO (4.19) and the virtual law $x_{vd}(k)$ (4.25), the following Lyapunov function is chosen as

$$\begin{aligned} L_1(k) = {} & \frac{1}{\mathscr{K}_1}s_1^{\mathrm{T}}(k)s_1(k) + \sum_{i=1}^{3}\frac{1}{\mathscr{P}_{1i}\rho_{1i}}\tilde{\theta}_{1i}^{\mathrm{T}}(k)\tilde{\theta}_{1i}(k) \\ & + \sum_{i=1}^{3}\frac{1}{\mathscr{P}_{0i}\rho_{0i}}\tilde{\theta}_{0i}^{\mathrm{T}}(k)\tilde{\theta}_{0i}(k) + \frac{1}{\hbar_1}\sum_{i=1}^{3}\tilde{\mathscr{D}}_{1i}^2(k) \end{aligned} \tag{4.31}$$

where $\mathscr{K}_1$ and $\hbar_1$ are positive constants.

Based on (4.31), the first difference of the Lyapunov function $L_1(k)$ can be written as

$$\begin{aligned} \Delta L_1(k) = {} & L_1(k+1) - L_1(k) \\ = {} & \frac{1}{\mathscr{K}_1}s_1^{\mathrm{T}}(k+1)s_1(k+1) - \frac{1}{\mathscr{K}_1}s_1^{\mathrm{T}}(k)s_1(k) + \sum_{i=1}^{3}\frac{1}{\mathscr{P}_{1i}\rho_{1i}}\tilde{\theta}_{1i}^{\mathrm{T}}(k+1)\tilde{\theta}_{1i}(k+1) \\ & - \frac{1}{\hbar_1}\sum_{i=1}^{3}\tilde{\mathscr{D}}_{1i}^2(k) + \frac{1}{\hbar_1}\sum_{i=1}^{3}\tilde{\mathscr{D}}_{1i}^2(k+1) - \sum_{i=1}^{3}\frac{1}{\mathscr{P}_{1i}\rho_{1i}}\tilde{\theta}_{1i}^{\mathrm{T}}(k)\tilde{\theta}_{1i}(k) \\ & + \sum_{i=1}^{3}\frac{1}{\mathscr{P}_{0i}\rho_{0i}}\tilde{\theta}_{0i}^{\mathrm{T}}(k+1)\tilde{\theta}_{0i}(k+1) - \sum_{i=1}^{3}\frac{1}{\mathscr{P}_{0i}\rho_{0i}}\tilde{\theta}_{0i}^{\mathrm{T}}(k)\tilde{\theta}_{0i}(k) \end{aligned} \tag{4.32}$$

From (4.26), one has

$$\begin{aligned} s_1^{\mathrm{T}}(k+1)s_1(k+1) = {} & s_1^{\mathrm{T}}(k+1)(\tilde{\mathscr{D}}_1(k) - \tilde{\theta}_1(k) + \bar{\lambda}(k)\bar{G}_1(k)\tilde{\theta}_0(k)) \\ & - s_1^{\mathrm{T}}(k+1)\mathscr{B}s_1(k) - s_1^{\mathrm{T}}(k+1)\bar{\lambda}(k)\bar{G}_1(k)\mathrm{E}_0(k) \\ & + s_1^{\mathrm{T}}(k+1)\bar{G}_1(k)(s_2(k) + \mathscr{Y}(k)) \end{aligned} \tag{4.33}$$

Considering (4.27)–(4.30), the following expressions can be obtained:

$$\begin{aligned} \sum_{i=1}^{3} \tfrac{1}{\mathscr{P}_{1i}\rho_{1i}} \tilde{\theta}_{1i}^{\mathrm{T}}(k+1)\tilde{\theta}_{1i}(k+1) = {} & \sum_{i=1}^{3} \tfrac{1}{\mathscr{P}_{1i}\rho_{1i}} \tilde{\theta}_{1i}^{\mathrm{T}}(k+1)\tilde{\theta}_{1i}(k) \\ + \sum_{i=1}^{3} \tfrac{\mathscr{Q}_{1i}}{\rho_{1i}} \tilde{\theta}_{1i}^{\mathrm{T}}(k+1)\phi_{1i}(k)s_{1i}(k+1) & - \sum_{i=1}^{3} \tfrac{\mathscr{R}_{1i}}{\rho_{1i}} \tilde{\theta}_{1i}^{\mathrm{T}}(k+1)\hat{\theta}_{1i}(k) \end{aligned} \tag{4.34}$$

$$\begin{aligned} \sum_{i=1}^{3} \tfrac{\mathscr{Q}_{1i}}{\rho_{1i}} \tilde{\theta}_{1i}^{\mathrm{T}}(k+1)\phi_{1i}(k)s_{1i}(k+1) = {} & \sum_{i=1}^{3} \tfrac{\mathscr{Q}_{1i}^2}{\rho_{1i}} \mathscr{P}_{1i}\phi_{1i}^{\mathrm{T}}(k)s_{1i}(k+1)\phi_{1i}(k)s_{1i}(k+1) \\ + \sum_{i=1}^{3} \tfrac{\mathscr{Q}_{1i}}{\rho_{1i}} \tilde{\theta}_{1i}^{\mathrm{T}}(k)\phi_{1i}(k)s_{1i}(k+1) & - \sum_{i=1}^{3} \tfrac{\mathscr{Q}_{1i}}{\rho_{1i}} \mathscr{P}_{1i}\mathscr{R}_{1i}\hat{\theta}_{1i}^{\mathrm{T}}(k)\phi_{1i}(k)s_{1i}(k+1) \end{aligned} \tag{4.35}$$

$$\begin{aligned} -\sum_{i=1}^{3} \tfrac{\mathscr{R}_{1i}}{\rho_{1i}} \tilde{\theta}_{1i}^{\mathrm{T}}(k+1)\hat{\theta}_{1i}(k) = {} & -\sum_{i=1}^{3} \tfrac{\mathscr{R}_{1i}}{\rho_{1i}} \tilde{\theta}_{1i}^{\mathrm{T}}(k)\hat{\theta}_{1i}(k) \\ -\sum_{i=1}^{3} \mathscr{Q}_{1i}\tfrac{\mathscr{R}_{1i}}{\rho_{1i}} \mathscr{P}_{1i}\phi_{1i}^{\mathrm{T}}(k)s_{1i}(k+1)\hat{\theta}_{1i}(k) & + \sum_{i=1}^{3} \tfrac{\mathscr{R}_{1i}^2}{\rho_{1i}} \mathscr{P}_{1i}\hat{\theta}_{1i}^{\mathrm{T}}(k)\hat{\theta}_{1i}(k) \end{aligned} \tag{4.36}$$

$$\begin{aligned} -\sum_{i=1}^{3} \tfrac{1}{\mathscr{P}_{1i}\rho_{1i}} \tilde{\theta}_{1i}^{\mathrm{T}}(k)\tilde{\theta}_{1i}(k) = {} & -\sum_{i=1}^{3} \tfrac{1}{\mathscr{P}_{1i}\rho_{1i}} \tilde{\theta}_{1i}^{\mathrm{T}}(k)\tilde{\theta}_{1i}(k+1) \\ + \sum_{i=1}^{3} \tfrac{\mathscr{Q}_{1i}}{\rho_{1i}} \tilde{\theta}_{1i}^{\mathrm{T}}(k)\phi_{1i}(k)s_{1i}(k+1) & - \sum_{i=1}^{3} \tfrac{\mathscr{R}_{1i}}{\rho_{1i}} \tilde{\theta}_{1i}^{\mathrm{T}}(k)\hat{\theta}_{1i}(k) \end{aligned} \tag{4.37}$$

$$\begin{aligned} \sum_{i=1}^{3} \tfrac{1}{\mathscr{P}_{0i}\rho_{0i}} \tilde{\theta}_{0i}^{\mathrm{T}}(k+1)\tilde{\theta}_{0i}(k+1) = {} & \sum_{i=1}^{3} \tfrac{1}{\mathscr{P}_{0i}\rho_{0i}} \tilde{\theta}_{0i}^{\mathrm{T}}(k+1)\tilde{\theta}_{0i}(k) \\ + \sum_{i=1}^{3} \tfrac{\mathscr{Q}_{1i}}{\rho_{0i}} \tilde{\theta}_{1i}^{\mathrm{T}}(k+1)\phi_{0i}(k)s_{0i}(k+1) & - \sum_{i=1}^{3} \tfrac{\mathscr{R}_{0i}}{\rho_{0i}} \tilde{\theta}_{0i}^{\mathrm{T}}(k+1)\hat{\theta}_{0i}(k) \end{aligned} \tag{4.38}$$

$$\sum_{i=1}^{3} \frac{\mathscr{Q}_{1i}}{\rho_{0i}} \tilde{\theta}_{0i}^{\mathrm{T}}(k+1)\phi_{0i}(k)s_{1i}(k+1) = \sum_{i=1}^{3} \frac{\mathscr{Q}_{1i}}{\rho_{1i}^{2}} \mathscr{P}_{0i}\phi_{0i}^{\mathrm{T}}(k)s_{1i}(k+1)\phi_{0i}(k)s_{1i}(k+1)$$
$$+\sum_{i=1}^{3} \frac{\mathscr{Q}_{1i}}{\rho_{0i}} \tilde{\theta}_{0i}^{\mathrm{T}}(k)\phi_{0i}(k)s_{1i}(k+1) - \sum_{i=1}^{3} \frac{\mathscr{Q}_{1i}}{\rho_{0i}} \mathscr{P}_{0i}\mathscr{R}_{1i}\hat{\theta}_{0i}^{\mathrm{T}}(k)\phi_{0i}(k)s_{1i}(k+1) \tag{4.39}$$

$$-\sum_{i=1}^{3} \frac{\mathscr{R}_{0i}}{\rho_{0i}} \tilde{\theta}_{0i}^{\mathrm{T}}(k+1)\hat{\theta}_{0i}(k) = -\sum_{i=1}^{3} \frac{\mathscr{R}_{0i}}{\rho_{0i}} \tilde{\theta}_{0i}^{\mathrm{T}}(k)\hat{\theta}_{0i}(k)$$
$$-\sum_{i=1}^{3} \mathscr{Q}_{1i}\frac{\mathscr{R}_{0i}}{\rho_{0i}} \mathscr{P}_{0i}\phi_{0i}^{\mathrm{T}}(k)s_{1i}(k+1)\hat{\theta}_{0i}(k) + \sum_{i=1}^{3} \frac{\mathscr{R}_{0i}^{2}}{\rho_{0i}} \mathscr{P}_{0i}\hat{\theta}_{0i}^{\mathrm{T}}(k)\hat{\theta}_{0i}(k) \tag{4.40}$$

$$-\sum_{i=1}^{3} \frac{1}{\mathscr{P}_{0i}\rho_{0i}} \tilde{\theta}_{0i}^{\mathrm{T}}(k)\tilde{\theta}_{0i}(k) = -\sum_{i=1}^{3} \frac{1}{\mathscr{P}_{0i}\rho_{0i}} \tilde{\theta}_{0i}^{\mathrm{T}}(k)\tilde{\theta}_{0i}(k+1)$$
$$+\sum_{i=1}^{3} \frac{\mathscr{Q}_{1i}}{\rho_{0i}} \tilde{\theta}_{0i}^{\mathrm{T}}(k)\phi_{0i}(k)s_{1i}(k+1) - \sum_{i=1}^{3} \frac{\mathscr{R}_{0i}}{\rho_{0i}} \tilde{\theta}_{0i}^{\mathrm{T}}(k)\hat{\theta}_{0i}(k) \tag{4.41}$$

According to (4.34)–(4.41), (4.32) can be transformed into the following form:

$$\Delta L_1(k) = \frac{1}{\mathscr{K}_1} s_1^{\mathrm{T}}(k+1)s_1(k+1) - \frac{1}{\mathscr{K}_1} s_1^{\mathrm{T}}(k)s_1(k) + 2\sum_{i=1}^{3} \frac{\mathscr{Q}_{1i}}{\rho_{1i}} \tilde{\theta}_{1i}^{\mathrm{T}}(k)\phi_{1i}(k)s_{1i}(k+1)$$
$$+\sum_{i=1}^{3} \frac{\mathscr{Q}_{1i}^{2}}{\rho_{1i}} \mathscr{P}_{1i}\phi_{1i}^{\mathrm{T}}(k)s_{1i}(k+1)\phi_{1i}(k)s_{1i}(k+1) - 2\sum_{i=1}^{3} \frac{\mathscr{R}_{1i}}{\rho_{1i}} \tilde{\theta}_{1i}^{\mathrm{T}}(k)\hat{\theta}_{1i}(k)$$
$$-2\sum_{i=1}^{3} \frac{\mathscr{Q}_{1i}}{\rho_{1i}} \mathscr{P}_{1i}\mathscr{R}_{1i}\hat{\theta}_{1i}^{\mathrm{T}}(k)\phi_{1i}(k)s_{1i}(k+1) + \sum_{i=1}^{3} \frac{\mathscr{R}_{1i}^{2}}{\rho_{1i}} \mathscr{P}_{1i}\hat{\theta}_{1i}^{\mathrm{T}}(k)\hat{\theta}_{1i}(k)$$
$$+\frac{1}{\hbar_1}\sum_{i=1}^{3} \tilde{\mathscr{D}}_{1i}^{2}(k+1) - \frac{1}{\hbar_1}\sum_{i=1}^{3} \tilde{\mathscr{D}}_{1i}^{2}(k) + 2\sum_{i=1}^{3} \frac{\mathscr{Q}_{1i}}{\rho_{0i}} \tilde{\theta}_{0i}^{\mathrm{T}}(k)\phi_{0i}(k)s_{1i}(k+1)$$
$$+\sum_{i=1}^{3} \frac{\mathscr{Q}_{1i}^{2}}{\rho_{0i}} \mathscr{P}_{0i}\phi_{0i}^{\mathrm{T}}(k)s_{1i}(k+1)\phi_{0i}(k)s_{1i}(k+1) - 2\sum_{i=1}^{3} \frac{\mathscr{R}_{0i}}{\rho_{0i}} \tilde{\theta}_{0i}^{\mathrm{T}}(k)\hat{\theta}_{0i}(k)$$
$$-2\sum_{i=1}^{3} \frac{\mathscr{Q}_{1i}}{\rho_{0i}} \mathscr{P}_{0i}\mathscr{R}_{0i}\hat{\theta}_{0i}^{\mathrm{T}}(k)\phi_{0i}(k)s_{1i}(k+1) + \sum_{i=1}^{3} \frac{\mathscr{R}_{0i}^{2}}{\rho_{0i}} \mathscr{P}_{0i}\hat{\theta}_{0i}^{\mathrm{T}}(k)\hat{\theta}_{0i}(k) \tag{4.42}$$

On the basis of (4.33), one can obtain

$$2\mathscr{Q}_1 s_1^{\mathrm{T}}(k+1)\tilde{\theta}_1(k) = 2\mathscr{Q}_1 s_1^{\mathrm{T}}(k+1)\bar{\lambda}(k)\bar{G}_1(k)\tilde{\theta}_0(k) + 2\mathscr{Q}_1 s_1^{\mathrm{T}}(k+1)\bar{G}_1(k)s_2(k)$$
$$+2\mathscr{Q}_1 s_1^{\mathrm{T}}(k+1)\bar{G}_1(k)\mathscr{Y}(k) - 2\mathscr{Q}_1 s_1^{\mathrm{T}}(k+1)s_1(k+1) - 2\mathscr{Q}_1 s_1^{\mathrm{T}}(k+1)\mathscr{B}s_1(k)$$
$$+2\mathscr{Q}_1 s_1^{\mathrm{T}}(k+1)\tilde{\mathscr{D}}_1(k) - 2\mathscr{Q}_1 s_1^{\mathrm{T}}(k+1)\bar{\lambda}(k)\bar{G}_1(k)\mathrm{E}_0(k) \tag{4.43}$$

where the matrix $\mathscr{Q}_1 = \mathrm{diag}[\mathscr{Q}_{11}, \mathscr{Q}_{12}, \mathscr{Q}_{13}]$.

From (4.9) and (4.19), one has

$$\sum_{i=1}^{3}\tilde{\mathscr{D}}_{1i}^{2}(k+1)=\sum_{i=1}^{3}\tilde{\mathscr{D}}_{1i}(k+1)\Delta\mathscr{D}_{1i}(k)+\sum_{i=1}^{3}\tilde{\mathscr{D}}_{1i}(k+1)\frac{1}{\rho_{1i}}\tilde{\theta}_{1i}^{T}(k)\phi_{1i}(k) \tag{4.44}$$

where $\Delta\mathscr{D}_{1i}(k)=\mathscr{D}_{1i}(k+1)-\mathscr{D}_{1i}(k)$.

Remark 4.3 Based on the NN approximation theory, the unknown approximation error $\varepsilon_{1i}(k+1)$ is bounded. In addition, according to Assumption 4.2, it can be obtained that the external disturbance $\bar{d}_1(k+1)$ is bounded. Thus, $\Delta\mathscr{D}_{1i}(k)$ is also bounded, and $\Delta\mathscr{D}_{1i}(k)$ satisfies that $|\Delta\mathscr{D}_{1i}(k)|\leq\mathscr{W}_{1i}$, $\mathscr{W}_{1i}$ is a positive constant.

Therefore, the following inequalities can be obtained:

$$2\mathscr{Q}_1 s_1^{\mathrm{T}}(k+1)\tilde{\mathscr{D}}_1(k)\leq\|\mathscr{Q}_1\|^2\tau_{11}\left\|\tilde{\mathscr{D}}_1(k)\right\|^2+\frac{1}{\tau_{11}}\|s_1(k+1)\|^2 \tag{4.45}$$

where τ_{11} is a positive constant.

$$\begin{aligned}&2\mathscr{Q}_1 s_1^{\mathrm{T}}(k+1)(\bar{G}_1(k)\mathscr{Y}(k)-\overline{\lambda}(k)\bar{G}_1(k)\mathrm{E}_0(k))\leq\\&\quad\|\mathscr{Q}_1\|^2\tau_{10}\left\|\bar{\mathrm{E}}_0\right\|^2+\tfrac{1}{\tau_{10}}\|s_1(k+1)\|^2\end{aligned} \tag{4.46}$$

where τ_{10} is a positive constant, and $\left\|\bar{G}_1(k)\mathscr{Y}(k)-\overline{\lambda}(k)\bar{G}_1(k)\mathrm{E}_0(k)\right\|\leq\bar{\mathrm{E}}_0$.

$$\sum_{i=1}^{3}\frac{\mathscr{Q}_{1i}^2}{\rho_{1i}}\mathscr{P}_{1i}\phi_{1i}^{\mathrm{T}}(k)s_{1i}(k+1)\phi_{1i}(k)s_{1i}(k+1)\leq\frac{\mathscr{Q}_{1\max}^2}{\rho_{1\min}}\mathscr{P}_{1\max}\bar{l}_1\|s_1(k+1)\|^2 \tag{4.47}$$

where $\mathscr{P}_{1\max}$ is the maximum value of $\mathscr{P}_{1i}$, $\mathscr{Q}_{1\max}$ is the maximum value of $\mathscr{Q}_{1i}$, $\phi_{1i}^{\mathrm{T}}(k)\phi_{1i}(k)\leq\bar{l}_1$, $\rho_{1\min}$ is the minimum value of ρ_{1i}, and $\bar{l}_1\geq 1$.

$$\begin{aligned}&2\sum_{i=1}^{3}\tfrac{\mathscr{Q}_{1i}}{\rho_{1i}}\mathscr{P}_{1i}\mathscr{R}_{1i}\hat{\theta}_{1i}^{\mathrm{T}}(k)\phi_{1i}(k)s_{1i}(k+1)\leq\\&\sum_{i=1}^{3}\tfrac{\mathscr{Q}_{1i}}{\rho_{1i}}\mathscr{P}_{1i}\mathscr{R}_{1i}\tau_{12}\left\|\hat{\theta}_{1i}(k)\right\|^2+\tfrac{\mathscr{Q}_{1\max}\mathscr{P}_{1\max}\mathscr{R}_{1\max}}{\tau_{12}\rho_{1min}}\bar{l}_1\|s_1(k+1)\|^2\end{aligned} \tag{4.48}$$

where $\mathscr{R}_{1\max}$ is the maximum value of $\mathscr{R}_{1i}$, and τ_{12} is a positive constant.

$$\begin{aligned}2\sum_{i=1}^{3}\frac{\mathscr{R}_{1i}}{\rho_{1i}}\tilde{\theta}_{1i}^{\mathrm{T}}(k)\hat{\theta}_{1i}(k)=&\sum_{i=1}^{3}\frac{\mathscr{R}_{1i}}{\rho_{1i}}\left\|\tilde{\theta}_{1i}(k)\right\|^2+\sum_{i=1}^{3}\frac{\mathscr{R}_{1i}}{\rho_{1i}}\left\|\hat{\theta}_{1i}(k)\right\|^2\\&-\sum_{i=1}^{3}\frac{\mathscr{R}_{1i}}{\rho_{1i}}\|\theta_{1i}^{*}(k)\|^2\end{aligned} \tag{4.49}$$

$$\sum_{i=1}^{3} \frac{\mathscr{R}_{1i}}{\rho_{1i}} \mathscr{P}_{1i} \mathscr{R}_{1i} \hat{\theta}_{1i}^{\mathrm{T}}(k) \hat{\theta}_{1i}(k) = \sum_{i=1}^{3} \frac{\mathscr{R}_{1i}^{2}}{\rho_{1i}} \mathscr{P}_{1i} \left\| \hat{\theta}_{1i}(k) \right\|^{2} \tag{4.50}$$

$$2\mathscr{Q}_{1} s_{1}^{\mathrm{T}}(k+1) \bar{G}_{1}(k) s_{2}(k) \leq \tau_{13} \left\| \mathscr{Q}_{1} \right\|^{2} \bar{g}_{1}^{2} \| s_{1}(k+1) \|^{2} + \frac{1}{\tau_{13}} \| s_{2}(k) \|^{2} \tag{4.51}$$

where τ_{13} is a positive constant.

$$\sum_{i=1}^{3} \tilde{\mathscr{D}}_{1i}(k+1) \Delta \mathscr{D}_{1i}(k) \leq \frac{\tau_{14}}{2} \sum_{i=1}^{3} \tilde{\mathscr{D}}_{1i}^{2}(k+1) + \frac{1}{2\tau_{14}} \sum_{i=1}^{3} \mathscr{W}_{1i}^{2} \tag{4.52}$$

where τ_{14} is a positive constant.

$$\sum_{i=1}^{3} \tilde{\mathscr{D}}_{1i}(k+1) \frac{1}{\rho_{1i}} \tilde{\theta}_{1i}^{T}(k) \phi_{1i}(k) \leq \frac{\tau_{15}}{2} \sum_{i=1}^{3} \tilde{\mathscr{D}}_{1i}^{2}(k+1) + \frac{\bar{l}_{1}}{2\rho_{1\min}^{2} \tau_{15}} \sum_{i=1}^{3} \left\| \tilde{\theta}_{1i}(k) \right\|^{2} \tag{4.53}$$

where τ_{15} is a positive constant.

$$\sum_{i=1}^{3} \frac{\mathscr{Q}_{1i}^{2}}{\rho_{0i}} \mathscr{P}_{0i} \phi_{0i}^{\mathrm{T}}(k) s_{1i}(k+1) \phi_{0i}(k) s_{1i}(k+1) \leq \frac{\mathscr{Q}_{1\max}^{2}}{\rho_{0\min}} \mathscr{P}_{0\max} \bar{l}_{0} \| s_{1}(k+1) \|^{2} \tag{4.54}$$

where $\mathscr{P}_{0\max}$ is the maximum value of $\mathscr{P}_{0i}$, $\phi_{0i}^{\mathrm{T}}(k)\phi_{0i}(k) \leq \bar{l}_0$, $\rho_{0\min}$ is the minimum value of ρ_{0i}, and $\bar{l}_0 \geq 1$.

$$\begin{aligned} 2\sum_{i=1}^{3} \frac{\mathscr{Q}_{1i}}{\rho_{0i}} \mathscr{P}_{0i} \mathscr{R}_{0i} \hat{\theta}_{0i}^{\mathrm{T}}(k) \phi_{0i}(k) s_{1i}(k+1) \leq \\ \sum_{i=1}^{3} \frac{\mathscr{Q}_{1i}}{\rho_{0i}} \mathscr{P}_{0i} \mathscr{R}_{0i} \bar{\tau}_{12} \left\| \hat{\theta}_{0i}(k) \right\|^{2} + \frac{\mathscr{Q}_{1\max} \mathscr{P}_{0\max} \mathscr{R}_{0\max}}{\bar{\tau}_{12} \rho_{0\min}} \bar{l}_{0} \| s_{1}(k+1) \|^{2} \end{aligned} \tag{4.55}$$

where $\mathscr{R}_{0\max}$ is the maximum value of $\mathscr{R}_{0i}$, and $\bar{\tau}_{12}$ is a positive constant.

$$\begin{aligned} 2\sum_{i=1}^{3} \frac{\mathscr{R}_{0i}}{\rho_{0i}} \tilde{\theta}_{0i}^{\mathrm{T}}(k) \hat{\theta}_{0i}(k) = \sum_{i=1}^{3} \frac{\mathscr{R}_{0i}}{\rho_{0i}} \left\| \tilde{\theta}_{0i}(k) \right\|^{2} + \sum_{i=1}^{3} \frac{\mathscr{R}_{0i}}{\rho_{0i}} \left\| \hat{\theta}_{0i}(k) \right\|^{2} \\ - \sum_{i=1}^{3} \frac{\mathscr{R}_{0i}}{\rho_{0i}} \left\| \theta_{0i}^{*}(k) \right\|^{2} \end{aligned} \tag{4.56}$$

$$\begin{aligned} 2\sum_{i=1}^{3} \frac{\mathscr{Q}_{1i}}{\rho_{0i}} \tilde{\theta}_{0i}^{\mathrm{T}}(k) \phi_{0i}(k) s_{1i}(k+1) \leq \mathscr{Q}_{1\max}^{2} \bar{\tau}_{13} \| s_{1}(k+1) \|^{2} \\ + \frac{\bar{l}_{0}}{\bar{\tau}_{13} \rho_{0\min}^{2}} \sum_{i=1}^{3} \left\| \tilde{\theta}_{0i}(k) \right\|^{2} \end{aligned} \tag{4.57}$$

where $\bar{\tau}_{13}$ is a positive constant.

$$2\mathscr{Q}_1 s_1^{\mathrm{T}}(k+1)\overline{\lambda}(k)\bar{G}_1(k)\tilde{\theta}_0(k) \le \bar{\tau}_{14}\|\mathscr{Q}_1\|^2\|s_1(k+1)\|^2 + \frac{\bar{l}_0\ell^2\bar{g}_1^2}{\bar{\tau}_{14}\rho_{0\min}^2}\sum_{i=1}^{3}\left\|\tilde{\theta}_{0i}(k)\right\|^2 \tag{4.58}$$

where $\bar{\tau}_{14}$ is a positive constant.

$$2\mathscr{Q}_1 s_1^{\mathrm{T}}(k+1)\mathscr{B}s_1(k) \le \bar{\tau}_{15}\|\mathscr{Q}_1\|^2\|s_1(k+1)\|^2 + \frac{\|B\|^2}{\bar{\tau}_{15}}\|s_1(k)\|^2 \tag{4.59}$$

where $\bar{\tau}_{15}$ is a positive constant.

According to (4.43)–(4.59), one has

$$\begin{aligned}
\Delta L_1(k) \le\; & \frac{1}{\mathscr{K}_1}s_1^{\mathrm{T}}(k+1)s_1(k+1) - \frac{1}{\mathscr{K}_1}s_1^{\mathrm{T}}(k)s_1(k) - 2\mathscr{Q}_1 s_1^{\mathrm{T}}(k+1)s_1(k+1)\\
& + \mathscr{Q}_{1i}^2\tau_{11}\left\|\tilde{\mathscr{D}}_1(k)\right\|^2 + \frac{1}{\tau_{11}}\|s_1(k+1)\|^2 - \sum_{i=1}^{3}\frac{\mathscr{R}_{1i}}{\rho_{1i}}\left\|\tilde{\theta}_{1i}(k)\right\|^2\\
& + \frac{\mathscr{Q}_{1\max}^2\mathscr{P}_{1\max}\bar{l}_1}{\rho_{1\min}}\|s_1(k+1)\|^2 + \frac{1}{\tau_{13}}\|s_2(k)\|^2 + \frac{1}{2\tau_{14}}\sum_{i=1}^{3}\mathscr{W}_{1i}^2\\
& + \sum_{i=1}^{3}\frac{\mathscr{Q}_{1i}\mathscr{P}_{1i}\mathscr{R}_{1i}\tau_{12}}{\rho_{1i}}\left\|\hat{\theta}_{1i}(k)\right\|^2 + \frac{\mathscr{Q}_{1\max}\mathscr{P}_{1\max}\mathscr{R}_{1\max}}{\rho_{1\min}\tau_{12}}\bar{l}_1\|s_1(k+1)\|^2\\
& - \sum_{i=1}^{3}\frac{\mathscr{R}_{1i}}{\rho_{1i}}\left\|\hat{\theta}_{1i}(k)\right\|^2 + \sum_{i=1}^{3}\frac{\mathscr{R}_{1i}}{\rho_{1i}}\|\theta_{1i}^*(k)\|^2 + \sum_{i=1}^{3}\frac{\mathscr{R}_{1i}^2}{\rho_{1i}}\mathscr{P}_{1i}\left\|\hat{\theta}_{1i}(k)\right\|^2\\
& + \tau_{13}\|\mathscr{Q}_1\|^2\bar{g}_1^2\|s_1(k+1)\|^2 + \frac{\tau_{14}}{2}\sum_{i=1}^{3}\tilde{\mathscr{D}}_{1i}^2(k+1) + \frac{\tau_{15}}{2}\sum_{i=1}^{3}\tilde{\mathscr{D}}_{1i}^2(k+1)\\
& + \frac{\bar{l}_1}{2\rho_{1i\min}^2\tau_{15}}\sum_{i=1}^{3}\left\|\tilde{\theta}_{1i}(k)\right\|^2 - \sum_{i=1}^{3}\tilde{\mathscr{D}}_{1i}^2(k+1) + \frac{1}{\hbar_1}\sum_{i=1}^{3}\tilde{\mathscr{D}}_{1i}^2(k+1)\\
& - \frac{1}{\hbar_1}\sum_{i=1}^{3}\tilde{\mathscr{D}}_{1i}^2(k) + \frac{\mathscr{Q}_{1\max}^2}{\rho_{0\min}}\mathscr{P}_{0\max}\bar{l}_0\|s_1(k+1)\|^2 + \frac{1}{\tau_{10}}\|s_1(k+1)\|^2\\
& + \sum_{i=1}^{3}\frac{\mathscr{Q}_{1i}}{\rho_{0i}}\mathscr{P}_{0i}\mathscr{R}_{0i}\bar{\tau}_{12}\left\|\hat{\theta}_{0i}(k)\right\|^2 + \frac{\mathscr{Q}_{1\max}\mathscr{P}_{0\max}\mathscr{R}_{0\max}}{\rho_{0\min}\bar{\tau}_{12}}\bar{l}_0\|s_1(k+1)\|^2\\
& - \sum_{i=1}^{3}\frac{\mathscr{R}_{0i}}{\rho_{0i}}\left\|\tilde{\theta}_{0i}(k)\right\|^2 - \sum_{i=1}^{3}\frac{\mathscr{R}_{0i}}{\rho_{0i}}\left\|\hat{\theta}_{0i}(k)\right\|^2 + \sum_{i=1}^{3}\frac{\mathscr{R}_{0i}}{\rho_{0i}}\|\theta_{0i}^*(k)\|^2
\end{aligned}$$

$$
\begin{aligned}
&+\|\mathscr{Q}_1\|^2\tau_{10}\left\|\bar{\mathrm{E}}_0\right\|^2+\sum_{i=1}^{3}\frac{\mathscr{R}_{0i}^2}{\rho_{0i}}\mathscr{P}_{0i}\left\|\hat{\theta}_{0i}(k)\right\|^2+\mathscr{Q}_{1\max}^2\bar{\tau}_{13}\|s_1(k+1)\|^2\\
&+\frac{\bar{l}_0}{\bar{\tau}_{13}\rho_{0\min}^2}\sum_{i=1}^{3}\left\|\tilde{\theta}_{0i}(k)\right\|^2+\bar{\tau}_{14}\|\mathscr{Q}_1\|^2\|s_1(k+1)\|^2+\frac{\|B\|^2}{\bar{\tau}_{15}}\|s_1(k)\|^2\\
&+\bar{\tau}_{15}\|\mathscr{Q}_1\|^2\|s_1(k+1)\|^2+\frac{\bar{l}_0\ell^2\bar{g}_1^2}{\bar{\tau}_{14}\rho_{0\min}^2}\sum_{i=1}^{3}\left\|\tilde{\theta}_{0i}(k)\right\|^2
\end{aligned}
\tag{4.60}
$$

Furthermore, (4.60) can be written as

$$
\begin{aligned}
\Delta L_1(k)\leq{}&-\kappa_{11}\|s_1(k+1)\|^2-\kappa_{12}\|s_1(k)\|^2-\sum_{i=1}^{3}\kappa_{13}\left\|\hat{\theta}_{1i}(k)\right\|^2\\
&-\kappa_{14}\sum_{i=1}^{3}\tilde{\mathscr{D}}_{1i}^2(k+1)-\kappa_{15}\sum_{i=1}^{3}\tilde{\mathscr{D}}_{1i}^2(k)+\kappa_{16}\|s_2(k)\|^2+\kappa_0\\
&-\kappa_{17}\sum_{i=1}^{3}\left\|\tilde{\theta}_{1i}(k)\right\|^2-\sum_{i=1}^{3}\kappa_{18}\left\|\hat{\theta}_{0i}(k)\right\|^2-\kappa_{19}\sum_{i=1}^{3}\left\|\tilde{\theta}_{0i}(k)\right\|^2
\end{aligned}
\tag{4.61}
$$

where $\kappa_{11}=2\mathscr{Q}_{1\min}-\frac{1}{\mathscr{K}_1}-\frac{1}{\tau_{11}}-\frac{\mathscr{Q}_{1\max}^2\mathscr{P}_{1\max}\bar{l}_1}{\rho_{1\min}}-\frac{\mathscr{Q}_{1\max}\mathscr{P}_{1\max}\mathscr{R}_{1\max}}{\rho_{1\min}\tau_{12}}\bar{l}_1-\tau_{13}\|\mathscr{Q}_1\|^2$ $\bar{g}_1^2-\frac{\mathscr{Q}_{1\max}^2}{\rho_{0\min}}\mathscr{P}_{0\max}\bar{l}_0-\frac{1}{\tau_{10}}-\frac{\mathscr{Q}_{1\max}\mathscr{P}_{0\max}\mathscr{R}_{0\max}}{\bar{\tau}_{12}\rho_{0\min}}-\mathscr{Q}_{1\max}^2\bar{\tau}_{13}-\bar{\tau}_{14}\|\mathscr{Q}_1\|^2-\bar{\tau}_{15}\|\mathscr{Q}_1\|^2$, $\kappa_{13}=\frac{\mathscr{R}_{1i}}{\rho_{1i}}-\frac{\mathscr{R}_{1i}^2}{\rho_{1i}}\mathscr{P}_{1i}-\frac{\mathscr{R}_{1i}}{\rho_{1i}}\mathscr{P}_{1i}\mathscr{Q}_{1i}\tau_{12}$, $\kappa_{12}=\frac{1}{\mathscr{K}_1}-\frac{\|B\|^2}{\bar{\tau}_{15}}$, $\kappa_{14}=1-\frac{\tau_{14}}{2}-\frac{\tau_{15}}{2}-\frac{1}{\hbar_1}$, $\kappa_{15}=\frac{1}{\hbar_1}-\frac{1}{2\tau_{16}}-\mathscr{Q}_{1i}^2\tau_{11}$, $\kappa_{16}=\frac{1}{\tau_{13}}$, $\kappa_{17}=\frac{\mathscr{R}_{1min}}{\rho_{1\max}}-\frac{\bar{l}_1}{2\rho_{1\min}^2\tau_{15}}$, $\kappa_{18}=\frac{\mathscr{R}_{0i}}{\rho_{0i}}-\frac{\mathscr{R}_{0i}^2}{\rho_{0i}}\mathscr{P}_{0i}-\frac{\mathscr{R}_{0i}}{\rho_{0i}}\mathscr{P}_{0i}\mathscr{Q}_{1i}\bar{\tau}_{12}$, $\kappa_0=\sum_{i=1}^{3}\frac{\mathscr{R}_{0i}}{\rho_{0i}}\left\|\theta_{0i}^*(k)\right\|^2+\|\mathscr{Q}_1\|^2\tau_{10}\left\|\bar{\mathrm{E}}_0\right\|^2+\frac{1}{2\tau_{14}}\sum_{i=1}^{3}\mathscr{W}_{1i}^2+\sum_{i=1}^{3}\frac{\mathscr{R}_{1i}}{\rho_{1i}}$ $\left\|\theta_{1i}^*(k)\right\|^2$, $\kappa_{19}=\frac{\mathscr{R}_{0\min}}{\rho_{0\max}}-\frac{\bar{l}_0\ell^2\bar{g}_1^2}{\bar{\tau}_{14}\rho_{0\min}^2}-\frac{\bar{l}_0}{\bar{\tau}_{13}\rho_{0\min}^2}$, $\mathscr{R}_{1min}$ is the minimum value of $\mathscr{R}_{1i}$, $\mathscr{R}_{0min}$ is the minimum value of $\mathscr{R}_{0i}$, $\mathscr{Q}_{1\min}$ is the minimum value of $\mathscr{Q}_1$, $\rho_{1\max}$ is the maximum value of ρ_{1i}, $\rho_{0\max}$ is the maximum value of ρ_{0i}, and $\bar{l}_0\geq 1$. and $\mathscr{R}_{0min}$ is the minimum value of $\mathscr{R}_{0i}$.

Step 2: According to (4.2) and $s_2(k)=x_2(k)-\mathrm{sat}(x_{vd}(k))$, one has

$$
s_2(k+1)=\bar{F}_2(k)+\Delta\bar{F}_2(k)+\bar{G}_2(k)u(k)+\bar{d}_2(k)-v_d(k+1) \tag{4.62}
$$

On the basis of Lemma 2.2, the NN is used to approximate $\rho_{2i}\Delta F_{2i}(k)(i=1,2,3)$ with $\Delta\bar{F}_{2i}(k)$ being the ith variable of $\Delta\bar{F}_2(k)$. Then, (4.62) can be written as

$$
s_2(k+1)=\bar{F}_2(k)+\Theta_2(k)+\bar{G}_2(k)u(k)+\bar{d}_2(k)-v_d(k+1)+\mathrm{E}_2(k) \tag{4.63}
$$

where $\Theta_2(k) = [\frac{1}{\rho_{21}}\theta_{21}^{*\mathrm{T}}(k)\phi_{21}(Z_2(k)), \frac{1}{\rho_{22}}\theta_{22}^{*\mathrm{T}}(k)\phi_{22}(Z_2(k)), \frac{1}{\rho_{23}}\theta_{23}^{*\mathrm{T}}(k)\phi_{23}(Z_2(k))]^{\mathrm{T}}$, $Z_2(k) = x(k), \mathrm{E}_2(k) = [\frac{1}{\rho_{21}}\varepsilon_{21}(k), \frac{1}{\rho_{22}}\varepsilon_{22}(k), \frac{1}{\rho_{23}}\varepsilon_{23}(k)]^{\mathrm{T}}$, ρ_{2i} is a designed constant, $\theta_{2i}^*(k)$ is the optimal weight vector, $\varepsilon_{2i}(k)$ is the minimum approximation error, $\phi_{2i}(\cdot)$ is the basis function vector. To facilitate control scheme design, define $\phi_{2i}(Z_2(k)) = \phi_{2i}(k)$.

Moreover, (4.63) can be described as

$$s_2(k+1) = \bar{F}_2(k) + \Theta_2(k) + \bar{G}_2(k)u(k) + \mathscr{D}_2(k) - v_d(k+1) \tag{4.64}$$

where $\mathscr{D}_2(k) = \mathrm{E}_2(k) + \bar{d}_2(k)$ is the complex disturbance.

In order to deal with the complex disturbance $\mathscr{D}_2(k)$ in (4.64), the NN-based nonlinear DTDO is designed according to (4.7), one has

$$\begin{cases} \eta_{2i}(k+1) = \bar{F}_{2i}(k) + \mathscr{V}_{2i}(k) \\ \mathscr{H}_{2i}(k+1) = -k_{2i}\hat{\theta}_{2i}^{\mathrm{T}}(k)\phi_{2i}(k) \\ \hat{\mathscr{D}}_{2i}(k) = x_{2i}(k) - \eta_{2i}(k) + \frac{1}{k_{2i}\rho_{2i}}\mathscr{H}_{2i}(k) \end{cases} \tag{4.65}$$

where $x_{2i}(k)$ is the ith variable of $x_2(k)$, $\bar{F}_{2i}(k)$ is the ith variable of $\bar{F}_2(k)$, $\hat{\mathscr{D}}_{2i}(k)$ is the output of DTDO, and $\hat{\mathscr{D}}_2(k) = [\hat{\mathscr{D}}_{21}(k), \hat{\mathscr{D}}_{22}(k), \hat{\mathscr{D}}_{23}(k)]^{\mathrm{T}}$. The disturbance estimation error $\tilde{\mathscr{D}}_{2i}(k) = \mathscr{D}_{2i}(k) - \hat{\mathscr{D}}_{2i}(k)$. $\mathscr{V}_{2i}(k)$ is the ith variable of $\mathscr{V}_2(k)$, and $\mathscr{V}_2(k) = \bar{G}_2(k)u(k)$. $\eta_{2i}(k)$ and $\mathscr{H}_{2i}(k)$ are the state variables of nonlinear DTDO, k_{2i} is the designed constant, $\hat{\theta}_{2i}(k)$ is the estimation of $\theta_{2i}^*(k)$.

To forecast $v_d(k+1)$, the following discrete-time tracking differentiator based on Lemma 2.3 is employed to estimate $v_d(k+1)$:

$$\begin{cases} \bar{\chi}_{11i}(k+1) = \bar{\chi}_{11i}(k) + h_{01i}\bar{\chi}_{12i}(k) \\ \bar{\chi}_{12i}(k+1) = \bar{\chi}_{12i}(k) - h_{01i}r_{01i}\bar{\chi}_{13i}(\bar{\chi}_{14i}(k), \delta_{01i}) \end{cases} \tag{4.66}$$

where $i = 1, 2, 3$, h_{01i} and r_{01i} are designed constants, and $\bar{\chi}_{11i}(k)$ and $\bar{\chi}_{12i}(k)$ are the state variables of discrete-time tracking differentiator.

According to (4.66)and Lemma 2.3, one has

$$v_d(k+1) = \bar{\chi}_{11}(k) + h_{01}\bar{\chi}_{12}(k) - \tilde{\chi} \tag{4.67}$$

where $\bar{\chi}_{11}(k) = [\bar{\chi}_{111}(k), \bar{\chi}_{112}(k), \bar{\chi}_{113}(k)]^{\mathrm{T}}$, $\bar{\chi}_{12}(k) = [\bar{\chi}_{121}(k), \bar{\chi}_{122}(k), \bar{\chi}_{123}(k)]^{\mathrm{T}}$, $h_{01} = \mathrm{diag}[h_{011}, h_{012}, h_{013}], r_{01} = \mathrm{diag}[r_{011}, r_{012}, r_{013}], \tilde{\chi} = [\tilde{\chi}_1, \tilde{\chi}_2, \tilde{\chi}_3]^{\mathrm{T}}$ is the estimated error vector, and $\tilde{\chi}$ satisfies that $\|\tilde{\chi}\| \leq \bar{\chi}$ based on Lemma 2.3, $\bar{\chi} > 0$ is a constant.

Remark 4.4 Based on the discrete-time form of tracking differentiator, the kth step information is used to approximate the $k+1$th step information of the virtual controller. According to the physical characteristics of the UAV system, the attitude angle signal and attitude angle speed in the UAV system should vary in two bounded ranges. In addition, the expected trace signal is bounded according to the actual situation. Therefore, according to the above description and the controllability of the UAV system, the variation range of the signal $v_d(k)$ is bounded. Based on the previous statement, the $k+1$th step information in the k step controller is avoided by using the discrete-time tracking differentiator. The approximation error is considered in the stability analysis of the closed loop system.

Substituting (4.67) into (4.64), it yields

$$
\begin{aligned}
s_2(k+1) = &\bar{F}_2(k) + \Theta_2(k) + \bar{G}_2(k)u(k) + \mathscr{D}_2(k) \\
&-\bar{\chi}_{11}(k) - h_{01}\bar{\chi}_{12}(k) + \tilde{\chi}
\end{aligned} \tag{4.68}
$$

Then, the discrete-time controller is designed by

$$
\begin{aligned}
u(k) = &-\bar{G}_2^{-1}(k)\bar{F}_2(k) - \bar{G}_2^{-1}(k)(\hat{\Theta}_2(k) + \hat{\mathscr{D}}_2(k)) \\
&+\bar{G}_1^{-1}(k)\bar{\chi}_{11}(k) + \bar{G}_1^{-1}(k)h_{01}\bar{\chi}_{12}(k)
\end{aligned} \tag{4.69}
$$

where $\hat{\Theta}_2(k) = [\frac{1}{\rho_{21}}\hat{\theta}_{21}^{\mathrm{T}}(k)\phi_{21}(k), \frac{1}{\rho_{22}}\hat{\theta}_{22}^{\mathrm{T}}(k)\phi_{22}(k), \frac{1}{\rho_{23}}\hat{\theta}_{23}^{\mathrm{T}}(k)\phi_{23}(k)]^{\mathrm{T}}$, and $\hat{\Theta}_2(k)$ is the estimation of $\Theta_2(k)$.

Based on (4.69) and (4.68), one has

$$
s_2^{\mathrm{T}}(k+1)s_2(k+1) = s_2^{\mathrm{T}}(k+1)(\tilde{\mathscr{D}}_2(k) - \tilde{\Theta}_2(k)) + s_1^{\mathrm{T}}(k+1)\tilde{\chi} \tag{4.70}
$$

where $\tilde{\Theta}_2(k) = [\frac{1}{\rho_{21}}\tilde{\theta}_{21}^{\mathrm{T}}(k)\phi_2(k), \frac{1}{\rho_{22}}\tilde{\theta}_{22}^{\mathrm{T}}(k)\phi_2(k), \frac{1}{\rho_{23}}\tilde{\theta}_{23}^{\mathrm{T}}(k)\phi_2(k)]^{\mathrm{T}}$ and the error variable $\tilde{\Theta}_2(k) = \hat{\Theta}_2(k) - \Theta_2(k)$.

In addition, the adaptive law $\hat{\theta}_{2i}(k)$ is designed by

$$
\hat{\theta}_{2i}(k+1) = \hat{\theta}_{2i}(k) + \mathscr{P}_{2i}\mathscr{Q}_{2i}\phi_{2i}(k)s_{2i}(k+1) - \mathscr{P}_{2i}\mathscr{R}_{2i}\hat{\theta}_{2i}(k) \tag{4.71}
$$

where $s_{2i}(k+1)$ is the ith variable of $s_2(k+1)$, $\mathscr{P}_{2i}$, $\mathscr{Q}_{2i}$ and $\mathscr{R}_{2i}$ are designed positive constants.

From (4.71), one can obtain

$$
\tilde{\theta}_{2i}(k+1) = \tilde{\theta}_{2i}(k) + \mathscr{P}_{2i}\mathscr{Q}_{2i}\phi_{2i}(k)s_{2i}(k+1) - \mathscr{P}_{2i}\mathscr{R}_{2i}\hat{\theta}_{2i}(k) \tag{4.72}
$$

where $\tilde{\theta}_{2i}(k+1) = \hat{\theta}_{2i}(k+1) - \theta_{2i}^*(k+1)$ and $\tilde{\theta}_{2i}(k) = \hat{\theta}_{2i}(k) - \theta_{2i}^*(k)$.

To prove the effectiveness of the nonlinear DTDO (4.65) and the controller $x_{vd}(k)$ (4.69), the following Lyapunov function is chosen as

$$
L_2(k) = \frac{1}{\mathscr{K}_2}s_2^{\mathrm{T}}(k)s_2(k) + \sum_{i=1}^{3}\frac{1}{\mathscr{P}_{2i}}\tilde{\theta}_{2i}^{\mathrm{T}}(k)\tilde{\theta}_{2i}(k) + \frac{1}{\bar{\hbar}_2}\sum_{i=1}^{3}\tilde{\mathscr{D}}_{2i}^2(k) \tag{4.73}
$$

where $\mathscr{K}_2$ and $\bar{\hbar}_2$ are positive constants.

Based on (4.73), the first difference of the Lyapunov function $L_2(k)$ can be written as

$$\begin{aligned}\Delta L_2(k) &= L_2(k+1) - L_2(k)\\ &= \frac{1}{\mathscr{K}_2} s_2^{\mathrm{T}}(k+1)s_2(k+1) - \frac{1}{\mathscr{K}_2} s_2^{\mathrm{T}}(k)s_2(k) + \sum_{i=1}^{3}\frac{1}{\mathscr{P}_{2i}}\tilde{\theta}_{2i}^{\mathrm{T}}(k+1)\tilde{\theta}_{2i}(k+1)\\ &\quad - \frac{1}{\bar{\hbar}_2}\sum_{i=1}^{3}\tilde{\mathscr{D}}_{2i}^2(k) + \frac{1}{\bar{\hbar}_2}\sum_{i=1}^{3}\tilde{\mathscr{D}}_{2i}^2(k+1) - \sum_{i=1}^{3}\frac{1}{\mathscr{P}_{2i}}\tilde{\theta}_{2i}^{\mathrm{T}}(k)\tilde{\theta}_{2i}(k) \end{aligned} \tag{4.74}$$

According to the analysis in Step 1, $\Delta L_2(k)$ can satisfy the following expression:

$$\begin{aligned}\Delta L_2(k) \le\ & \frac{1}{\mathscr{K}_2} s_2^{\mathrm{T}}(k+1)s_2(k+1) - \frac{1}{\mathscr{K}_2} s_2^{\mathrm{T}}(k)s_2(k) - 2\mathscr{Q}_2 s_2^{\mathrm{T}}(k+1)s_2(k+1)\\ & + \|\mathscr{Q}_2\|^2\tau_{21}\left\|\tilde{\mathscr{D}}_2(k)\right\|^2 + \frac{1}{\tau_{21}}\|s_2(k+1)\|^2 - \sum_{i=1}^{3}\frac{\mathscr{R}_{2i}}{\rho_{2i}}\left\|\tilde{\theta}_{2i}(k)\right\|^2\\ & + \frac{\mathscr{Q}_{2\max}^2\mathscr{P}_{2\max}\bar{l}_2}{\rho_{2\min}}\|s_2(k+1)\|^2 + \frac{1}{\tau_{23}}\bar{\chi}^2 + \sum_{i=1}^{3}\frac{\mathscr{Q}_{2i}}{\mathscr{P}_{2i}\mathscr{R}_{2i}\tau_{22}\rho_{2i}}\left\|\hat{\theta}_{2i}(k)\right\|^2\\ & + \frac{1}{2\tau_{24}}\sum_{i=1}^{3}\mathscr{W}_{2i}^2 + \frac{\mathscr{Q}_{2\max}\mathscr{P}_{2\max}\mathscr{R}_{2\max}}{\tau_{22}\rho_{2\min}}\bar{l}_2\|s_2(k+1)\|^2 - \sum_{i=1}^{3}\frac{\mathscr{R}_{2i}}{\rho_{2i}}\left\|\hat{\theta}_{2i}(k)\right\|^2\\ & + \sum_{i=1}^{3}\frac{\mathscr{R}_{2i}}{\rho_{2i}}\left\|\theta_{2i}^*(k)\right\|^2 + \sum_{i=1}^{3}\frac{\mathscr{R}_{2i}^2}{\rho_{2i}}\mathscr{P}_{2i}\left\|\hat{\theta}_{2i}(k)\right\|^2 + \tau_{23}\|\mathscr{Q}_2\|^2\|s_2(k+1)\|^2\\ & + \frac{\tau_{24}}{2}\sum_{i=1}^{3}\tilde{\mathscr{D}}_{2i}^2(k+1) + \frac{\tau_{25}}{2}\sum_{i=1}^{3}\tilde{\mathscr{D}}_{2i}^2(k+1) + \frac{\bar{l}_2}{2\rho_{2\min}^2\tau_{25}}\sum_{i=1}^{3}\left\|\tilde{\theta}_{2i}(k)\right\|^2\\ & - \sum_{i=1}^{3}\tilde{\mathscr{D}}_{2i}^2(k+1) + \frac{1}{\bar{\hbar}_2}\sum_{i=1}^{3}\tilde{\mathscr{D}}_{2i}^2(k+1) - \frac{1}{\bar{\hbar}_2}\sum_{i=1}^{3}\tilde{\mathscr{D}}_{2i}^2(k) \end{aligned} \tag{4.75}$$

where τ_{21}, τ_{22}, τ_{23}, τ_{24} and τ_{25} are positive constants, $\mathscr{P}_{2\max}$ is the maximum value of $\mathscr{P}_{2i}$, $\mathscr{Q}_{2\max}$ is the maximum value of $\mathscr{Q}_{2i}$, $\mathscr{R}_{2\max}$ is the maximum value of $\mathscr{R}_{2i}$, $\phi_{2i}^{\mathrm{T}}(k)\phi_{2i}(k) \le \bar{l}_2$, $\bar{l}_2 \ge 1$, $|\Delta\mathscr{D}_{2i}(k)| \le \mathscr{W}_{2i}$, and $\mathscr{W}_{2i}$ is a positive constant.

Furthermore, (4.75) can be written as

$$\begin{aligned}\Delta L_2(k) \le & -\kappa_{21}\|s_2(k+1)\|^2 - \kappa_{22}\|s_2(k)\|^2 - \sum_{i=1}^{3}\kappa_{23}\left\|\hat{\theta}_{2i}(k)\right\|^2\\ & -\kappa_{24}\sum_{i=1}^{3}\tilde{\mathscr{D}}_{2i}^2(k+1) - \kappa_{25}\sum_{i=1}^{3}\tilde{\mathscr{D}}_{2i}^2(k) - \kappa_{26}\sum_{i=1}^{3}\left\|\tilde{\theta}_{2i}(k)\right\|^2 + \kappa_{27} \end{aligned} \tag{4.76}$$

where $\kappa_{21} = 2\mathscr{Q}_{2\min} - \frac{1}{\mathscr{K}_2} - \frac{1}{\tau_{21}} - \frac{\mathscr{Q}^2_{2\max}\mathscr{P}_{2\max}\bar{l}_2}{\rho_{2\min}} - \frac{\mathscr{Q}_{2\max}\mathscr{P}_{2\max}\mathscr{R}_{2\max}}{\tau_{22}\rho_{2\min}}\bar{l}_2 - \tau_{23}\|\mathscr{Q}_2\|^2$, $\kappa_{22} = \frac{1}{\mathscr{K}_2}$, $\kappa_{23} = \frac{\mathscr{R}_{2i}}{\rho_{2i}} - \frac{\mathscr{R}^2_{2i}}{\rho_{2i}}\mathscr{P}_{2i} - \frac{\mathscr{R}_{2i}}{\rho_{2i}}\mathscr{P}_{2i}\mathscr{Q}_{2i}\tau_{22}$, $\kappa_{24} = 1 - \frac{\tau_{24}}{2} - \frac{\tau_{25}}{2} - \frac{1}{\hbar_2}$, $\kappa_{25} = \frac{1}{\hbar_2} - \|\mathscr{Q}_2\|^2\tau_{21}$, $\kappa_{26} = \frac{\mathscr{R}_{2\min}}{\rho_{2\max}} - \frac{\bar{l}_2}{2\rho^2_{2\min}\tau_{25}}$, $\kappa_{27} = \frac{1}{\tau_{23}}\bar{\chi}^2 + \frac{1}{2\tau_{24}}\sum_{i=1}^{3}\mathscr{W}^2_{2i} + \sum_{i=1}^{3}\frac{\mathscr{R}^2_{2i}}{\rho_{2i}}\left\|\theta^*_{2i}(k)\right\|^2$, $\mathscr{Q}_{2\min}$ is the minimum value of $\mathscr{Q}_2$, $\rho_{2\max}$ is the maximum value of ρ_{2i}, $\rho_{0\min}$ is the minimum value of ρ_{0i}, $\mathscr{R}_{2\min}$ is the minimum value of $\mathscr{R}_{2i}$, $\mathscr{R}_{2\max}$ is the maximum value of $\mathscr{R}_{2i}$, and the matrix $\mathscr{Q}_2 = \text{diag}[\mathscr{Q}_{21}, \mathscr{Q}_{22}, \mathscr{Q}_{23}]$.

For the attitude dynamic model of UAV (4.2) with wind disturbances and system uncertainties, the above designed discrete-time adaptive NN control scheme based on the designed DTDO can be summarized as follows:

Theorem 4.1 *For the uncertain discrete-time attitude dynamic model of UAV with external wind disturbances (4.2), the external disturbances are estimated by using the designed NN-based DTDO (4.19) and (4.65). On the basis of the designed control signal (4.25), the adaptive law of weight estimation (4.27), (4.29) and (4.71), and the developed discrete-time controller (4.69), the tracking error and disturbance estimation errors can be guaranteed to be bounded, and all signals in the closed-loop system can also be guaranteed to be bounded.*

Proof For the entire closed-loop system, the following Lyapunov function is selected as

$$L(k) = \sum_{\bar{j}=1}^{2} L_{\bar{j}}(k) \tag{4.77}$$

According to (4.61) and (4.76), one has

$$\begin{aligned}
\Delta L(k) &= \sum_{\bar{j}=1}^{2} L_{\bar{j}}(k+1) - \sum_{\bar{j}=1}^{2} L_{\bar{j}}(k) \\
&\leq -\kappa_{11}\|s_1(k+1)\|^2 - \kappa_{12}\|s_1(k)\|^2 - \sum_{i=1}^{3}\kappa_{13}\left\|\hat{\theta}_{1i}(k)\right\|^2 - \sum_{i=1}^{3}\kappa_{14}\tilde{\mathscr{D}}^2_{1i}(k+1) \\
&\quad - \sum_{i=1}^{3}\kappa_{15}\tilde{\mathscr{D}}^2_{1i}(k) + \kappa_{16}\|s_2(k)\|^2 - \kappa_{17}\sum_{i=1}^{3}\left\|\tilde{\theta}_{1i}(k)\right\|^2 - \kappa_{21}\|s_2(k+1)\|^2 \\
&\quad - \sum_{i=1}^{3}\kappa_{23}\left\|\hat{\theta}_{2i}(k)\right\|^2 - \sum_{i=1}^{3}\kappa_{24}\tilde{\mathscr{D}}^2_{2i}(k+1) - \sum_{i=1}^{3}\kappa_{25}\tilde{\mathscr{D}}^2_{2i}(k) - \kappa_{22}\|s_2(k)\|^2 \\
&\quad - \kappa_{26}\sum_{i=1}^{3}\left\|\tilde{\theta}_{2i}(k)\right\|^2 + \kappa_{27} - \sum_{i=1}^{3}\kappa_{18}\left\|\hat{\theta}_{0i}(k)\right\|^2 \\
&\quad - \kappa_{19}\sum_{i=1}^{3}\left\|\tilde{\theta}_{0i}(k)\right\|^2 + \kappa_0
\end{aligned} \tag{4.78}$$

Based on (4.78), if the appropriate control parameters can be selected to ensure that $\kappa_0 > 0, \kappa_{11} > 0, \kappa_{12} > 0, \kappa_{13} > 0, \kappa_{14} > 0, \kappa_{15} > 0, \kappa_{17} > 0, \kappa_{18} > 0, \kappa_{19} > 0, \kappa_{22} - \kappa_{16} > 0, \kappa_{21} > 0, \kappa_{23} > 0, \kappa_{24} > 0, \kappa_{25} > 0, \kappa_{26} > 0$ and $\kappa_{27} > 0$, the tracking error $s_1(k)$ and the disturbance estimation errors $\tilde{\mathscr{D}}_1(k)$ and $\tilde{\mathscr{D}}_2(k)$ are bounded. ◇

4.4 Simulation Study

In this section, to illustrate the performance of the designed DTDO-based discrete-time flight control scheme, the simulations of the UAV system with wind disturbance (4.2) are investigated.

In the simulation of uncertain UAV attitude dynamics system (4.2), only the effects of gust on the control of UAV system are considered, and the specific form of gust is shown in (2.40). In this numerical simulation, the maximum external wind speed is assumed as $\bar{V}_{Wm} = [20, 10, 5]^{\mathrm{T}}\mathrm{m/sec}$, $t_m = 3\mathrm{sec}$, the disturbances $\bar{d}_{11} = \bar{d}_{21} = \bar{d}_{23} = 0.7\sin(0.5t) - 0.3\cos(0.2t)$, where m denotes the meter and sec denotes the second. The discrete-time sampling period ΔT is chosen as $\Delta T = 0.01$. The appropriate proportional gain coefficients are selected, and the initial attitude values of the UAV system are set as $\beta_0 = 0.2\mathrm{deg}$, $\alpha_0 = 0\mathrm{deg}$, $\mu_0 = 0\mathrm{deg}$ and $p_0 = q_0 = r_0 = 0\mathrm{deg/sec}$. The desired attitude angle signal x_d of the UAV system is assumed as $\beta_d = 0\mathrm{deg}$, $\alpha_d = 4\mathrm{erf}(0.5t)\mathrm{deg}$ and $\mu_d = 3\mathrm{erf}(0.5 - t)\mathrm{deg}$. The control parameters of DTDO are designed as $k_{11} = 0.1, k_{12} = 0.002, k_{13} = 0.001, \rho_{11} = \rho_{12} = \rho_{13} = 1, k_{21} = 0.011, k_{22} = 0.012, k_{23} = 0.1$ and $\rho_{21} = \rho_{22} = \rho_{23} = 1$. The control parameters in the adaptive law are set as $\mathscr{P}_{11} = \mathscr{P}_{12} = \mathscr{P}_{13} = 0.001, \mathscr{Q}_{11} = \mathscr{Q}_{12} = \mathscr{Q}_{13} = 1, \mathscr{R}_{11} = \mathscr{R}_{12} = \mathscr{R}_{13} = 0.01, \mathscr{P}_{21} = \mathscr{P}_{22} = \mathscr{P}_{23} = 0.001, \mathscr{Q}_{21} = \mathscr{Q}_{22} = \mathscr{Q}_{23} = 1$ and $\mathscr{R}_{21} = \mathscr{R}_{22} = \mathscr{R}_{23} = 0.01$. The control parameters of the tracking differentiator in the discrete-time form are selected as $h_{01} = \mathrm{diag}[20, 20, 20]$ and $r_{01} = \mathrm{diag}[20, 20, 20]$. In addition, $\hat{\bar{d}}_{11}, \hat{\bar{d}}_{12}, \hat{\bar{d}}_{13}, \hat{\bar{d}}_{21}, \hat{\bar{d}}_{22}$ and $\hat{\bar{d}}_{23}$ are defined as the estimations of $\bar{d}_{11}, \bar{d}_{12}, \bar{d}_{13}, \bar{d}_{21}, \bar{d}_{22}$ and $\bar{d}_{23}$ respectively, the disturbance estimation error is defined as $\tilde{\bar{d}}_{11} = \bar{d}_{11} - \hat{\bar{d}}_{11}, \tilde{\bar{d}}_{12} = \bar{d}_{12} - \hat{\bar{d}}_{12}, \tilde{\bar{d}}_{13} = \bar{d}_{13} - \hat{\bar{d}}_{13}, \tilde{\bar{d}}_{21} = \bar{d}_{21} - \hat{\bar{d}}_{21}, \tilde{\bar{d}}_{22} = \bar{d}_{22} - \hat{\bar{d}}_{22}$ and $\tilde{\bar{d}}_{23} = \bar{d}_{23} - \hat{\bar{d}}_{23}$, and the tracking errors are defined as $e_\beta = \beta - \beta_d$, $e_\alpha = \alpha - \alpha_d$ and $e_\mu = \mu - \mu_d$.

According to the above designed control parameters, the numerical simulation results are shown in the Figs. 4.1, 4.2, 4.3, 4.4, 4.5 and 4.6. The tracking control results are presented in Fig. 4.1. As can be seen from Fig. 4.1, the output signal $y(k)$ can quickly track the reference signal $x_d(k)$. The tracking errors e_β, e_α and e_μ are shown in Fig. 4.2, and the tracking errors can quickly converge. The estimated performance response of DTDO and the estimated errors response of DTDO are shown in the Figs. 4.3, 4.4, 4.5 and 4.6. According to Figs. 4.3, 4.4, 4.5 and 4.6, the designed nonlinear DTDO based on the NN can estimate the external disturbances. Therefore, according to the above numerical simulation results, the discrete-time adaptive NN control method based on the DTDO can obtain the effective control performance of the UAV attitude dynamic system.

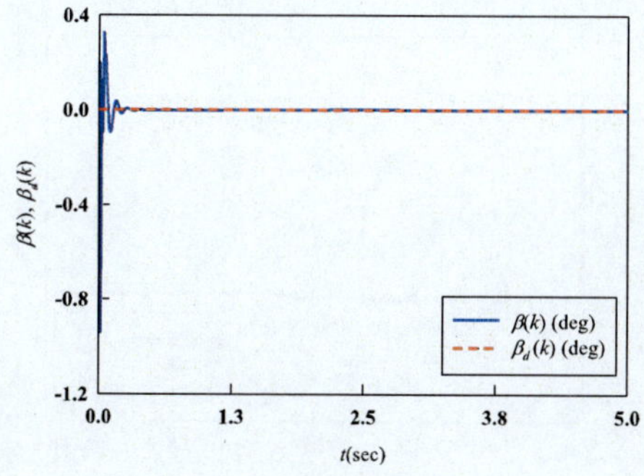

(a) β and β_d

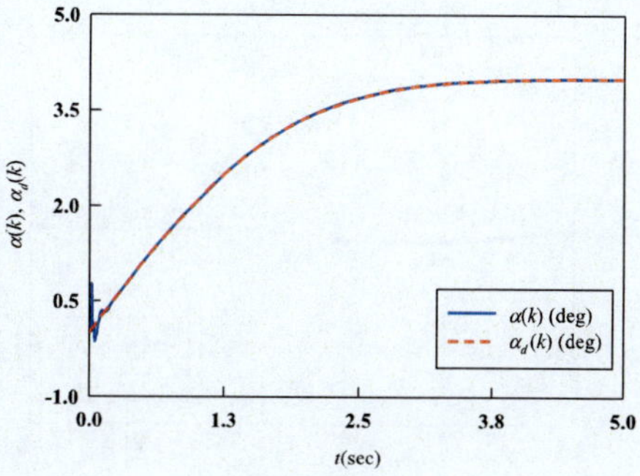

(b) α and α_d

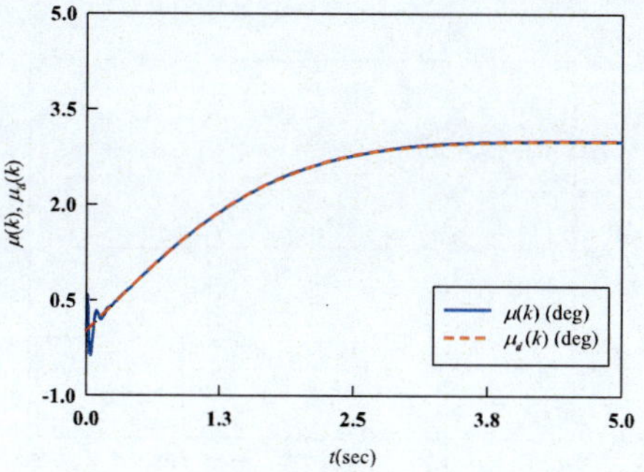

(c) μ and μ_d

Fig. 4.1 Attitude angle tracking response

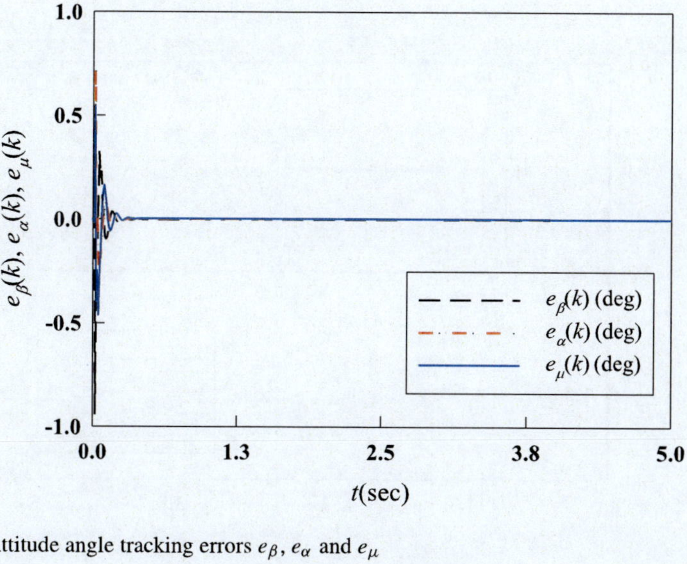

Fig. 4.2 Attitude angle tracking errors e_β, e_α and e_μ

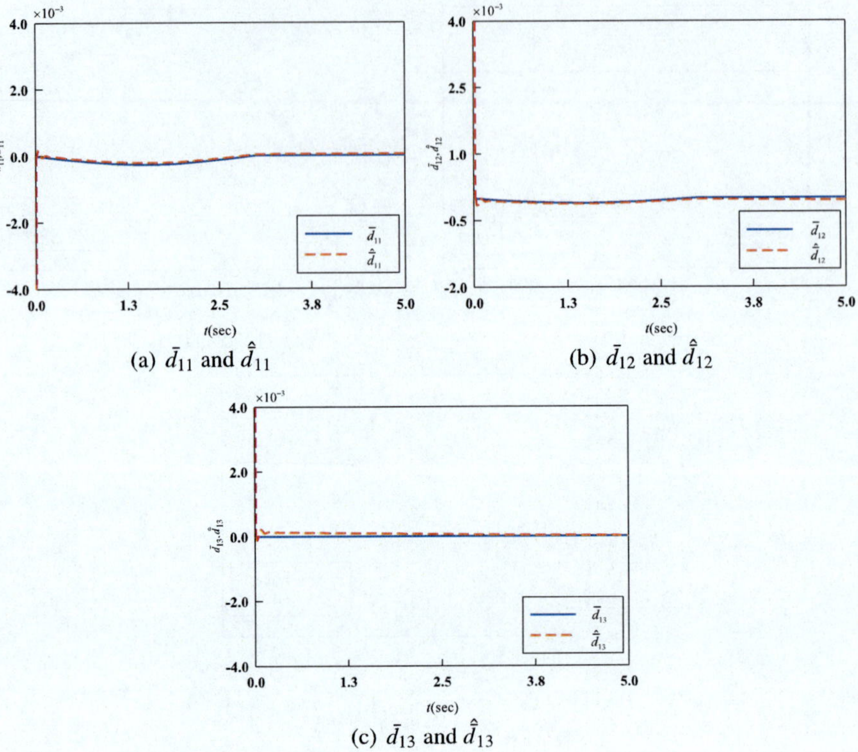

(a) $\bar{d}_{11}$ and $\hat{\bar{d}}_{11}$ (b) $\bar{d}_{12}$ and $\hat{\bar{d}}_{12}$

(c) $\bar{d}_{13}$ and $\hat{\bar{d}}_{13}$

Fig. 4.3 Estimation performance of DTDO

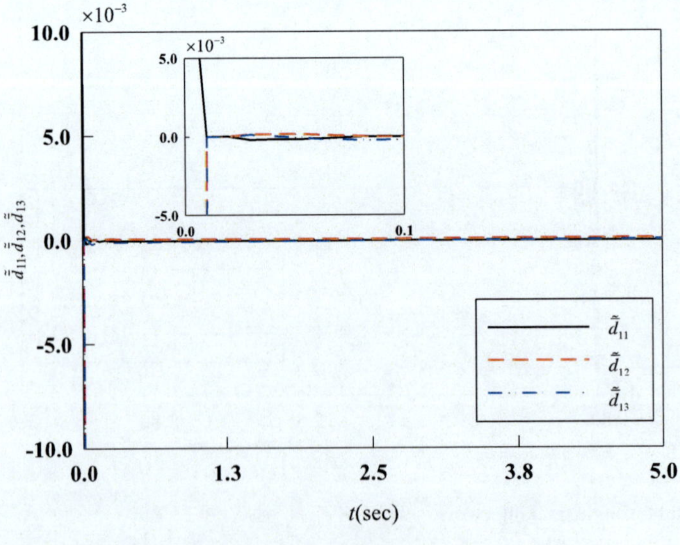

Fig. 4.4 Disturbance estimation errors $\tilde{\bar{d}}_{11}$, $\tilde{\bar{d}}_{12}$ and $\tilde{\bar{d}}_{13}$

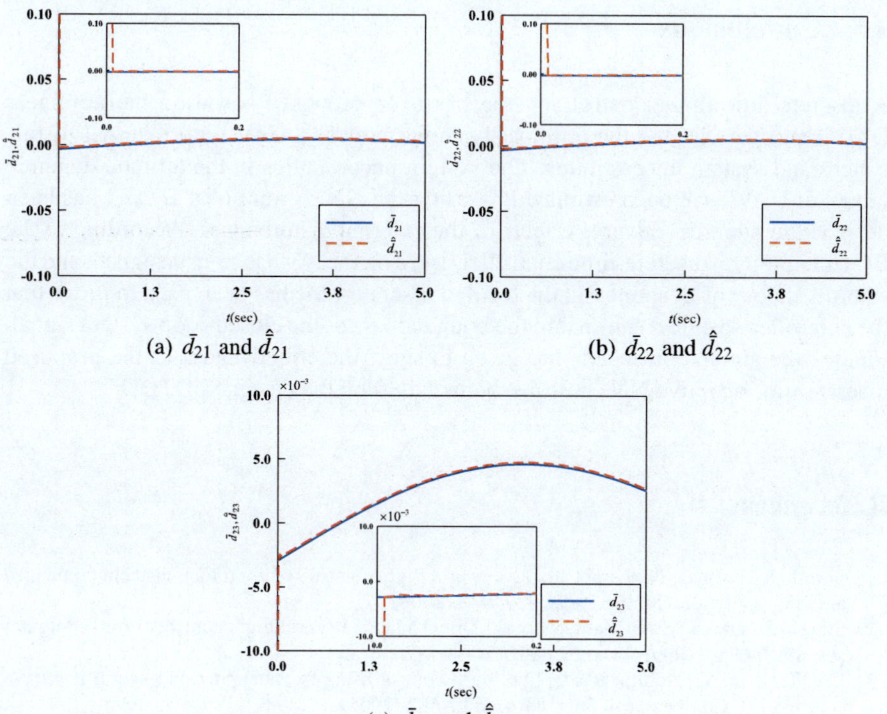

(a) $\bar{d}_{21}$ and $\hat{\bar{d}}_{21}$

(b) $\bar{d}_{22}$ and $\hat{\bar{d}}_{22}$

(c) $\bar{d}_{23}$ and $\hat{\bar{d}}_{23}$

Fig. 4.5 Estimation performance of DTDO

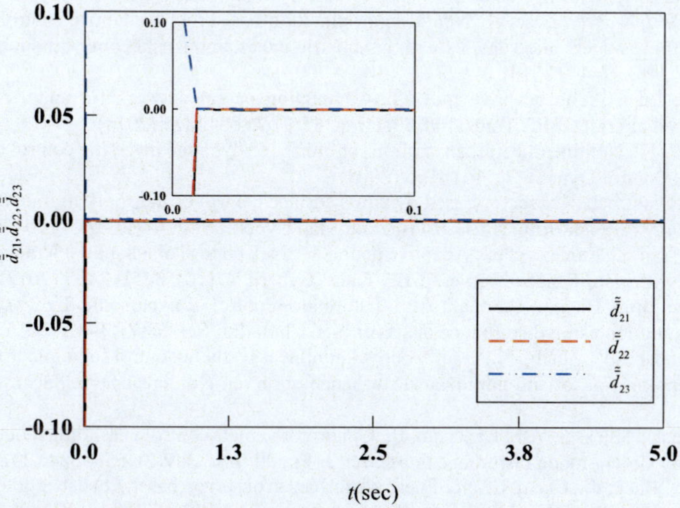

Fig. 4.6 Disturbance estimation errors $\tilde{\bar{d}}_{21}$, $\tilde{\bar{d}}_{22}$ and $\tilde{\bar{d}}_{23}$

4.5 Conclusions

A discrete-time flight control scheme has been proposed based on the nonlinear DTDO and the NN for the attitude dynamic model of UAV with external disturbances and system uncertainties. The system uncertainties in the attitude dynamic model of UAV have been estimated by using the NN. A nonlinear DTDO has been designed to suppress adverse effects of the external disturbances. According to the BC technology, a discrete-time controller based on the NN has been designed, and the stability theory of Lyapunov in the form of discrete-time has been used to prove that the controller designed can ensure the boundedness of the closed-loop system signal. Finally, the simulation results are given to show the effectiveness of the proposed discrete-time adaptive NN control scheme based on the designed DTDO.

References

1. Suresh, S., Kannan, N.: Direct adaptive neural flight control system for an unstable unmanned aircraft. Appl. Soft Comput. **8**(2), 937–948 (2008)
2. Suresh, S., Omkar, S.N., Mani, V., et al.: Direct adaptive neural flight controller for F-8 fighter aircraft. J. Guid. Control Dyn. **29**(2), 454–464 (2006)
3. Lee, T., Kim, Y.: Nonlinear adaptive flight control using backstepping and neural networks controller. J. Guid. Control Dyn. **24**(4), 675–682 (2001)
4. Lei, X., Ge, S.S., Fang, J.: Adaptive neural network control of small unmanned aerial rotorcraft. J. Intell. Robot. Syst. **75**(2), 331–341 (2014)
5. Zhang, C., Hu, H., Wang, J.: An adaptive neural network approach to the tracking control of micro aerial vehicles in constrained space. Int. J. Syst. Sci. **48**(1), 84–94 (2017)
6. Bu, X., Wu, X., Wei, D., et al.: Neural-approximation-based robust adaptive control of flexible air-breathing hypersonic vehicles with parametric uncertainties and control input constraints. Inf. Sci. **346**, 29–43 (2016)
7. Lei, X., Lu, P.: The adaptive radial basis function neural network for small rotary-wing unmanned aircraft. IEEE Trans. Ind. Electron. **61**(9), 4808–4815 (2014)
8. Chen, W.-H.: Nonlinear disturbance observer-enhanced dynamic inversion control of missiles. J. Guid. Control Dyn. **26**(1), 161–166 (2003)
9. Xu, B.: Disturbance observer-based dynamic surface control of transport aircraft with continuous heavy cargo airdrop. IEEE Trans. Syst. Man Cybern. Syst. **47**(1), 161–170 (2017)
10. He, W., Yan, Z., Sun, C., et al.: Adaptive neural network control of a flapping wing micro aerial vehicle with disturbance observer. IEEE Trans. Cybern. **47**(10), 3452–3465 (2017)
11. Chen, M., Ren, B., Wu, Q., et al.: Anti-disturbance control of hypersonic flight vehicles with input saturation using disturbance observer. Sci. China Inf. Sci. **58**(7), 1–12 (2015)
12. Chen, F., Lei, W., Zhang, K., et al.: A novel nonlinear resilient control for a quadrotor uav via backstepping control and nonlinear disturbance observer. Nonlinear Dyn. **85**(2), 1281–1295 (2016)
13. Besnard, L., Shtessel, Y.B., Landrum, B.: Quadrotor vehicle control via sliding mode controller driven by sliding mode disturbance observer. J. Frankl. Inst. **349**(2), 658–684 (2012)
14. Lee, K., Back, J., Choy, I.: Nonlinear disturbance observer based robust attitude tracking controller for quadrotor UAVs. Int. J. Control Autom. Syst. **12**(6), 1266–1275 (2014)
15. Wang, H., Chen, M.: Trajectory tracking control for an indoor quadrotor UAV based on the disturbance observer. Trans. Inst. Meas. Control **38**(6), 675–692 (2016)
16. Yang, J., Li, S., Sun, C., et al.: Nonlinear-disturbance-observer-based robust flight control for airbreathing hypersonic vehicles. IEEE Trans. Aerosp. Electron. Syst. **49**(2), 1263–1275 (2013)

17. Wu, G., Meng, X.: Nonlinear disturbance observer based robust backstepping control for a flexible air-breathing hypersonic vehicle. Aerosp. Sci. Technol. **54**, 174–182 (2016)
18. Sun, H., Li, S., Yang, J., et al.: Non-linear disturbance observer-based back-stepping control for airbreathing hypersonic vehicles with mismatched disturbances. IET Control Theory Appl. **8**(17), 1852–1865 (2014)
19. Chen, M., Yu, J.: Disturbance observer-based adaptive sliding mode control for near-space vehicles. Nonlinear Dyn. **82**(4), 1671–1682 (2015)
20. Lu, H., Liu, C., Guo, L., et al.: Flight control design for small-scale helicopter using disturbance-observer-based backstepping. J. Guid. Control Dyn. **38**(11), 2235–2240 (2015)
21. Han, Y., Li, P., Zheng, Z.: A non-decoupled backstepping control for fixed-wing UAVs with multivariable fixed-time sliding mode disturbance observer. In: Transactions of the Institute of Measurement and Control (2018). https://doi.org/10.1177/0142331218793178
22. Smith, J., Su, J., Liu, C., et al.: Disturbance observer based control with anti-windup applied to a small fixed wing UAV for disturbance rejection. J. Intell. Robot. Syst. **88**(2–4), 329–346 (2017)
23. Osa, Y., Mabuchi, T., Uchikado, S.: Synthesis of discrete time adaptive flight control system using nonlinear model matching. IEEE Int. Symp. Ind. Electron. **1**, 58–63 (2001)
24. Xiong, J.-J., Zhang, G.: Discrete-time sliding mode control for a quadrotor UAV. Optik **127**(8), 3718–3722 (2016)
25. Jiang, B., Chowdhury, F.N.: Fault estimation and accommodation for linear MIMO discrete-time systems. IEEE Trans. Control Syst. Technol. **13**(3), 493–499 (2005)
26. Xu, B., Sun, F., Yang, C., et al.: Adaptive discrete-time controller design with neural network for hypersonic flight vehicle via back-stepping. Int. J. Control **84**(9), 1543–1552 (2011)
27. Xu, B., Wang, D., Sun, F., et al.: Direct neural discrete control of hypersonic flight vehicle. Nonlinear Dyn. **70**(1), 269–278 (2012)
28. Xu, B., Zhang, Y.: Neural discrete back-stepping control of hypersonic flight vehicle with equivalent prediction model. Neurocomputing **154**, 337–346 (2015)
29. Shin, D.-H., Kim, Y.: Nonlinear discrete-time reconfigurable flight control law using neural networks. IEEE Trans. Control Syst. Technol. **14**(3), 408–422 (2006)
30. Mareels, I.M., Penfold, H., Evans, R.J.: Controlling nonlinear time-varying systems via euler approximations. Automatica **28**(4), 681–696 (1992)
31. Yan, X., Chen, M., Feng, G. et al.: Fuzzy robust constrained control for nonlinear systems with input saturation and external disturbances. In: IEEE Transactions on Fuzzy Systems (2019). https://doi.org/10.1109/TFUZZ.2019.2952794

Chapter 5
Discrete-Time NN Attitude Tracking Control for UAV System with Disturbance and Input Saturation

5.1 Introduction

According to the physical characteristics of the aerodynamic rudder surface of the fixed-wing UAV, the rudder deflection angle, angular velocity and engine thrust of the fixed-wing UAV are all limited [1]. In recent years, some research works on considering input saturation in flight control have been reported [2–8]. The above research results proposed some corresponding control schemes for the continuous time systems with input saturation, and effectively inhibited the negative effects of input saturation on the control performance of the systems. However, so far, there are few researches on actuator constraint (that is, input saturation) in some research results of discrete-time flight control methods [9–15]. In addition, if the designed UAV control method does not consider the influence of the input saturation, when the deflection angle of the aerodynamic rudder surface reaches the limit boundary, it will no longer change, which will affect the control performance of UAV system and even destroy the stability of UAV system. Therefore, for the research of discrete-time flight control methods of UAV system, robust discrete-time constraint control should be further considered in the design of UAV control scheme. According to the above mentioned problems and based on the research in Chap. 4, a NN tracking control scheme is proposed based on a designed DTDO and an auxiliary system for the uncertain UAV attitude dynamic model with external disturbance and input saturation in this chapter.

5.2 Problem Statement

For actual control systems, the input constraints of the system are often unavoidable. Therefore, the input saturation problem is considered in the control study for the UAV system (2.41) in this section. Furthermore, to design a discrete-time control scheme

S. Shao et al., *Robust Discrete-Time Flight Control of UAV with External Disturbances*, Studies in Systems, Decision and Control 317,
https://doi.org/10.1007/978-3-030-57957-9_5

using a discrete-time model of Euler approximation [16], the continuous time model of the UAV is converted to an approximated discrete-time model subject to system uncertainties and external disturbances. The discrete-time nonlinear dynamics form is described as

$$\begin{cases} x_1(k+1) = \bar{F}_1(k) + \Delta\bar{F}_1(k) + \bar{G}_1(k)x_2(k) + \bar{d}_1(k) \\ x_2(k+1) = \bar{F}_2(k) + \Delta\bar{F}_2(k) + \bar{G}_2(k)\mathrm{sat}(u(k)) + \bar{d}_2(k) \\ y(k) = x_1(k) \end{cases} \tag{5.1}$$

where $\bar{F}_1(k) = \Delta T F_1(x_1(k)) + x_1(k)$, $\bar{G}_1(k) = \Delta T G_1(x_1(k))$, $\bar{F}_2(k) = \Delta T F_2(x(k)) + x_2(k)$, $\bar{G}_2(k) = \Delta T G_2(x(k))$, ΔT is a sampling period, $\Delta\bar{F}_1(k)$ are $\Delta\bar{F}_2(k)$ are unknown system uncertainties including the errors of Euler approximation. In addition, $\bar{d}_1(k) = [\bar{d}_{11}, \bar{d}_{12}, \bar{d}_{13}]^{\mathrm{T}}$ and $\bar{d}_2(k) = [\bar{d}_{21}, \bar{d}_{22}, \bar{d}_{23}]^{\mathrm{T}}$ are disturbances, the wind disturbances $\bar{d}_{11}$, $\bar{d}_{12}$, and $\bar{d}_{13}$ have been given in Chap. 3, $\bar{d}_2(k)$ are the external unknown time-varying disturbances. The saturation function $\mathrm{sat}(u(k))$ is described as $\mathrm{sat}(u(k)) = [\mathrm{sat}(u_1(k)), \mathrm{sat}(u_2(k)), \mathrm{sat}(u_3(k))]^{\mathrm{T}} \in \Re^3$. According to (2.54), $\mathrm{sat}(u_i(k))$ can be written as

$$\mathrm{sat}(u_i(k)) = \begin{cases} \bar{\upsilon}_{mi}\,\mathrm{sign}(u_i(k)), & |u_i(k)| \geq \bar{\upsilon}_{mi}, \\ u_i(k), & |u_i(k)| < \bar{\upsilon}_{mi}. \end{cases} \tag{5.2}$$

where $\bar{\upsilon}_{mi}$ denotes the bound of $\mathrm{sat}(u_i(k))$, $\mathrm{sign}(\cdot)$ denotes a sign function and $i = 1, 2, 3$.

In this chapter, an adaptive neural control method is proposed for the uncertain attitude dynamics model with consideration of the input saturation and external disturbances by utilizing a designed DTDO and the discrete-time tracking differentiator. The control purpose is to develop a DTDO-based neural controller such that: (1) the output signal $y(k)$ tracks a known and bounded signal $x_{1d}(k) = [\beta_d, \alpha_d, \mu_d]^{\mathrm{T}}$ to a bounded compact set; (2) the boundedness of all the signals is guaranteed in the closed-loop system. The developed control method can be showed by a block diagram in Fig. 5.1 for better demonstration of the design concept.

To demonstrate the design procedure towards the above-mentioned control objective, some assumptions are presented first:

Assumption 5.1 Based on the physical properties of the UAV system (2.41), the control coefficient matrix $\bar{G}_1(k)$ is invertible, and $\bar{G}_1(k)$ statifies that $\underline{g}_1 \leq \|\bar{G}_1(k)\| \leq \bar{g}_1$, where $\underline{g}_1 > 0$ and $\bar{g}_1 > 0$ are constants.

Assumption 5.2 For external disturbances $\bar{d}_{\bar{j}}(k)$ in the uncertain UAV attitude dynamic model, $\bar{d}_{\bar{j}}(k)$ and $\bar{d}_{\bar{j}}(k+1)$ are assumed to be bounded disturbances, and $\bar{j} = 1, 2$.

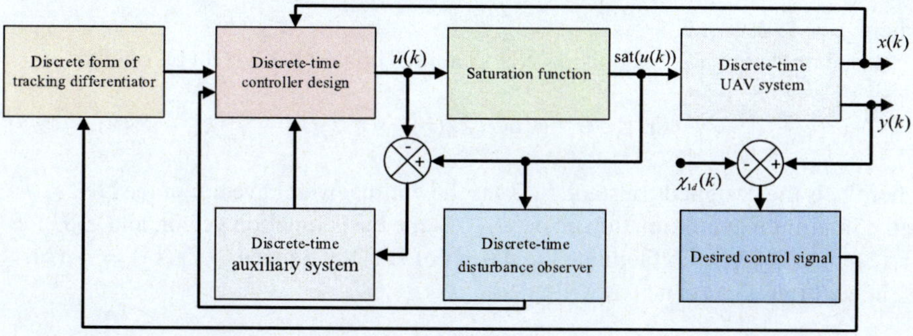

Fig. 5.1 The block diagram of the developed control strategy

5.3 Disturbance Observer-Based Tracking Control Scheme

In this section, the NN-based DTDO design method is first presented, and then an auxiliary system is designed to compensate for the influence of input saturation on the dynamic model control of discrete-time UAV system. Finally, according to the NN-based DTDO, the structured auxiliary system, the discrete-time tracking differentiator and the BC technology, a kind of NN-based discrete-time adaptive controller is designed, and the signals boundedness of the closed-loop system is proven by using Lyapunov stability theory.

5.3.1 Design of NN-Based DTDO

This section focuses on the discrete-time uncertain UAV system with external disturbances and input saturation (5.1), a kind of NN-based DTDO is designed. Firstly, according to (5.1), one has

$$\bar{x}(k+1) = \bar{F}(k) + \Delta\bar{F}(k) + \bar{v}(k) + \bar{d}(k) \tag{5.3}$$

where $\bar{x}(k+1) = [x_1^{\mathrm{T}}(k+1), x_2^{\mathrm{T}}(k+1)]^{\mathrm{T}}$ is the state vector, $\bar{F}(k) = [\bar{F}_1^{\mathrm{T}}(k), \bar{F}_2^{\mathrm{T}}(k)]^{\mathrm{T}}$ is the known nonlinear function vector, $\Delta\bar{F}(k) = [\Delta\bar{F}_1^{\mathrm{T}}(k), \Delta\bar{F}_2^{\mathrm{T}}(k)]^{\mathrm{T}}$ is the unknown nonlinear function vector, $\bar{d}(k) = [\bar{d}_1^{\mathrm{T}}(k), \bar{d}_2^{\mathrm{T}}(k)]^{\mathrm{T}}$ is the unknown disturbance vector, and $\bar{v}(k) = [(\bar{G}_1(k)x_2(k))^{\mathrm{T}}, (\bar{G}_2(k)\mathrm{sat}(u(k)))^{\mathrm{T}}]^{\mathrm{T}}$. Furthermore, one can define that $\bar{x}(k+1) = [\bar{x}_1(k+1), \ldots, \bar{x}_6(k+1)]^{\mathrm{T}} \in \Re^6$, $\bar{F}(k) = [\bar{f}_1(k), \ldots, \bar{f}_6(k)]^{\mathrm{T}} \in \Re^6$, $\Delta\bar{F}(k) = [\Delta\bar{f}_1(k), \ldots, \Delta\bar{f}_6(k)]^{\mathrm{T}} \in \Re^6$, $\bar{v}(k) = [\bar{v}_1(k), \ldots, \bar{v}_6(k)]^{\mathrm{T}} \in \Re^6$, $\bar{d}(k) = [\underline{d}_1(k), \ldots, \underline{d}_6(k)]^{\mathrm{T}} \in \Re^6$.

According to (5.3), the jth variable of $\bar{x}$ can be written as

$$\bar{x}_j(k+1) = \bar{f}_j(k) + \Delta\bar{f}_j(k) + \bar{v}_j(k) + \underline{d}_j(k) \tag{5.4}$$

where $j = 1, 2, \ldots, 6$.

Based on Lemma 2.2 and using NN to approximate $(1/\zeta_j)\Delta \bar{f}_j(k)$, one has

$$\bar{x}_j(k+1) = \bar{f}_j(k) + \zeta_j \theta_j^{*T}(k)\phi_j(Z_0(k)) + \zeta_j \varepsilon_j(k) + \bar{v}_j(k) + \underline{d}_j(k) \quad (5.5)$$

where ζ_j is the designed constant, $\theta_j^*(k)$ is the optimal weight vector of the NN, $\varepsilon_j(k)$ is the minimum approximation error, $\phi_j(\cdot)$ is the basis function vector, and $Z_0(k) = [\bar{x}_1(k), \ldots, \bar{x}_j(k)]^{\mathrm{T}}$. To facilitate the design of DTDO, define $\phi_j(Z_0(k)) = \phi_j(k)$.

In addition, (5.5) can be rewritten as

$$\bar{x}_j(k+1) = \bar{f}_j(k) + \zeta_j \theta_j^{*T}(k)\phi_j(k) + \bar{v}_j(k) + D_j(k) \quad (5.6)$$

where $D_j(k) = \zeta_j \varepsilon_j(k) + \underline{d}_j(k)$ is the complex disturbance.

On the basis of the approximation theory of the NN, the estimation error $\varepsilon_j(k+1)$ is bounded. Moreover, based on Assumption 5.2, it can be obtained that the disturbance $\underline{d}_j(k+1)$ is bounded. Thus, the complex disturbance $D_j(k+1) = \zeta_j \varepsilon_j(k+1) + \underline{d}_j(k+1)$ is also bounded. Therefore, there exists an unknown positive constant $\bar{\delta}_j$ which satisfies the following inequality:

$$|D_j(k+1) - D_j(k)| \leq \bar{\delta}_j \quad (5.7)$$

To design the NN-based DTDO, the following auxiliary variable is defined:

$$\mathscr{P}_j(k) = D_j(k) - \bar{x}_j(k) \quad (5.8)$$

Combining (5.6) and (5.8), $\mathscr{P}_i(k+1)$ can be written as

$$\mathscr{P}_j(k+1) = D_j(k+1) - (\bar{f}_j(k) + \zeta_j \theta_j^{*T}(k)\phi_j(k) + D_j(k) + \bar{v}_j(k)) \quad (5.9)$$

Then, the estimation of auxiliary variable $\hat{\mathscr{P}}_j(k)$ is described as

$$\hat{\mathscr{P}}_j(k+1) = -(\bar{f}_j(k) + \zeta_j \hat{\theta}_j^{T}(k)\phi_j(k) + \bar{v}_j(k)) \quad (5.10)$$

where $\hat{\mathscr{P}}_j(k)$ is the estimation of $\mathscr{P}_j(k)$, $\hat{\theta}_j(k)$ is the estimation of $\theta_j^*(k)$.

According to (5.8), the disturbance $D_j(k)$ can be estimated by

$$\hat{D}_j(k) = \hat{\mathscr{P}}_j(k) + \bar{x}_j(k) \quad (5.11)$$

where $\hat{D}_j(k)$ is the estimation of $D_j(k)$.

Defining $\tilde{D}_i(k) = D_i(k) - \hat{D}_i(k)$ and $\tilde{\mathscr{P}}_j(k) = \mathscr{P}_j(k) - \hat{\mathscr{P}}_j(k)$, and considering (5.8) and (5.11), one can obtain

$$\tilde{\mathscr{P}}_j(k) = \tilde{D}_j(k) \quad (5.12)$$

From (5.9) and (5.10), one has

$$\tilde{\mathscr{P}}_j(k+1) = D_j(k+1) - D_j(k) + \zeta_j \tilde{\theta}_j^T(k)\phi_j(k) \tag{5.13}$$

where $\tilde{\theta}_j(k) = \hat{\theta}_j(k) - \theta_j^*(k)$.

A theorem on the design of the NN-based DTDO is given as follows:

Theorem 5.1 *To compensate for the negative effects of the external disturbances in the uncertain discrete-time UAV system (5.1), an NN-based DTDO is designed as (5.10) and (5.11). The estimation error is bounded between the output of the proposed DTDO and the external disturbance.*

The proof of Theorem 5.1 and the adaptive law of $\hat{\theta}_j(k)$ will be given in next subsection.

5.3.2 Design of Control Based on Auxiliary System

To restrain the adverse effects of the input saturation in (5.1), the following auxiliary system is given [17]:

$$\begin{cases} \hbar_1(k+1) = -\mathscr{C}_1\hbar_1(k) + \mathscr{G}_1\bar{G}_1(k)\hbar_2(k) \\ \hbar_2(k+1) = -\mathscr{C}_2\hbar_2(k) + \bar{G}_2(k)\Delta v(k) \end{cases} \tag{5.14}$$

where $\hbar_{\bar{j}}(k) \in \Re^3 (\bar{j} = 1, 2)$ are the variables, $\mathscr{C}_{\bar{j}} = \mathscr{C}_{\bar{j}}^{\mathrm{T}} > 0$ are design diagonal matrices, $\mathscr{G}_1 \in \Re^{3\times 3}$ is a design diagonal matrix with $\mathscr{G}_1^{\mathrm{T}} = \mathscr{G}_1 > 0$, and $\Delta v(k) = \mathrm{sat}(u(k)) - u(k)$.

Remark 5.1 In the present world, the control energy cannot be supplied infinitely by the actuator for the flight control systems. Thus, $\Delta v(k)$ can be assumed to be bounded, and $\Delta v(k)$ satisfies $\|\Delta v(k)\| \le \nu$ based on the system controllability, where $\nu > 0$ is a constant [1, 17, 18].

From $\Delta v(k) = \mathrm{sat}(u(k)) - u(k)$, (5.1) can be written as

$$\begin{cases} x_1(k+1) = \bar{F}_1(k) + \Delta\bar{F}_1(k) + \bar{G}_1(k)x_2(k) + \bar{d}_1(k) \\ x_2(k+1) = \bar{F}_2(k) + \Delta\bar{F}_2(k) + \bar{G}_2(k)u(k) + \bar{G}_2(k)\Delta v(k) + \bar{d}_2(k) \\ y(k) = x_1(k) \end{cases} \tag{5.15}$$

According to (5.15), the discrete-time controller will be designed to guarantee that the output signal $y(k)$ tracks a known and bounded signal $x_{1d}(k)$ to a bounded compact set. Based on the backstepping technique and the NN-based DTDO, the following design method of discrete-time tracking controller is presented:

Step 1: Defining error variables as $z_1(k) = x_1(k) - x_{1d}(k) - \hbar_1(k)$ and $z_2(k) = x_2(k) - \mathrm{sat}(x_{vd}(k)) - \hbar_2(k)$, where $x_{vd}(k)$ denotes the designed signal, and based on (5.15), one can obtain

$$\begin{aligned} z_1(k+1) &= x_1(k+1) - x_{1d}(k+1) - \hbar_1(k+1) \\ &= \bar{F}_1(k) + \Delta\bar{F}_1(k) + \bar{G}_1(k)(z_2(k) + \mathrm{sat}(x_{vd}(k))) \\ &\quad + \bar{G}_1(k)\hbar_2(k) + \bar{d}_1(k) - x_{1d}(k+1) - \hbar_1(k+1) \end{aligned} \tag{5.16}$$

For the saturation function $\mathrm{sat}(x_{vd}(k))$, the handling method in Chap. 4 is used. Then, (5.16) can be written as

$$\begin{aligned} z_1(k+1) = {} & \bar{F}_1(k) + \Delta\bar{F}_1(k) + \bar{G}_1(k)(z_2(k) + \mathscr{Y}(k) \\ & + \lambda\!\!\!^{-}(k)x_{vd}(k)) + \bar{G}_1(k)\hbar_2(k) + \bar{d}_1(k) \\ & - x_{1d}(k+1) - \hbar_1(k+1) \end{aligned} \tag{5.17}$$

where $\mathscr{Y}(k)$ is the approximation error, $\lambda\!\!\!^{-}(k)$ is the unknown and bounded diagonal matrix, $\|\lambda\!\!\!^{-}(k)\| \le \bar{\ell}$, and $\bar{\ell}$ is a positive constant.

According to Lemma 2.2, by using the NN to approximate $(1/\zeta_{1i})\Delta\bar{F}_{1i}(k)(i = 1, 2, 3)$ with $\Delta\bar{F}_{1i}(k)$ being the ith variable of $\Delta\bar{F}_1(k)$, it yields

$$\begin{aligned} z_1(k+1) = {} & \bar{F}_1(k) + \Xi_1(k) + \bar{G}_1(k)(z_2(k) + \mathscr{Y}(k) + \lambda\!\!\!^{-}(k)x_{vd}(k)) \\ & + \bar{G}_1(k)\hbar_2(k) + \bar{d}_1(k) - x_{1d}(k+1) \\ & + \mathrm{E}_1(k) - \hbar_1(k+1) \end{aligned} \tag{5.18}$$

where $\Xi_1(k) = [\zeta_{11}\theta_{11}^{*T}(k)\phi_{11}(Z_1(k)), \zeta_{12}\theta_{12}^{*T}(k)\phi_{12}(Z_1(k)), \zeta_{13}\theta_{13}^{*T}(k)\phi_{13}(Z_1(k))]^{\mathrm{T}}$, $Z_1(k) = x_1(k)$, $\mathrm{E}_1(k) = [\zeta_{11}\varepsilon_{11}(k), \zeta_{12}\varepsilon_{12}(k), \zeta_{13}\varepsilon_{13}(k)]^{\mathrm{T}}$, $\theta_{1i}^*(k)$ is the optimal weight vector of the NN, $\varepsilon_{1i}(k)$ is the minimum approximation error, $\phi_{1i}(\cdot)$ is the basis function vector. $\Delta\bar{F}_{1i}(k)$ is the ith variable of $\Delta\bar{F}_1(k)$, ζ_{1i} is a positive constant. To facilitate the design of control scheme, define $\phi_{1i}(Z_1(k)) = \phi_{1i}(k)$ with $i = 1, 2, 3$.

Furthermore, one can have

$$\begin{aligned} z_1(k+1) = {} & \bar{F}_1(k) + \Xi_1(k) + \bar{G}_1(k)(z_2(k) + \mathscr{Y}(k) + \lambda\!\!\!^{-}(k)x_{vd}(k)) \\ & + \bar{G}_1(k)\hbar_2(k) + \mathscr{D}_1(k) - x_{1d}(k+1) - \hbar_1(k+1) \end{aligned} \tag{5.19}$$

where $\mathscr{D}_1(k) = \mathrm{E}_1(k) + \bar{d}_1(k)$ denotes the compound disturbance.

From (5.14) and (5.19), it yields

$$\begin{aligned} z_1(k+1) = {} & \bar{F}_1(k) + \Xi_1(k) + \bar{G}_1(k)(z_2(k) + \mathscr{Y}(k) + \lambda\!\!\!^{-}(k)x_{vd}(k)) \\ & + \mathscr{D}_1(k) - x_{1d}(k+1) + \mathscr{C}_1\hbar_1(k) \\ & + (\mathscr{I} - \mathscr{G}_1)\bar{G}_1(k)\hbar_2(k) \end{aligned} \tag{5.20}$$

where $\mathscr{I} \in \Re^{3\times3}$ is the identity matrix.

In order to deal with the disturbance $\mathscr{D}_1(k)$ in (5.20), the DTDO is designed according to (5.10) and (5.11), which is given by

$$\begin{cases} \hat{\mathscr{P}}_{1i}(k+1) = -(\bar{F}_{1i}(k) + \zeta_{1i}\hat{\theta}_{1i}^T(k)\phi_{1i}(k) + \mathscr{V}_{1i}(k)) \\ \hat{\mathscr{D}}_{1i}(k) = \hat{\mathscr{P}}_{1i}(k) + x_{1i}(k) \end{cases} \tag{5.21}$$

where $i = 1, 2, 3$, $x_{1i}(k)$ denotes the ith variable of $x_1(k)$, $\hat{\mathscr{D}}_{1i}(k)$ is the output of DTDO, and $\hat{\mathscr{D}}_1(k) = [\hat{\mathscr{D}}_{11}(k), \hat{\mathscr{D}}_{12}(k), \hat{\mathscr{D}}_{13}(k)]^{\mathrm{T}}$. $\hat{\mathscr{P}}_{1i}(k)$ denotes the variable of the DTDO. The disturbance estimation error is defined as $\tilde{\mathscr{D}}_{1i}(k) = \mathscr{D}_{1i}(k) - \hat{\mathscr{D}}_{1i}(k)$. $\mathscr{V}_{1i}(k)$ denotes the ith variable of $\mathscr{V}_1(k)$ and $\mathscr{V}_1(k) = \bar{G}_1(k)x_2(k)$. Furthermore, according to (5.7), one can obtain that $|\mathscr{D}_{1i}(k+1) - \mathscr{D}_{1i}(k)| \leq \bar{\delta}_{1i}$, and $\bar{\delta}_{1i}$ is a positive constant.

According to the designed method of control signal $x_{vd}(k)$ in Chap. 4, Lemma 2.2 and (5.20), the NN is used to approximate the uncertainty $(1/\zeta_{0i})\Delta\bar{\mathscr{F}}_{1i}(k)(i = 1, 2, 3)$ with $\Delta\bar{\mathscr{F}}_{1i}(k)$ being the ith variable of $\Delta\bar{\mathscr{F}}_1(k) = -(\lambda(k)\bar{G}_1(k))^{-1}(\bar{F}_1(k) + \hat{\Xi}_1(k) + \hat{\mathscr{D}}_1(k) + \mathscr{C}_1\hbar_1(k) - x_{1d}(k+1)) + \bar{G}_1^{-1}(k)(\bar{F}_1(k) + \hat{\Xi}_1(k) + \hat{\mathscr{D}}_1(k) + \mathscr{C}_1\hbar_1(k) - x_{1d}(k+1))$, the control signal $x_{vd}(k)$ can be written as

$$\begin{aligned} x_{vd}(k) = &-\bar{G}_1^{-1}(k)(\bar{F}_1(k) + \hat{\Xi}_1(k) + \hat{\mathscr{D}}_1(k) + \mathscr{C}_1\hbar_1(k)) \\ &+\bar{G}_1^{-1}(k)x_{1d}(k+1) + \Xi_0(k) + \mathrm{E}_0(k) \end{aligned} \tag{5.22}$$

where $\Xi_0(k) = [\zeta_{01}\theta_{01}^{*\mathrm{T}}(k)\phi_{01}(Z_0(k)), \zeta_{02}\theta_{02}^{*\mathrm{T}}(k)\phi_{02}(Z_0(k)), \zeta_{03}\theta_{03}^{*\mathrm{T}}(k)\phi_{03}(Z_0(k))]^{\mathrm{T}}$, $\hat{\Xi}_1(k) = [\zeta_{11}\hat{\theta}_{11}^{\mathrm{T}}(k)\phi_{11}(k), \zeta_{12}\hat{\theta}_{12}^{\mathrm{T}}(k)\phi_{12}(k), \zeta_{13}\hat{\theta}_{13}^{\mathrm{T}}(k)\phi_{13}(k)]^{\mathrm{T}}$, $\hat{\mathscr{D}}_1(k)$ is the estimation of $\mathscr{D}_1(k)$, $\mathrm{E}_0(k) = [\zeta_{01}\varepsilon_{01}(k), \zeta_{02}\varepsilon_{02}(k), \zeta_{03}\varepsilon_{03}(k)]^{\mathrm{T}}$, $\hat{\Xi}_1(k)$ is the estimation of $\Xi_1(k)$, $Z_0(k) = [x_1^{\mathrm{T}}(k), \hbar_1{}^{\mathrm{T}}(k), \hat{\Xi}_1^{\mathrm{T}}(k), \hat{\mathscr{D}}_1^{\mathrm{T}}(k)]^{\mathrm{T}}$, ζ_{0i} is the designed constant, $\theta_{0i}^*(k)$ is the optimal weight vector of the NN, $\varepsilon_{0i}(k)$ is the minimum approximation error, $\phi_{0i}(\cdot)$ is the basis function vector. To facilitate the design of control scheme, define $\phi_{0i}(Z_0(k)) = \phi_{0i}(k)$.

Then, the control signal $x_{vd}(k)$ is designed as

$$\begin{aligned} x_{vd}(k) = &-\bar{G}_1^{-1}(k)(\bar{F}_1(k) + \hat{\Xi}_1(k) + \hat{\mathscr{D}}_1(k) + \mathscr{C}_1\hbar_1(k)) \\ &+\bar{G}_1^{-1}(k)x_{1d}(k+1) + \hat{\Xi}_0(k) \end{aligned} \tag{5.23}$$

where $\hat{\Xi}_0(k) = [\zeta_{01}\hat{\theta}_{01}^{\mathrm{T}}(k)\phi_{01}(k), \zeta_{02}\hat{\theta}_{02}^{\mathrm{T}}(k)\phi_{02}(k), \zeta_{03}\hat{\theta}_{03}^{\mathrm{T}}(k)\phi_{03}(k)]^{\mathrm{T}}$ and $\hat{\Xi}_0(k)$ is the estimation of $\Xi_0(k)$.

Substituting (5.23) into (5.22), one can obtain

$$\begin{aligned} z_1(k+1) = &\tilde{\mathscr{D}}_1(k) - \tilde{\Xi}_1(k)) + \lambda(k)\bar{G}_1(k)\tilde{\Xi}_0(k) + \bar{G}_1(k)z_2(k) \\ &+(\mathscr{I} - \mathscr{G}_1)\bar{G}_1(k)\hbar_2(k) - \lambda(k)\bar{G}_1(k)\mathrm{E}_0(k) + \bar{G}_1(k)\mathscr{Y}(k) \end{aligned} \tag{5.24}$$

where $\tilde{\Xi}_1(k) = [\zeta_{11}\tilde{\theta}_{11}^{\mathrm{T}}(k)\phi_{11}(k), \zeta_{12}\tilde{\theta}_{12}^{\mathrm{T}}(k)\phi_{12}(k), \zeta_{13}\tilde{\theta}_{13}^{\mathrm{T}}(k)\phi_{13}(k)]^{\mathrm{T}}$, $\tilde{\Xi}_1(k) = \hat{\Xi}_1(k) - \Xi_1(k)$, $\tilde{\Xi}_0(k) = [\zeta_{01}\tilde{\theta}_{01}^{\mathrm{T}}(k)\phi_{01}(k), \zeta_{02}\tilde{\theta}_{02}^{\mathrm{T}}(k)\phi_{02}(k), \zeta_{03}\tilde{\theta}_{03}^{\mathrm{T}}(k)\phi_{03}(k)]^{\mathrm{T}}$, and $\tilde{\Xi}_0(k) = \hat{\Xi}_0(k) - \Xi_0(k)$.

In addition, the adaptive law of $\hat{\theta}_{1i}(k)$ is designed by

$$\hat{\theta}_{1i}(k+1) = \hat{\theta}_{1i}(k) + h_{1i}\sigma_{1i}\phi_{1i}(k)z_{1i}(k+1) - h_{1i}\varsigma_{1i}\hat{\theta}_{1i}(k) \tag{5.25}$$

where $z_{1i}(k+1)$ denotes the ith variable of $z_1(k+1)$, h_{1i}, σ_{1i} and ς_{1i} are designed constants.

From (5.25), one has

$$\tilde{\theta}_{1i}(k+1) = \tilde{\theta}_{1i}(k) + h_{1i}\sigma_{1i}\phi_{1i}(k)z_{1i}(k+1) - h_{1i}\varsigma_{1i}\hat{\theta}_{1i}(k) \tag{5.26}$$

where $\tilde{\theta}_{1i}(k+1) = \hat{\theta}_{1i}(k+1) - \theta_{1i}^*(k+1)$ and $\tilde{\theta}_{1i}(k) = \hat{\theta}_{1i}(k) - \theta_{1i}^*(k)$.

Furthermore, the adaptive law of $\hat{\theta}_{0i}(k)$ is designed by

$$\hat{\theta}_{0i}(k+1) = \hat{\theta}_{0i}(k) + h_{0i}\sigma_{1i}\phi_{0i}(k)z_{1i}(k+1) - h_{0i}\varsigma_{0i}\hat{\theta}_{0i}(k) \tag{5.27}$$

where $z_{1i}(k+1)$ denotes the ith variable of $z_1(k+1)$, h_{0i} and ς_{0i} are designed constants.

From (5.27), one can obtain

$$\tilde{\theta}_{0i}(k+1) = \tilde{\theta}_{0i}(k) + h_{0i}\sigma_{1i}\phi_{0i}(k)z_{1i}(k+1) - h_{0i}\varsigma_{0i}\hat{\theta}_{0i}(k) \tag{5.28}$$

where $\tilde{\theta}_{0i}(k+1) = \hat{\theta}_{0i}(k+1) - \theta_{0i}^*(k+1)$ and $\tilde{\theta}_{0i}(k) = \hat{\theta}_{0i}(k) - \theta_{0i}^*(k)$.

To analyze the boundedness of the tracking error, the Lyapunov function is selected as follows:

$$\begin{aligned} L_1(k) = {} & \eta_1 z_1^{\mathrm{T}}(k)z_1(k) + \sum_{i=1}^{3}\frac{\zeta_{1i}}{h_{1i}}\tilde{\theta}_{1i}^{\mathrm{T}}(k)\tilde{\theta}_{1i}(k) + \bar{\hbar}_1\hbar_1^{\mathrm{T}}(k)\hbar_1(k) \\ & + \sum_{i=1}^{3}\tilde{\mathscr{D}}_{1i}^2(k) + \sum_{i=1}^{3}\frac{\zeta_{0i}}{h_{0i}}\tilde{\theta}_{0i}^{\mathrm{T}}(k)\tilde{\theta}_{0i}(k) \end{aligned} \tag{5.29}$$

where η_1 is the minimum value of $\sigma_{1i}(i = 1, 2, 3)$, $\bar{\hbar}_1$ is the value of $\|\mathscr{C}_1\|$.

According to (5.29), one has

$$\begin{aligned} \Delta L_1(k) = {} & L_1(k+1) - L_1(k) \\ = {} & \eta_1 z_1^{\mathrm{T}}(k+1)z_1(k+1) - \eta_1 z_1^{\mathrm{T}}(k)z_1(k) + \sum_{i=1}^{3}\frac{\zeta_{1i}}{h_{1i}}\tilde{\theta}_{1i}^{\mathrm{T}}(k+1)\tilde{\theta}_{1i}(k+1) \\ & - \sum_{i=1}^{3}\tilde{\mathscr{D}}_{1i}^2(k) + \bar{\hbar}_1\hbar_1^{\mathrm{T}}(k+1)\hbar_1(k+1) - \bar{\hbar}_1\hbar_1^{\mathrm{T}}(k)\hbar_1(k) \end{aligned}$$

$$+\sum_{i=1}^{3}\tilde{\mathscr{D}}_{1i}^{2}(k+1)-\sum_{i=1}^{3}\frac{\zeta_{1i}}{h_{1i}}\tilde{\theta}_{1i}^{\mathrm{T}}(k)\tilde{\theta}_{1i}(k)-\sum_{i=1}^{3}\frac{\zeta_{0i}}{h_{0i}}\tilde{\theta}_{0i}^{\mathrm{T}}(k)\tilde{\theta}_{0i}(k)$$
$$+\sum_{i=1}^{3}\frac{\zeta_{0i}}{h_{0i}}\tilde{\theta}_{0i}^{\mathrm{T}}(k+1)\tilde{\theta}_{0i}(k+1) \tag{5.30}$$

Based on (5.20) and (5.24), it yields

$$\begin{aligned}z_1^{\mathrm{T}}(k+1)z_1(k+1) = {} & z_1^{\mathrm{T}}(k+1)(\tilde{\mathscr{D}}_1(k)-\tilde{\Xi}_1(k))+z_1^{\mathrm{T}}(k+1)\bar{G}_1(k)z_2(k)\\ & +z_1^{\mathrm{T}}(k+1)(\mathscr{I}-\mathscr{G}_1)\bar{G}_1(k)\hbar_2(k)+z_1^{\mathrm{T}}(k+1)\lambda(k)\bar{G}_1(k)\tilde{\Xi}_0(k)\\ & +z_1^{\mathrm{T}}(k+1)(\bar{G}_1(k)\mathscr{Y}(k)-\lambda(k)\bar{G}_1(k)\mathrm{E}_0(k))\end{aligned} \tag{5.31}$$

According to (5.25)–(5.28), one can obtain

$$\begin{aligned}\sum_{i=1}^{3}\tfrac{\zeta_{1i}}{h_{1i}}\tilde{\theta}_{1i}^{\mathrm{T}}(k+1)\tilde{\theta}_{1i}(k+1) = {} & \sum_{i=1}^{3}\tfrac{\zeta_{1i}}{h_{1i}}\tilde{\theta}_{1i}^{\mathrm{T}}(k+1)\tilde{\theta}_{1i}(k)\\ +\sum_{i=1}^{3}\zeta_{1i}\sigma_{1i}\tilde{\theta}_{1i}^{\mathrm{T}}(k+1)\phi_{1i}(k)z_{1i}(k+1) & -\sum_{i=1}^{3}\zeta_{1i}\varsigma_{1i}\tilde{\theta}_{1i}^{\mathrm{T}}(k+1)\hat{\theta}_{1i}(k)\end{aligned} \tag{5.32}$$

$$\begin{aligned}\sum_{i=1}^{3}\sigma_{1i}\zeta_{1i}\tilde{\theta}_{1i}^{\mathrm{T}}(k+1)\phi_{1i}(k)z_{1i}(k+1) = {} & \sum_{i=1}^{3}\sigma_{1i}\zeta_{1i}\tilde{\theta}_{1i}^{\mathrm{T}}(k)\phi_{1i}(k)z_{1i}(k+1)\\ & -\sum_{i=1}^{3}\sigma_{1i}\zeta_{1i}h_{1i}\varsigma_{1i}\hat{\theta}_{1i}^{\mathrm{T}}(k)\phi_{1i}(k)z_{1i}(k+1)\\ & +\sum_{i=1}^{3}\sigma_{1i}^{2}\zeta_{1i}h_{1i}\phi_{1i}^{\mathrm{T}}(k)z_{1i}(k+1)\phi_{1i}(k)z_{1i}(k+1)\end{aligned} \tag{5.33}$$

$$\begin{aligned}-\sum_{i=1}^{3}\zeta_{1i}\varsigma_{1i}\tilde{\theta}_{1i}^{\mathrm{T}}(k+1)\hat{\theta}_{1i}(k) = {} & -\sum_{i=1}^{3}\zeta_{1i}\varsigma_{1i}\tilde{\theta}_{1i}^{\mathrm{T}}(k)\hat{\theta}_{1i}(k)\\ -\sum_{i=1}^{3}\zeta_{1i}\sigma_{1i}\varsigma_{1i}h_{1i}\phi_{1i}^{\mathrm{T}}(k)z_{1i}(k+1)\hat{\theta}_{1i}(k) & +\sum_{i=1}^{3}\zeta_{1i}\varsigma_{1i}^{2}h_{1i}\hat{\theta}_{1i}^{\mathrm{T}}(k)\hat{\theta}_{1i}(k)\end{aligned} \tag{5.34}$$

$$\begin{aligned}-\sum_{i=1}^{3}\tfrac{\zeta_{1i}}{h_{1i}}\tilde{\theta}_{1i}^{\mathrm{T}}(k)\tilde{\theta}_{1i}(k) = {} & -\sum_{i=1}^{3}\tfrac{\zeta_{1i}}{h_{1i}}\tilde{\theta}_{1i}^{\mathrm{T}}(k)\tilde{\theta}_{1i}(k+1)\\ +\sum_{i=1}^{3}\zeta_{1i}\sigma_{1i}\tilde{\theta}_{1i}^{\mathrm{T}}(k)\phi_{1i}(k)z_{1i}(k+1) & -\sum_{i=1}^{3}\zeta_{1i}\varsigma_{1i}\tilde{\theta}_{1i}^{\mathrm{T}}(k)\hat{\theta}_{1i}(k)\end{aligned} \tag{5.35}$$

$$
\begin{aligned}
&\sum_{i=1}^{3}\tfrac{\zeta_{0i}}{h_{0i}}\tilde{\theta}_{0i}^{\mathrm{T}}(k+1)\tilde{\theta}_{0i}(k+1)=\sum_{i=1}^{3}\tfrac{\zeta_{0i}}{h_{0i}}\tilde{\theta}_{0i}^{\mathrm{T}}(k+1)\tilde{\theta}_{0i}(k)\\
&+\sum_{i=1}^{3}\zeta_{0i}\sigma_{1i}\tilde{\theta}_{0i}^{\mathrm{T}}(k+1)\phi_{0i}(k)z_{1i}(k+1)-\sum_{i=1}^{3}\zeta_{0i}\varsigma_{0i}\tilde{\theta}_{0i}^{\mathrm{T}}(k+1)\hat{\theta}_{0i}(k)
\end{aligned}
\tag{5.36}
$$

$$
\begin{aligned}
\sum_{i=1}^{3}\zeta_{0i}\sigma_{1i}\tilde{\theta}_{0i}^{\mathrm{T}}(k+1)\phi_{0i}(k)z_{1i}(k+1)&=\sum_{i=1}^{3}\zeta_{0i}\sigma_{1i}\tilde{\theta}_{0i}^{\mathrm{T}}(k)\phi_{0i}(k)z_{1i}(k+1)\\
&-\sum_{i=1}^{3}\zeta_{0i}\sigma_{1i}h_{0i}\varsigma_{0i}\hat{\theta}_{0i}^{\mathrm{T}}(k)\phi_{0i}(k)z_{1i}(k+1)\\
&+\sum_{i=1}^{3}\zeta_{0i}\sigma_{1i}^{2}h_{0i}\phi_{0i}^{\mathrm{T}}(k)z_{0i}(k+1)\phi_{0i}(k)z_{1i}(k+1)
\end{aligned}
\tag{5.37}
$$

$$
\begin{aligned}
&-\sum_{i=1}^{3}\zeta_{0i}\varsigma_{0i}\tilde{\theta}_{0i}^{\mathrm{T}}(k+1)\hat{\theta}_{0i}(k)=-\sum_{i=1}^{3}\zeta_{0i}\varsigma_{0i}\tilde{\theta}_{0i}^{\mathrm{T}}(k)\hat{\theta}_{0i}(k)\\
&-\sum_{i=1}^{3}\zeta_{0i}\sigma_{1i}\varsigma_{0i}h_{0i}\phi_{0i}^{\mathrm{T}}(k)z_{1i}(k+1)\hat{\theta}_{0i}(k)+\sum_{i=1}^{3}\zeta_{0i}\varsigma_{0i}^{2}h_{0i}\hat{\theta}_{0i}^{\mathrm{T}}(k)\hat{\theta}_{0i}(k)
\end{aligned}
\tag{5.38}
$$

$$
\begin{aligned}
&-\sum_{i=1}^{3}\tfrac{\zeta_{0i}}{h_{0i}}\tilde{\theta}_{0i}^{\mathrm{T}}(k)\tilde{\theta}_{0i}(k)=-\sum_{i=1}^{3}\tfrac{\zeta_{0i}}{h_{0i}}\tilde{\theta}_{0i}^{\mathrm{T}}(k)\tilde{\theta}_{0i}(k+1) \qquad cc\\
&+\sum_{i=1}^{3}\zeta_{0i}\sigma_{1i}\tilde{\theta}_{0i}^{\mathrm{T}}(k)\phi_{0i}(k)z_{1i}(k+1)-\sum_{i=1}^{3}\zeta_{0i}\varsigma_{0i}\tilde{\theta}_{0i}^{\mathrm{T}}(k)\hat{\theta}_{0i}(k)
\end{aligned}
\tag{5.39}
$$

From (5.32)–(5.39), (5.30) can be described as

$$
\begin{aligned}
\Delta L_1(k)&=\eta_1 z_1^{\mathrm{T}}(k+1)z_1(k+1)-\eta_1 z_1^{\mathrm{T}}(k)z_1(k)+2\sum_{i=1}^{3}\zeta_{1i}\sigma_{1i}\tilde{\theta}_{1i}^{\mathrm{T}}(k)\phi_{1i}(k)z_{1i}(k+1)\\
&+\sum_{i=1}^{3}\zeta_{1i}\sigma_{1i}^{2}h_{1i}\phi_{1i}^{\mathrm{T}}(k)z_{1i}(k+1)\phi_{1i}(k)z_{1i}(k+1)-2\sum_{i=1}^{3}\zeta_{0i}\varsigma_{1i}\tilde{\theta}_{1i}^{\mathrm{T}}(k)\hat{\theta}_{1i}(k)\\
&-2\sum_{i=1}^{3}\zeta_{1i}\sigma_{1i}h_{1i}\varsigma_{1i}\hat{\theta}_{1i}^{\mathrm{T}}(k)\phi_{1i}(k)z_{1i}(k+1)+\sum_{i=1}^{3}\zeta_{1i}\varsigma_{1i}^{2}h_{1i}\hat{\theta}_{1i}^{\mathrm{T}}(k)\hat{\theta}_{1i}(k)\\
&+\bar{\hbar}_1\hbar_1^{\mathrm{T}}(k+1)\hbar_1(k+1)-\bar{\hbar}_1\hbar_1^{\mathrm{T}}(k)\hbar_1(k)+\sum_{i=1}^{3}\tilde{\mathscr{D}}_{1i}^{2}(k+1)\\
&-\sum_{i=1}^{3}\tilde{\mathscr{D}}_{1i}^{2}(k)+2\sum_{i=1}^{3}\zeta_{0i}\sigma_{1i}\tilde{\theta}_{0i}^{\mathrm{T}}(k)\phi_{0i}(k)z_{1i}(k+1)\\
&+\sum_{i=1}^{3}\zeta_{0i}\sigma_{1i}^{2}h_{0i}\phi_{0i}^{\mathrm{T}}(k)z_{1i}(k+1)\phi_{0i}(k)z_{1i}(k+1)-2\sum_{i=1}^{3}\zeta_{0i}\varsigma_{0i}\tilde{\theta}_{0i}^{\mathrm{T}}(k)\hat{\theta}_{0i}(k)\\
&-2\sum_{i=1}^{3}\zeta_{0i}\sigma_{1i}h_{0i}\varsigma_{0i}\hat{\theta}_{0i}^{\mathrm{T}}(k)\phi_{0i}(k)z_{1i}(k+1)+\sum_{i=1}^{3}\zeta_{0i}\varsigma_{0i}^{2}h_{0i}\hat{\theta}_{0i}^{\mathrm{T}}(k)\hat{\theta}_{0i}(k)
\end{aligned}
\tag{5.40}
$$

According to (5.14) and (5.31), one has

$$
\begin{aligned}
2\sigma_1 z_1^{\mathrm{T}}(k+1)\tilde{\Xi}_1(k) &= 2\sigma_1 z_1^{\mathrm{T}}(k+1)\lambda(k)\bar{G}_1(k)\tilde{\Xi}_0(k) + 2\sigma_1 z_1^{\mathrm{T}}(k+1)\bar{G}_1(k)z_2(k) \\
&+2\sigma_1 z_1^{\mathrm{T}}(k+1)\tilde{\mathscr{D}}_1(k) + 2\sigma_1 z_1^{\mathrm{T}}(k+1)(\mathscr{I} - \mathscr{G}_1)\bar{G}_1(k)\hbar_2(k) \\
&+2\sigma_1 z_1^{\mathrm{T}}(k+1)(\bar{G}_1(k)\mathscr{Y}(k) - \lambda(k)\bar{G}_1(k)\mathrm{E}_0(k)) - 2\sigma_1 z_1^{\mathrm{T}}(k+1)z_1(k+1)
\end{aligned}
\tag{5.41}
$$

where the matrix $\sigma_1 = \mathrm{diag}[\sigma_{11}, \sigma_{12}, \sigma_{13}]$.

$$
\hbar_1^{\mathrm{T}}(k+1)\hbar_1(k+1) = \hbar_1^{\mathrm{T}}(k+1)\mathscr{G}_1\bar{G}_1(k)\hbar_2(k) - \hbar_1^{\mathrm{T}}(k+1)\mathscr{C}_1\hbar_1(k) \tag{5.42}
$$

Moreover, some inequalities based on the above analysis can be achieved as follows:

$$
2\sigma_1 z_1^{\mathrm{T}}(k+1)\tilde{\mathscr{D}}_1(k) \le \|\sigma_1\|^2\rho_{11}\left\|\tilde{\mathscr{D}}_1(k)\right\|^2 + \frac{1}{\rho_{11}}\|z_1(k+1)\|^2 \tag{5.43}
$$

where ρ_{11} is a positive constant.

$$
2\sigma_1 z_1^{\mathrm{T}}(k+1)(\bar{G}_1(k)\mathscr{Y}(k) + \mathrm{E}_0(k)) \le \|\sigma_1\|^2\rho_{10}\left\|\bar{\mathrm{E}}_0\right\|^2 + \frac{1}{\rho_{10}}\|z_1(k+1)\|^2 \tag{5.44}
$$

where ρ_{10} is a positive constant, and $\left\|\bar{G}_1(k)\mathscr{Y}(k) - \lambda(k)\bar{G}_1(k)\mathrm{E}_0(k)\right\| \le \bar{\mathrm{E}}_0$.

$$
\begin{aligned}
&\sum_{i=1}^{3}\zeta_{1i}\sigma_{1i}^2 h_{1i}\phi_{1i}^{\mathrm{T}}(k)z_{1i}(k+1)\phi_{1i}(k)z_{1i}(k+1) \le \\
&\qquad \zeta_{1\max}\sigma_{1\max}^2 h_{1\max}\bar{l}_1\|z_1(k+1)\|^2
\end{aligned}
\tag{5.45}
$$

where $h_{1\max}$ denotes the maximum value of h_{1i}, $\zeta_{1\max}$ denotes the maximum value of ζ_{1i}, $\sigma_{1\max}$ denotes the maximum value of σ_{1i}, $\phi_{1i}^{\mathrm{T}}(k)\phi_{1i}(k) \le \bar{l}_1$, and $\bar{l}_1 \ge 1$.

$$
\begin{aligned}
2\sum_{i=1}^{3}\zeta_{1i}\sigma_{1i}h_{1i}\varsigma_{1i}\hat{\theta}_{1i}^{\mathrm{T}}(k)\phi_{1i}(k)z_{1i}(k+1) &\le \sum_{i=1}^{3}\sigma_{1i}h_{1i}\varsigma_{1i}\rho_{12}\left\|\hat{\theta}_{1i}(k)\right\|^2 \\
&+\tfrac{\zeta_{1\max}}{\rho_{12}}\sigma_{1\max}h_{1\max}\varsigma_{1\max}\bar{l}_1\|z_1(k+1)\|^2
\end{aligned}
\tag{5.46}
$$

where $\varsigma_{1\max}$ denotes the maximum value of ς_{1i}, and ρ_{12} is a positive constant.

$$
\begin{aligned}
2\sum_{i=1}^{3}\zeta_{1i}\varsigma_{1i}\tilde{\theta}_{1i}^{\mathrm{T}}(k)\hat{\theta}_{1i}(k) &= \sum_{i=1}^{3}\zeta_{1i}\varsigma_{1i}\left\|\tilde{\theta}_{1i}(k)\right\|^2 + \sum_{i=1}^{3}\zeta_{1i}\varsigma_{1i}\left\|\hat{\theta}_{1i}(k)\right\|^2 \\
&- \sum_{i=1}^{3}\zeta_{1i}\varsigma_{1i}\left\|\theta_{1i}^{*}(k)\right\|^2
\end{aligned}
\tag{5.47}
$$

$$\sum_{i=1}^{3} \zeta_{0i}\sigma_{1i}^2 h_{0i}\phi_{0i}^{\mathrm{T}}(k)z_{1i}(k+1)\phi_{0i}(k)z_{1i}(k+1) \leq \zeta_{0\max}\sigma_{1\max}^2 h_{0\max}\bar{l}_0\|z_1(k+1)\|^2 \tag{5.48}$$

where $h_{0\max}$ denotes the maximum value of h_{0i}, $\phi_{0i}^{\mathrm{T}}(k)\phi_{0i}(k) \leq \bar{l}_0$, $\zeta_{0\max}$ denotes the maximum value of ζ_{0i}, and $\bar{l}_0 \geq 1$.

$$2\sum_{i=1}^{3} \zeta_{0i}\sigma_{1i}h_{0i}\varsigma_{0i}\hat{\theta}_{0i}^{\mathrm{T}}(k)\phi_{0i}(k)z_{1i}(k+1) \leq \sum_{i=1}^{3} \zeta_{0i}\sigma_{1i}h_{0i}\varsigma_{0i}\bar{\rho}_{12}\left\|\hat{\theta}_{0i}(k)\right\|^2 + \frac{\zeta_{0i}}{\bar{\rho}_{12}}\sigma_{1\max}h_{0\max}\varsigma_{0\max}\bar{l}_0\|z_1(k+1)\|^2 \tag{5.49}$$

where $\varsigma_{0\max}$ denotes the maximum value of ς_{0i}, and $\bar{\rho}_{12}$ is a positive constant.

$$2\sum_{i=1}^{3} \zeta_{0i}\varsigma_{0i}\tilde{\theta}_{0i}^{\mathrm{T}}(k)\hat{\theta}_{0i}(k) = \sum_{i=1}^{3} \zeta_{0i}\varsigma_{0i}\left\|\tilde{\theta}_{0i}(k)\right\|^2 + \sum_{i=1}^{3} \zeta_{0i}\varsigma_{0i}\left\|\hat{\theta}_{0i}(k)\right\|^2 - \sum_{i=1}^{3} \zeta_{0i}\varsigma_{0i}\left\|\theta_{0i}^{*}(k)\right\|^2 \tag{5.50}$$

$$2\sigma_1 z_1^{\mathrm{T}}(k+1)\bar{G}_1(k)z_2(k) \leq \rho_{13}\|\sigma_1\|^2\bar{g}_1^2\|z_1(k+1)\|^2 + \frac{1}{\rho_{13}}\|z_2(k)\|^2 \tag{5.51}$$

where ρ_{13} is a positive constant.

$$2\sigma_1 z_1^{\mathrm{T}}(k+1)(\mathscr{I} - \mathscr{G}_1)\bar{G}_1(k)\hbar_2(k) \leq \frac{1}{\rho_{14}}\|\hbar_2(k)\|^2 + \|\sigma_1\|^2\rho_{14}\|z_1(k+1)\|^2\|(\mathscr{I} - \mathscr{G}_1)\|^2\bar{g}_1^2 \tag{5.52}$$

where ρ_{14} is a positive constant.

$$\hbar_1^{\mathrm{T}}(k+1)\mathscr{C}_1\hbar_1(k) \leq \frac{\rho_{15}}{2}\|\hbar_1(k+1)\|^2 + \frac{1}{2\rho_{15}}\|\mathscr{C}_1\|^2\|\hbar_1(k)\|^2 \tag{5.53}$$

where ρ_{15} is a positive constant.

$$\hbar_1^{\mathrm{T}}(k+1)\mathscr{G}_1\bar{G}_1(k)\hbar_2(k) \leq \frac{1}{2\rho_{16}}\|\mathscr{G}_1\|^2\bar{g}_1^2\|\hbar_1(k)\|^2 + \frac{\rho_{16}}{2}\|\hbar_1(k+1)\|^2 \tag{5.54}$$

where ρ_{16} is a positive constant.

$$2\sigma_1 z_1^{\mathrm{T}}(k+1)\aleph(k)\bar{G}_1(k)\tilde{\Xi}_0(k) \leq \rho_{17}\|\sigma_1\|^2\|z_1(k+1)\|^2 + \frac{\bar{l}_0\ell^2\bar{g}_1^2}{\rho_{17}}\sum_{i=1}^{3}\left\|\tilde{\theta}_{0i}(k)\right\|^2 \tag{5.55}$$

where ρ_{17} is a positive constant.

$$2\sum_{i=1}^{3}\zeta_{0i}\sigma_{1i}\tilde{\theta}_{0i}^{\mathrm{T}}(k)\phi_{0i}(k)z_{1i}(k+1) \le \rho_{18}\sigma_{1\max}^{2}\|z_1(k+1)\|^2 + \sum_{i=1}^{3}\frac{\zeta_{0i}^{2}\bar{l}_0}{\rho_{18}}\left\|\tilde{\theta}_{0i}(k)\right\|^2 \tag{5.56}$$

where ρ_{18} is a positive constant.

Moreover, according to (5.8)–(5.13) and (5.21), one can obtain that $\tilde{\mathscr{D}}_{1i} = \tilde{\mathscr{P}}_{1i}$ and

$$\tilde{\mathscr{D}}_{1i}^{2}(k+1) - \tilde{\mathscr{D}}_{1i}^{2}(k) \le 2\bar{\delta}_{1i}^{2} + 2\zeta_{1i}^{2}\bar{l}_1\left\|\tilde{\theta}_{1i}(k)\right\|^2 - \tilde{\mathscr{D}}_{1i}^{2}(k) \tag{5.57}$$

From (5.41)–(5.57), it yields

$$\begin{aligned}
\Delta L_1(k) \le{}& \eta_1 z_1^{\mathrm{T}}(k+1)z_1(k+1) - \eta_1 z_1^{\mathrm{T}}(k)z_1(k) + \|\sigma_1\|^2\rho_{11}\left\|\tilde{\mathscr{D}}_1(k)\right\|^2 \\
&-2\sigma_1 z_1^{\mathrm{T}}(k+1)z_1(k+1) + \tfrac{1}{\rho_{11}}\|z_1(k+1)\|^2 - \sum_{i=1}^{3}\zeta_{1i}\varsigma_{1i}\left\|\tilde{\theta}_{1i}(k)\right\|^2 \\
&+\zeta_{1\max}\sigma_{1\max}^{2}h_{1\max}\bar{l}_1\|z_1(k+1)\|^2 + \tfrac{1}{\rho_{13}}\|z_2(k)\|^2 + \tfrac{1}{\rho_{14}}\|\hbar_2(k)\|^2 \\
&+\sum_{i=1}^{3}\zeta_{1i}\sigma_{1i}h_{1i}\varsigma_{1i}\rho_{12}\left\|\hat{\theta}_{1i}(k)\right\|^2 + \tfrac{\zeta_{1\max}\sigma_{1\max}h_{1\max}\varsigma_{1\max}}{\rho_{12}}\bar{l}_1\|z_1(k+1)\|^2 \\
&-\sum_{i=1}^{3}\zeta_{1i}\varsigma_{1i}\left\|\hat{\theta}_{1i}(k)\right\|^2 + \sum_{i=1}^{3}\zeta_{1i}\varsigma_{1i}\left\|\theta_{1i}^{*}(k)\right\|^2 + \sum_{i=1}^{3}\zeta_{1i}\varsigma_{1i}^{2}h_{1i}\left\|\hat{\theta}_{1i}(k)\right\|^2 \\
&+\rho_{13}\|\sigma_1\|^2\bar{g}_1^2\|z_1(k+1)\|^2 + \|\sigma_1\|^2\rho_{14}\|z_1(k+1)\|^2\|(\mathscr{I}-\mathscr{G}_1)\|^2\bar{g}_1^2 \\
&+\tfrac{\rho_{13}}{2}\|h_1(k+1)\|^2 + \tfrac{1}{2\rho_{15}}\|\mathscr{C}_1\|^2\|\hbar_1(k)\|^2 + \tfrac{1}{2\rho_{16}}\|\mathscr{G}_1\|^2\bar{g}_1^2\|\hbar_1(k)\|^2 \\
&+\tfrac{\rho_{16}}{2}\|\hbar_1(k+1)\|^2 - \hbar_1^{\mathrm{T}}(k+1)\hbar_1(k+1) - \bar{\hbar}_1\hbar_1^{\mathrm{T}}(k)\hbar_1(k) \\
&+\bar{\hbar}_1\hbar_1^{\mathrm{T}}(k+1)\hbar_1(k+1) + \sum_{i=1}^{3}(2\bar{\delta}_{1i}^{2} + 2\zeta_{1i}^{2}\bar{l}_1\left\|\tilde{\theta}_{1i}(k)\right\|^2 - \tilde{D}_{1i}^{2}(k)) \\
&+\zeta_{0\max}\sigma_{1\max}^{2}h_{0\max}\bar{l}_0\|z_1(k+1)\|^2 + \sum_{i=1}^{3}\zeta_{0i}\sigma_{1i}h_{0i}\varsigma_{0i}\bar{\rho}_{12}\left\|\hat{\theta}_{0i}(k)\right\|^2 \\
&+\tfrac{\zeta_{0\max}\sigma_{1\max}h_{0\max}\varsigma_{0\max}}{\bar{\rho}_{12}}\bar{l}_0\|z_1(k+1)\|^2 - \sum_{i=1}^{3}\zeta_{0i}\varsigma_{0i}\left\|\tilde{\theta}_{0i}(k)\right\|^2 \\
&-\sum_{i=1}^{3}\zeta_{0i}\varsigma_{0i}\left\|\hat{\theta}_{0i}(k)\right\|^2 + \sum_{i=1}^{3}\zeta_{0i}\varsigma_{0i}\left\|\theta_{0i}^{*}(k)\right\|^2 + \sum_{i=1}^{3}\zeta_{0i}\varsigma_{0i}^{2}h_{0i}\left\|\hat{\theta}_{0i}(k)\right\|^2 \\
&+\|\sigma_1\|^2\rho_{10}\left\|\bar{\mathrm{E}}_0\right\|^2 + \tfrac{1}{\rho_{10}}\|z_1(k+1)\|^2 + \rho_{17}\|\sigma_1\|^2\|z_1(k+1)\|^2 \\
&+\tfrac{\bar{l}_0\ell^2\bar{g}_1^2}{\rho_{17}}\sum_{i=1}^{3}\left\|\tilde{\theta}_{0i}(k)\right\|^2 + \rho_{18}\sigma_{1\max}^{2}\|z_1(k+1)\|^2 + \sum_{i=1}^{3}\tfrac{\zeta_{0i}^{2}\bar{l}_0}{\rho_{18}}\left\|\tilde{\theta}_{0i}(k)\right\|^2
\end{aligned} \tag{5.58}$$

Furthermore, (5.58) can be written as

$$
\begin{aligned}
\Delta L_1(k) \leq & -\kappa_{11}\|z_1(k+1)\|^2 - \kappa_{12}\|z_1(k)\|^2 - \sum_{i=1}^{3}\kappa_{13}\left\|\hat{\theta}_{1i}(k)\right\|^2 \\
& -\kappa_{14}\|\hbar_1(k+1)\|^2 - \kappa_{15}\|\hbar_1(k)\|^2 - \sum_{i=1}^{3}\kappa_{16}\tilde{\mathscr{D}}_{1i}^2(k) \\
& +\kappa_{17}\|z_2(k)\|^2 + \kappa_{18}\|\hbar_2(k)\|^2 - \sum_{i=1}^{3}\kappa_{19}\left\|\tilde{\theta}_{1i}(k)\right\|^2 + \kappa_1 \\
& - \sum_{i=1}^{3}\kappa_{110}\left\|\hat{\theta}_{0i}(k)\right\|^2 - \sum_{i=1}^{3}\kappa_{111}\left\|\tilde{\theta}_{0i}(k)\right\|^2
\end{aligned}
\tag{5.59}
$$

where $\kappa_{12} = \eta_1, \kappa_{11} = 2\sigma_{1\min} - \eta_1 - \frac{1}{\rho_{11}} - \zeta_{1\max}\sigma_{1\max}^2 h_{1\max}\bar{l}_1 - \frac{\zeta_{1\max}\sigma_{1\max}h_{1\max}\varsigma_{1\max}}{\rho_{12}}\bar{l}_1 - \rho_{13}\|\sigma_1\|^2\bar{g}_1^2 - \|\sigma_1\|^2\rho_{14}\left\|\mathscr{I} - \zeta_{0\max}\sigma_{1\max}^2 h_{0\max}\bar{l}_0 - \mathscr{G}_1\right\|^2\bar{g}_1^2 - \frac{\zeta_{0\max}\sigma_{1\max}h_{0\max}\varsigma_{0\max}}{\bar{\rho}_{12}}\bar{l}_0 - \frac{1}{\rho_{10}} - \rho_{17}\|\sigma_1\|^2 - \rho_{18}\sigma_{1\max}^2$, $\kappa_{13} = \zeta_{1i}\varsigma_{1i} - \zeta_{1i}\varsigma_{1i}^2 h_{1i} - \zeta_{1i}\sigma_{1i}h_{1i}\varsigma_{1i}\rho_{12}$, $\kappa_{14} = 1 - \frac{\rho_{15}}{2} - \frac{\rho_{16}}{2} - \bar{\hbar}_1$, $\kappa_{15} = \bar{\hbar}_1 - \frac{1}{2\rho_{15}}\|\mathscr{C}_1\|^2 - \frac{1}{2\rho_{16}}\|\mathscr{G}_1\|^2\bar{g}_1^2$, $\kappa_{16} = 1 - \|\sigma_1\|^2\rho_{11}$, $\kappa_{17} = \frac{1}{\rho_{13}}$, $\kappa_{18} = \frac{1}{\rho_{14}}$, $\kappa_{19} = \zeta_{1i}\varsigma_{1i} - 2\zeta_{1i}\zeta_{1i}^2\bar{l}_1$, $\kappa_{110} = \zeta_{0i}\varsigma_{0i} - \zeta_{0i}\varsigma_{0i}^2 h_{0i} - \zeta_{0i}\sigma_{1i}h_{0i}\varsigma_{0i}\bar{\rho}_{12}$, $\kappa_{111} = \zeta_{0i}\varsigma_{0i} - \frac{\bar{l}_0\ell^2\bar{g}_1^2}{\rho_{17}} - \frac{\zeta_{0i}^2\bar{l}_0}{\rho_{18}}$ and $\kappa_1 = \sum_{i=1}^{3}\zeta_{1i}\varsigma_{1i}\left\|\theta_{1i}^*(k)\right\|^2 + \sum_{i=1}^{3}\zeta_{0i}\varsigma_{0i}\left\|\theta_{0i}^*(k)\right\|^2 + \sum_{i=1}^{3}2\bar{\delta}_{1i}^2 + \|\sigma_1\|^2\rho_{10}\left\|\bar{\mathrm{E}}_0\right\|^2$.

Step 2: Combining (5.15) and $z_2(k) = x_2(k) - \mathrm{sat}(x_{vd}(k)) - \hbar_2(k)$, it yields

$$
\begin{aligned}
z_2(k+1) = & \bar{F}_2(k) + \Delta\bar{F}_2(k) + \bar{G}_2(k)u(k) + \bar{d}_2(k) \\
& + \bar{G}_2(k)\Delta v(k) - v_d(k+1) - \hbar_2(k+1)
\end{aligned}
\tag{5.60}
$$

On the basis of Lemma 2.2, the NN is used to approximate $(1/\zeta_{2i})\Delta\bar{F}_{2i}(k)$. Then, (5.60) can be written as

$$
\begin{aligned}
z_2(k+1) = & \bar{F}_2(k) + \Xi_2(k) + \bar{G}_2(k)u(k) + \bar{d}_2(k) \\
& + \bar{G}_2(k)\Delta v(k) - v_d(k+1) + \mathrm{E}_2(k) - \hbar_2(k+1)
\end{aligned}
\tag{5.61}
$$

where $\Xi_2(k) = [\zeta_{21}\theta_{21}^{*T}(k)\phi_{21}(Z_2(k)), \zeta_{22}\theta_{22}^{*T}(k)\phi_{22}(Z_2(k)), \zeta_{23}\theta_{23}^{*T}(k)\phi_{23}(Z_2(k))]^{\mathrm{T}}$, $Z_2(k) = x(k)$, $\mathrm{E}_2(k) = [\zeta_{21}\varepsilon_{21}(k), \zeta_{22}\varepsilon_{22}(k), \zeta_{23}\varepsilon_{23}(k)]^{\mathrm{T}}$, $\theta_{2i}^*(k)$ is the optimal weight vector, $\bar{F}_{2i}(k)$ is the ith variable of $\bar{F}_2(k)$, ζ_{2i} is a positive constant, $\varepsilon_{2i}(k)$ is the minimum approximation error, $\phi_{2i}(\cdot)$ is the basis function vector. For convenience, $\phi_{2i}(Z_2(k)) = \phi_{2i}(k)$ is defined.

Furthermore, (5.61) can be written as

$$
\begin{aligned}
z_2(k+1) = & \bar{F}_2(k) + \Xi_2(k) + \bar{G}_2(k)u(k) + \mathscr{D}_2(k) \\
& + \bar{G}_2(k)\Delta v(k) - v_d(k+1) - \hbar_2(k+1)
\end{aligned}
\tag{5.62}
$$

where $\mathscr{D}_2(k) = \mathrm{E}_2(k) + \bar{d}_2(k)$ denotes the compound disturbance.

From (5.14) and (5.62), one can obtain

$$z_2(k+1) = \bar{F}_2(k) + \Xi_2(k) + \bar{G}_2(k)u(k) + \mathscr{D}_2(k) - x_{vd}(k+1) + \mathscr{C}_2\hbar_2(k) \tag{5.63}$$

In order to deal with the disturbance $\mathscr{D}_2(k)$ in (5.63), the DTDO is designed according to (5.10) and (5.11), which is given by

$$\begin{cases} \hat{\mathscr{P}}_{2i}(k+1) = -(\bar{F}_{2i}(k) + \zeta_{2i}\hat{\theta}_{2i}^T(k)\phi_{2i}(k) + \mathscr{V}_{2i}(k)) \\ \hat{\mathscr{D}}_{2i}(k) = \hat{\mathscr{P}}_{2i}(k) + x_{2i}(k) \end{cases} \tag{5.64}$$

where $i = 1, 2, 3$, $x_{2i}(k)$ denotes the ith variable of $x_2(k)$, $\hat{\mathscr{D}}_{2i}(k)$ is the output of DTDO with $\hat{\mathscr{D}}_2(k) = [\hat{\mathscr{D}}_{21}(k), \hat{\mathscr{D}}_{22}(k), \hat{\mathscr{D}}_{23}(k)]^{\mathrm{T}}$. $\hat{\mathscr{P}}_{2i}(k)$ denotes the variable of DTDO. The disturbance estimation error is defined as $\tilde{\mathscr{D}}_{2i}(k) = \mathscr{D}_{2i}(k) - \hat{\mathscr{D}}_{2i}(k)$. $\mathscr{V}_{2i}(k)$ denotes the ith variable of $\mathscr{V}_2(k)$, and $\mathscr{V}_2(k) = \bar{G}_2(k)\mathrm{sat}(u(k))$. Furthermore, according to (5.7), one can obtain that $|\mathscr{D}_{2i}(k+1) - \mathscr{D}_{2i}(k)| \leq \bar{\delta}_{2i}$, and $\bar{\delta}_{2i}$ is a positive constant.

To forecast $v_d(k+1)$, the following discrete-time tracking differentiator based on Lemma 2.3 is employed to estimate $v_d(k+1)$:

$$\begin{cases} \bar{\chi}_{11i}(k+1) = \bar{\chi}_{11i}(k) + h_{01i}\bar{\chi}_{12i}(k) \\ \bar{\chi}_{12i}(k+1) = \bar{\chi}_{12i}(k) - h_{01i}r_{01i}\bar{\chi}_{13i}(\bar{\chi}_{14i}(k), \delta_{01i}) \end{cases} \tag{5.65}$$

where $i = 1, 2, 3$, h_{01i} and r_{01i} are design constants, $\bar{\chi}_{11i}(k)$ and $\bar{\chi}_{12i}(k)$ are the state variables of the discrete-time tracking differentiator.

From (5.65) and Lemma 2.3, it yields

$$v_d(k+1) = \bar{\chi}_{11}(k) + h_{01}\bar{\chi}_{12}(k) - \tilde{\chi} \tag{5.66}$$

where $\bar{\chi}_{11}(k) = [\bar{\chi}_{111}(k), \bar{\chi}_{112}(k), \bar{\chi}_{113}(k)]^{\mathrm{T}}$, $\bar{\chi}_{12}(k) = [\bar{\chi}_{121}(k), \bar{\chi}_{122}(k), \bar{\chi}_{123}(k)]^{\mathrm{T}}$, $h_{01} = \mathrm{diag}[h_{011}, h_{012}, h_{013}]$, diag[$\cdot$]denotes a diagonal matrix, $\tilde{\chi} = [\tilde{\chi}_1, \tilde{\chi}_2, \tilde{\chi}_3]^{\mathrm{T}}$ is the estimation error vector and $\|\tilde{\chi}\| \leq \bar{\chi}$ with $\bar{\chi} > 0$.

Substituting (5.66) into (5.63), one can have

$$\begin{aligned} z_2(k+1) = {} & \bar{F}_2(k) + \Xi_2(k) + \bar{G}_2(k)u(k) + \mathscr{D}_2(k) \\ & -\bar{\chi}_{11}(k) - h_{01}\bar{\chi}_{12}(k) + \tilde{\chi} + \mathscr{C}_2\hbar_2(k) \end{aligned} \tag{5.67}$$

The controller is designed as

$$\begin{aligned} u(k) = {} & -\bar{G}_2^{-1}(k)\bar{F}_2(k) - \bar{G}_2^{-1}(k)(\hat{\Xi}_2(k) + \hat{\mathscr{D}}_2(k)) \\ & -\bar{G}_1^{-1}(k)\mathscr{C}_2\hbar_2(k) + \bar{G}_1^{-1}(k)\bar{\chi}_{11}(k) + \bar{G}_1^{-1}(k)h_{01}\bar{\chi}_{12}(k) \end{aligned} \tag{5.68}$$

where $\hat{\Xi}_2(k) = [\zeta_{21}\hat{\theta}_{21}^{\mathrm{T}}(k)\phi_{21}(k), \zeta_{22}\hat{\theta}_{22}^{\mathrm{T}}(k)\phi_{22}(k), \zeta_{23}\hat{\theta}_{23}^{\mathrm{T}}(k)\phi_{23}(k)]^{\mathrm{T}}$, and $\hat{\Xi}_2(k)$ denotes the estimation of $\Xi_2(k)$.

Invoking (5.67) and (5.68), one can obtain

$$z_2^{\mathrm{T}}(k+1)z_2(k+1) = z_2^{\mathrm{T}}(k+1)(\tilde{\mathscr{D}}_2(k) - \tilde{\Xi}_2(k)) + z_1^{\mathrm{T}}(k+1)\tilde{\chi} \tag{5.69}$$

where $\tilde{\Xi}_2(k) = [\zeta_{21}\tilde{\theta}_{21}^{\mathrm{T}}(k)\phi_{21}(k), \zeta_{22}\tilde{\theta}_{22}^{\mathrm{T}}(k)\phi_{22}(k), \zeta_{23}\tilde{\theta}_{23}^{\mathrm{T}}(k)\phi_{23}(k)]^{\mathrm{T}}$ and $\tilde{\Xi}_2(k) = \hat{\Xi}_2(k) - \Xi_2(k)$.

Moreover, the adaptive law of $\hat{\theta}_{2i}(k)$ is designed by

$$\hat{\theta}_{2i}(k+1) = \hat{\theta}_{2i}(k) + h_{2i}\sigma_{2i}\phi_{2i}(k)z_{2i}(k+1) - h_{2i}\varsigma_{2i}\hat{\theta}_{2i}(k) \tag{5.70}$$

where $z_{2i}(k+1)$ is the ith element of $z_2(k+1)$, h_{2i}, σ_{2i} and ς_{2i} are positive design constants.

From (5.70), it yields

$$\tilde{\theta}_{2i}(k+1) = \tilde{\theta}_{2i}(k) + h_{2i}\sigma_{2i}\phi_{2i}(k)z_{2i}(k+1) - h_{2i}\varsigma_{2i}\hat{\theta}_{2i}(k) \tag{5.71}$$

where $\tilde{\theta}_{2i}(k+1) = \hat{\theta}_{2i}(k+1) - \theta_{2i}^*(k+1)$ and $\tilde{\theta}_{2i}(k) = \hat{\theta}_{2i}(k) - \theta_{2i}^*(k)$.

To proof the effectiveness of the nonlinear DTDO (5.64) and the controller $u(k)$ (5.68), the following Lyapunov function is chosen as

$$L_2(k) = \eta_2 z_2^{\mathrm{T}}(k)z_2(k) + \sum_{i=1}^{3}\frac{\zeta_{2i}}{h_{2i}}\tilde{\theta}_{2i}^{\mathrm{T}}(k)\tilde{\theta}_{2i}(k) + \bar{\hbar}_2\hbar_2^{\mathrm{T}}(k)\hbar_2(k) + \sum_{i=1}^{3}\tilde{\mathscr{D}}_{2i}^2(k) \tag{5.72}$$

where η_2 is the minimum value of $\sigma_{2i}(i = 1, 2, 3)$, and $\bar{\hbar}_2$ is the value of $\|\mathscr{C}_2\|$.

Based on (5.72), the first difference of Lyapunov function $L_2(k)$ can be expressed as

$$\begin{aligned}\Delta L_2(k) = L_2(k+1) - L_2(k) = \eta_2 z_2^{\mathrm{T}}(k+1)z_2(k+1) - \eta_2 z_2^{\mathrm{T}}(k)z_2(k)\\ +\sum_{i=1}^{3}\tfrac{\zeta_{2i}}{h_{2i}}\tilde{\theta}_{2i}^{\mathrm{T}}(k+1)\tilde{\theta}_{2i}(k+1) - \sum_{i=1}^{3}\tilde{\mathscr{D}}_{2i}^2(k) + \bar{\hbar}_2\hbar_2^{\mathrm{T}}(k+1)\hbar_2(k+1)\\ -\bar{\hbar}_2\hbar_2^{\mathrm{T}}(k)\hbar_2(k) + \sum_{i=1}^{3}\tilde{\mathscr{D}}_{2i}^2(k+1) - \sum_{i=1}^{3}\tfrac{\zeta_{2i}}{h_{2i}}\tilde{\theta}_{2i}^{\mathrm{T}}(k)\tilde{\theta}_{2i}(k)\end{aligned} \tag{5.73}$$

In Step 1, the detailed formula analysis is shown. Thus, $\Delta L_2(k)$ can satisfy that

$$\begin{aligned}\Delta L_2(k) \le{}& \eta_2 z_2^{\mathrm{T}}(k+1)z_2(k+1) - \eta_2 z_2^{\mathrm{T}}(k)z_2(k) - 2\sigma_2 z_2^{\mathrm{T}}(k+1)z_2(k+1)\\ &+\|\sigma_2\|^2\rho_{21}\left\|\tilde{\mathscr{D}}_2(k)\right\|^2 + \frac{1}{\rho_{21}}\|z_2(k+1)\|^2 - \sum_{i=1}^{3}\zeta_{2i}\varsigma_{2i}\left\|\tilde{\theta}_{2i}(k)\right\|^2\\ &+\zeta_{2\max}\sigma_{2\max}^2 h_{2\max}\bar{l}_2\|z_2(k+1)\|^2 + \sum_{i=1}^{3}\zeta_{2i}\sigma_{2i}h_{2i}\varsigma_{2i}\rho_{22}\left\|\hat{\theta}_{2i}(k)\right\|^2\end{aligned}$$

$$
\begin{aligned}
&+\frac{\zeta_{2\max}\sigma_{2\max}h_{2\max}\varsigma_{2\max}}{\rho_{22}}\bar{l}_2\|z_2(k+1)\|^2-\sum_{i=1}^{3}\zeta_{2i}\varsigma_{2i}\left\|\hat{\theta}_{2i}(k)\right\|^2\\
&+\sum_{i=1}^{3}\zeta_{2i}\varsigma_{2i}^2h_{2i}\left\|\hat{\theta}_{2i}(k)\right\|^2+\rho_{23}\|\sigma_2\|^2\|z_2(k+1)\|^2+\frac{\rho_{24}}{2}\|\hbar_2(k+1)\|^2\\
&+\frac{1}{2\rho_{24}}\|\mathscr{C}_2\|^2\|\hbar_2(k)\|^2+\frac{\rho_{25}}{2}\|\hbar_2(k+1)\|^2-\hbar_2^{\mathrm{T}}(k+1)\hbar_2(k+1)\\
&-\bar{\hbar}_2\hbar_2^{\mathrm{T}}(k)\hbar_2(k)+\frac{1}{2\rho_{25}}\bar{g}_2^2\nu^2+\bar{\hbar}_2\hbar_2^{\mathrm{T}}(k+1)\hbar_2(k+1)+\frac{1}{\rho_{23}}\bar{\chi}^2\\
&+\sum_{i=1}^{3}(2\bar{\delta}_{2i}^2+2\zeta_{2i}^2\bar{l}_2\left\|\tilde{\theta}_{2i}(k)\right\|^2-\tilde{D}_{2i}^2(k))+\sum_{i=1}^{3}\zeta_{2i}\varsigma_{2i}\left\|\theta_{2i}^*(k)\right\|^2
\end{aligned}
\tag{5.74}
$$

where ρ_{21}, ρ_{22}, ρ_{23}, ρ_{24} and ρ_{25} are positive constants, $\zeta_{2\max}$ denotes the maximum value of ζ_{2i}, $h_{2\max}$ denotes the maximum value of h_{2i}, $\sigma_{2\max}$ is the maximum value of σ_{2i}, $\varsigma_{2\max}$ is the maximum value of ς_{2i}, $\phi_{2i}^{\mathrm{T}}(k)\phi_{2i}(k)\le\bar{l}_2$, and $\bar{l}_2\ge 1$.

Furthermore, (5.74) can be written as

$$
\begin{aligned}
\Delta L_2(k)\le&-\kappa_{21}\|z_2(k+1)\|^2-\kappa_{22}\|z_2(k)\|^2-\sum_{i=1}^{3}\kappa_{23}\left\|\hat{\theta}_{2i}(k)\right\|^2-\kappa_{24}\|\hbar_2(k+1)\|^2\\
&-\kappa_{25}\|\hbar_2(k)\|^2-\sum_{i=1}^{3}\kappa_{26}\tilde{\mathscr{D}}_{2i}^2(k)-\sum_{i=1}^{3}\kappa_{27}\left\|\tilde{\theta}_{2i}(k)\right\|^2+\kappa_2
\end{aligned}
\tag{5.75}
$$

where $\kappa_{21}=2\sigma_{2\min}-\eta_2-\frac{1}{\rho_{21}}-\zeta_{2\max}\sigma_{2\max}^2h_{2\max}\bar{l}_2-\frac{\zeta_{2\max}\sigma_{2\max}h_{2\max}\varsigma_{2\max}}{\rho_{22}}\bar{l}_2-\rho_{23}\|\sigma_2\|^2$, $\kappa_{22}=\eta_2$, $\kappa_{23}=\zeta_{2i}\varsigma_{2i}-\zeta_{2i}\varsigma_{2i}^2h_{2i}-\zeta_{2i}\sigma_{2i}h_{2i}\varsigma_{2i}\rho_{22}$, $\kappa_{24}=1-\frac{\rho_{24}}{2}-\frac{\rho_{25}}{2}-\bar{\hbar}_2$, $\kappa_{25}=\bar{\hbar}_2-\frac{1}{2\rho_{24}}\|\mathscr{C}_2\|^2$, $\kappa_{26}=1-\|\sigma_2\|^2\rho_{21}$, $\kappa_{27}=\zeta_{2i}\varsigma_{2i}-2\zeta_{2i}^2\bar{l}_2$, $\kappa_2=\sum_{i=1}^{3}\zeta_{2i}\varsigma_{2i}\left\|\theta_{2i}^*(k)\right\|^2+\sum_{i=1}^{3}2\bar{\delta}_{2i}^2+\frac{1}{\rho_{23}}\bar{\chi}^2+\frac{1}{2\rho_{25}}\bar{g}_2^2\nu^2$, $\sigma_{2\min}$ is the minimum value of σ_2, and the matrix $\sigma_2=\mathrm{diag}[\sigma_{21},\sigma_{22},\sigma_{23}]$.

The above adaptive neural control scheme based on the DTDO for the uncertain discrete-time UAV system (5.1) with external disturbance and input saturation can be summarized as the following theorem:

Theorem 5.2 *Consider the uncertain discrete-time UAV system (5.1), if the design parameters and matrices are chosen appropriately, by constructing the actual controller $u(k)$ (5.68), the control signal $x_{vd}(k)$ (5.23), the NN-based DTDOs $\hat{\mathscr{D}}_1(k)$ (5.21) and $\hat{\mathscr{D}}_2(k)$ (5.64), and the adaptive laws $\hat{\theta}_{1i}(k)$ (5.25), $\hat{\theta}_{0i}(k)$ (5.27) and $\hat{\theta}_{2i}(k)$ (5.70), then the proposed NN control method can guarantee that the error between $x_1(k)$ and $x_{1d}(k)$ is bounded and the estimation errors of disturbances are bounded.*

Proof The entire Lyapunov function is chosen as

$$L(k) = \sum_{\bar{j}=1}^{2} L_{\bar{j}}(k) \tag{5.76}$$

According to (5.59) and (5.75), one has

$$\begin{aligned}
\Delta L(k) &= \sum_{\bar{j}=1}^{2} L_{\bar{j}}(k+1) - \sum_{\bar{j}=1}^{2} L_{\bar{j}}(k) \\
&\leq -\kappa_{11}\|z_1(k+1)\|^2 - \kappa_{12}\|z_1(k)\|^2 - \sum_{i=1}^{3} \kappa_{13}\left\|\hat{\theta}_{1i}(k)\right\|^2 - \kappa_{14}\|\hbar_1(k+1)\|^2 \\
&\quad -\kappa_{25}\|\hbar_2(k)\|^2 - \kappa_{15}\|\hbar_1(k)\|^2 - \sum_{i=1}^{3} \kappa_{16}\tilde{\mathscr{D}}_{1i}^2(k) + \kappa_{17}\|z_2(k)\|^2 \\
&\quad +\kappa_{18}\|\hbar_2(k)\|^2 + \kappa_1 + \kappa_2 - \kappa_{21}\|z_2(k+1)\|^2 - \sum_{i=1}^{3} \kappa_{23}\left\|\hat{\theta}_{2i}(k)\right\|^2 \\
&\quad -\kappa_{24}\|\hbar_2(k+1)\|^2 - \sum_{i=1}^{3} \kappa_{26}\tilde{\mathscr{D}}_{2i}^2(k) - \kappa_{22}\|z_2(k)\|^2 - \sum_{i=1}^{3} \kappa_{19}\left\|\tilde{\theta}_{1i}(k)\right\|^2 \\
&\quad - \sum_{i=1}^{3} \kappa_{27}\left\|\tilde{\theta}_{2i}(k)\right\|^2 - \sum_{i=1}^{3} \kappa_{110}\left\|\hat{\theta}_{0i}(k)\right\|^2 - \sum_{i=1}^{3} \kappa_{111}\left\|\tilde{\theta}_{0i}(k)\right\|^2
\end{aligned} \tag{5.77}$$

From (5.77), if $\kappa_{11} > 0$, $\kappa_{12} > 0$, $\kappa_{13} > 0$, $\kappa_{14} > 0$, $\kappa_{15} > 0$, $\kappa_{16} > 0$, $\kappa_{19} > 0$, $\kappa_{22} - \kappa_{17} > 0$, $\kappa_{25} - \kappa_{18} > 0$, $\kappa_{110} > 0$, $\kappa_{111} > 0$, $\kappa_{21} > 0$, $\kappa_{23} > 0$, $\kappa_{24} > 0$, $\kappa_{25} > 0$, $\kappa_{26} > 0$ and $\kappa_{27} > 0$ are satisfied by choosing appropriately parameters and matrices, one can obtain that the error $z_1(k)$ is bounded. Furthermore, the conclusion that the estimation errors $\tilde{\mathscr{D}}_{1i}(k)$ and $\tilde{\mathscr{D}}_{2i}(k)$ are bounded from (5.77). This concludes the proof. ◇

Remark 5.2 According to the traditional method of selecting parameters, relatively small parameter values should be selected in the designed adaptive laws, e.g., the parameters h_{01}, h_{02}, h_{03}, h_{11}, h_{12}, h_{13}, h_{21}, h_{22}, h_{23}, ς_{01}, ς_{02}, ς_{03}, ς_{11}, ς_{12}, ς_{13}, ς_{21}, ς_{22} and ς_{23} should be chosen smaller. Moreover, the control parameters in the DTDO and the auxiliary system ζ_{11}, ζ_{12}, ζ_{13}, ζ_{21}, ζ_{22}, ζ_{23}, $\mathscr{C}_1$ and $\mathscr{C}_2$ are also selected smaller. In addition, the parameters σ_{11}, σ_{12}, σ_{13}, σ_{21}, σ_{22} and σ_{23} should be chosen larger to guarantee that $\kappa_{11} > 0, \kappa_{12} > 0, \kappa_{13} > 0, \kappa_{14} > 0, \kappa_{15} > 0, \kappa_{16} > 0, \kappa_{19} > 0$, $\kappa_{110} > 0$, $\kappa_{111} > 0$, $\kappa_{22} - \kappa_{17} > 0$, $\kappa_{25} - \kappa_{18} > 0$, $\kappa_{21} > 0$, $\kappa_{23} > 0$, $\kappa_{24} > 0$, $\kappa_{25} > 0$, $\kappa_{26} > 0$ and $\kappa_{27} > 0$.

Remark 5.3 In this chapter, the system uncertainties are tackled based on the NN and an NN-based DTDO is presented to compensate for the adverse impacts of unknown disturbances. In general, it is advantageous that the approximation error of the NN converges faster. However, the control approach is studied based on the entire system, and the control objective of this paper is to develop a DTDO-based neural controller to ensure that the signal tracking errors are bounded and the signals in entire system are bounded. Therefore, there is no requirement on the convergence rate of the two kinds of different approximation based on the research objective of this chapter.

5.4 Simulation Study

This section shows the effectiveness of discrete-time adaptive NN control scheme based on the DTDO and an auxiliary system through numerical simulation analysis. In the simulation of uncertain UAV attitude dynamics system, only the effects of gust on the control of UAV system are considered, and the specific form of gust is shown in (2.40). In this numerical simulation, the the maximum external wind speed is assumed as $\bar{V}_{Wm} = [30, 20, 10]^{\mathrm{T}}\mathrm{m/sec}$, and $t_m = 3\mathrm{sec}$. The external disturbance is set as $\bar{d}_2 = \Delta T[0.2\sin(0.1t) - 0.2\cos(0.4t), 0.1\sin(0.2t) - 0.2\cos(0.5t), 0.3\sin(0.3t) - 0.3\cos(0.5t)]^{\mathrm{T}}$. The appropriate proportional gain coefficients are selected, and the sampling period is selected as $\Delta T = 0.01$. The saturation levels of control inputs are $\bar{v}_{m1} = 21.5\mathrm{deg}$ $\bar{v}_{m2} = 25\mathrm{deg}$ and $\bar{v}_{m3} = 30\mathrm{deg}$. The initial values are chosen as $\beta_{10} = 0\mathrm{deg}$, $\alpha_{10} = 0.3\mathrm{deg}$, $\mu_{10} = 0\mathrm{deg}$ and $p_{20} = q_{20} = r_{20} = 0\mathrm{deg}/s$. The flight altitude is 3000m. The desired tracking signals are given as $\beta_d = 0\mathrm{deg}$, $\alpha_d = 2\mathrm{erf}(0.4t)\mathrm{deg}$ and $\mu_d = 3\mathrm{erf}(0.4t)\mathrm{deg}$, where $\mathrm{erf}(\cdot)\mathrm{deg}$ denotes the Gaussian error function. The control parameters in the auxiliary system are set as $\mathscr{C}_1 = \mathrm{diag}[0.002, 0.002, 0.002]$, $\mathscr{C}_2 = \mathrm{diag}[0.002, 0.002, 0.002]$ and $\mathscr{G}_1 = \mathrm{diag}[1, 1, 1]$. The parameters in the adaptive laws and the DTDO are designed as $h_{11} = h_{12} = h_{13} = 0.005$, $\sigma_{11} = \sigma_{12} = \sigma_{13} = 0.1$, and $\varsigma_{21} = \varsigma_{22} = \varsigma_{23} = 0.01$. The control parameters of the discrete-time tracking differentiator are selected as $h_{01} = \mathrm{diag}[0.05, 0.05, 0.05]$ and $r_{01} = \mathrm{diag}[20, 20, 20]$. Furthermore, tracking errors are defined as $e_\beta = \beta - \beta_d$, $e_\alpha = \alpha - \alpha_d$, $e_\mu = \mu - \mu_d$, and the following definitions are that $\hat{\bar{d}}_{11}$, $\hat{\bar{d}}_{12}$, $\hat{\bar{d}}_{13}$, $\hat{\bar{d}}_{21}$, $\hat{\bar{d}}_{22}$ and $\hat{\bar{d}}_{23}$ are the estimations of $\bar{d}_{11}$, $\bar{d}_{12}$, $\bar{d}_{13}$, $\bar{d}_{21}$, $\bar{d}_{22}$ and $\bar{d}_{23}$. Moreover, the disturbance estimation errors are defined as $\tilde{\bar{d}}_{11} = \bar{d}_{11} - \hat{\bar{d}}_{11}$, $\tilde{\bar{d}}_{12} = \bar{d}_{12} - \hat{\bar{d}}_{12}$, $\tilde{\bar{d}}_{13} = \bar{d}_{13} - \hat{\bar{d}}_{13}$, $\tilde{\bar{d}}_{21} = \bar{d}_{21} - \hat{\bar{d}}_{21}$, $\tilde{\bar{d}}_{22} = \bar{d}_{22} - \hat{\bar{d}}_{22}$ and $\tilde{\bar{d}}_{23} = \bar{d}_{23} - \hat{\bar{d}}_{23}$.

Based on the above simulation parameters, the simulations are presented in Figs. 5.2, 5.3, 5.4, 5.5, 5.6, 5.7, 5.8 and 5.9. According to the tracking results in Figs. 5.2 and 5.3, the desired attitude angles can be tracked in bounded under the proposed DTDO-based adaptive controller. Simultaneously, the attitude angular rates are stable from Fig. 5.4 once the desired signals are tracked by the output signals. Then, the control inputs are given in Fig. 5.5. It is shown that the negative impacts of input saturation can be restrained using the proposed adaptive neural control scheme. The estimation performance of the designed DTDO is shown in Figs. 5.6, 5.7, 5.8 and 5.9. Based on Figs. 5.6, 5.7, 5.8 and 5.9, the designed DTDO can estimate the unknown disturbances well. Thus, the designed DTDO is effective to restrain the

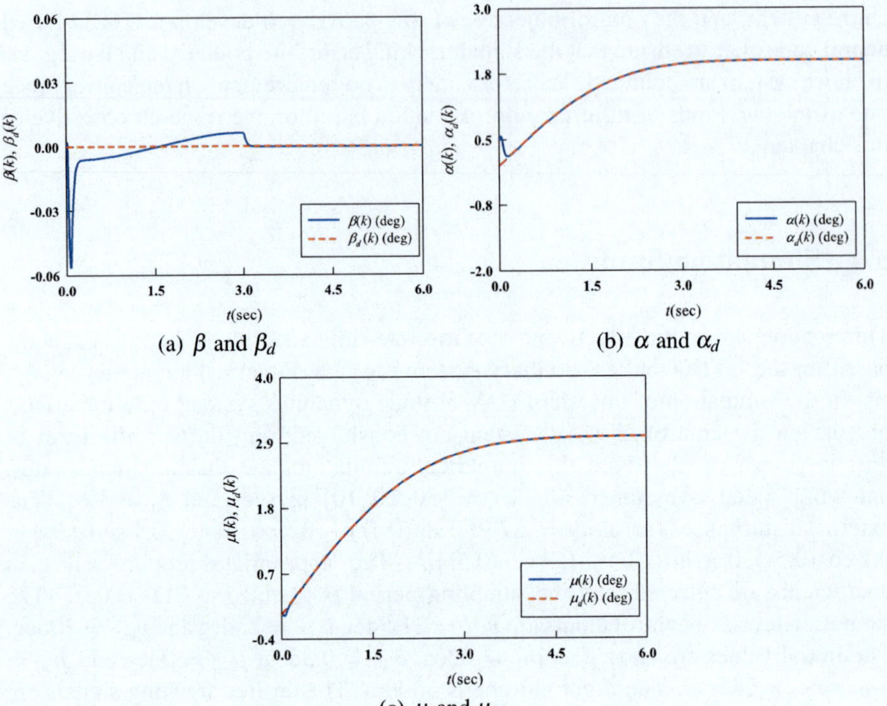

(a) β and β_d (b) α and α_d (c) μ and μ_d

Fig. 5.2 The tracking results

Fig. 5.3 The tracking errors $e_\beta(k)$, $e_\alpha(k)$ and $e_\mu(k)$

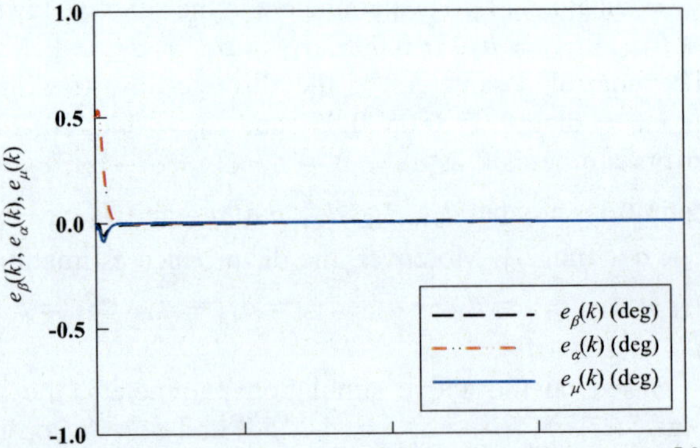

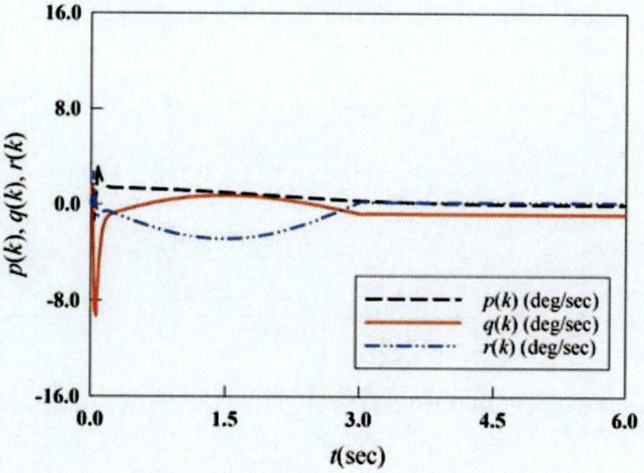

Fig. 5.4 The attitude angle rates $p(k)$, $q(k)$ and $r(k)$

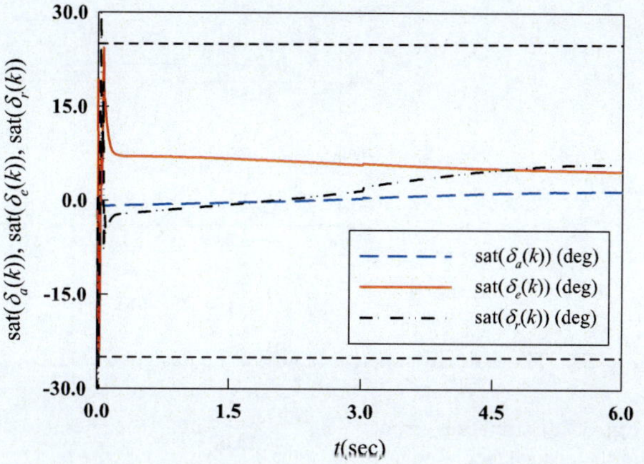

Fig. 5.5 Control inputs $\mathrm{sat}(\delta_a(k))$, $\mathrm{sat}(\delta_e(k))$ and $\mathrm{sat}(\delta_r(k))$

adverse effects of disturbances. The above simulation studies can show that the proposed DTDO-based adaptive neural control method is effective for the UAV system.

To further illustrate the validity of the designed adaptive neural control scheme, its performance has been compared with a SMC method and a PID control method via handling uncertainties and disturbances in the UAV system. The two comparisons are as follows:

Case1: In the design of the controller, the sliding mode term is directly used to replace the designed DO and the NN to handle the uncertainties. Other simulation conditions are identical to the proposed DTDO-based ANC method. The response results of the attitude tracking are presented in Fig. 5.10, and the control inputs are given in Fig. 5.11. Thus, the output signal x_1 can follow the desired signal x_{1d} well using appropriate SMC parameters with satisfactory performance. However, there is a relatively strong chattering of the control input signals as can be observed from Fig. 5.10.

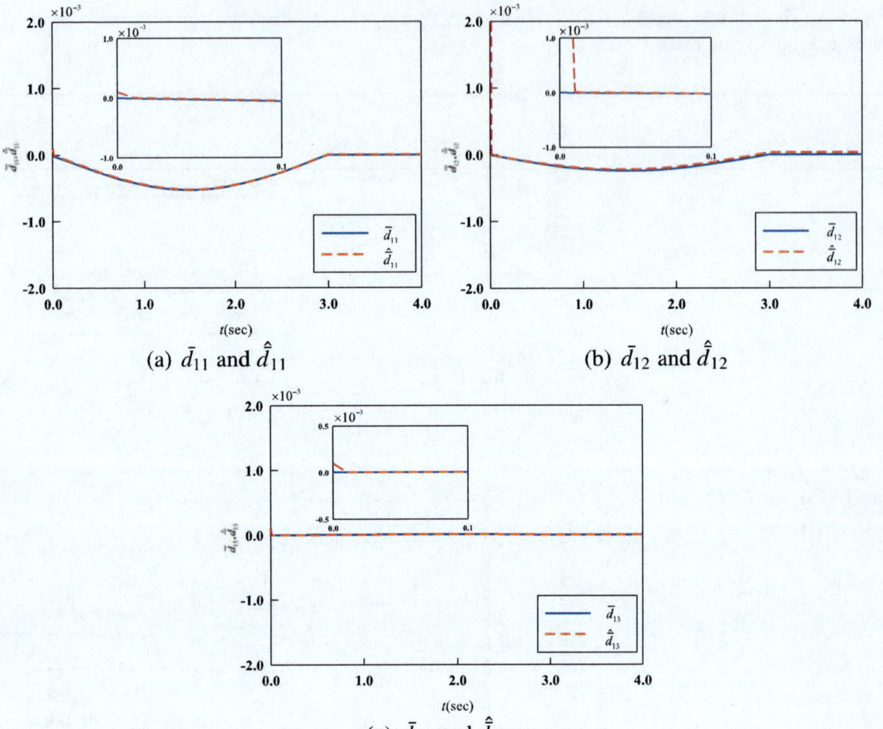

(a) $\bar{d}_{11}$ and $\hat{\bar{d}}_{11}$ (b) $\bar{d}_{12}$ and $\hat{\bar{d}}_{12}$

(c) $\bar{d}_{13}$ and $\hat{\bar{d}}_{13}$

Fig. 5.6 The estimation performance of the designed DTDO

Fig. 5.7 Estimation errors of the disturbance observer $\tilde{\bar{d}}_{11}$, $\tilde{\bar{d}}_{12}$ and $\tilde{\bar{d}}_{13}$

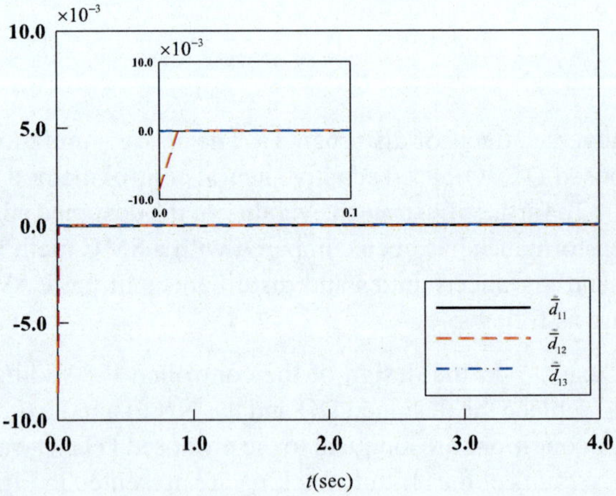

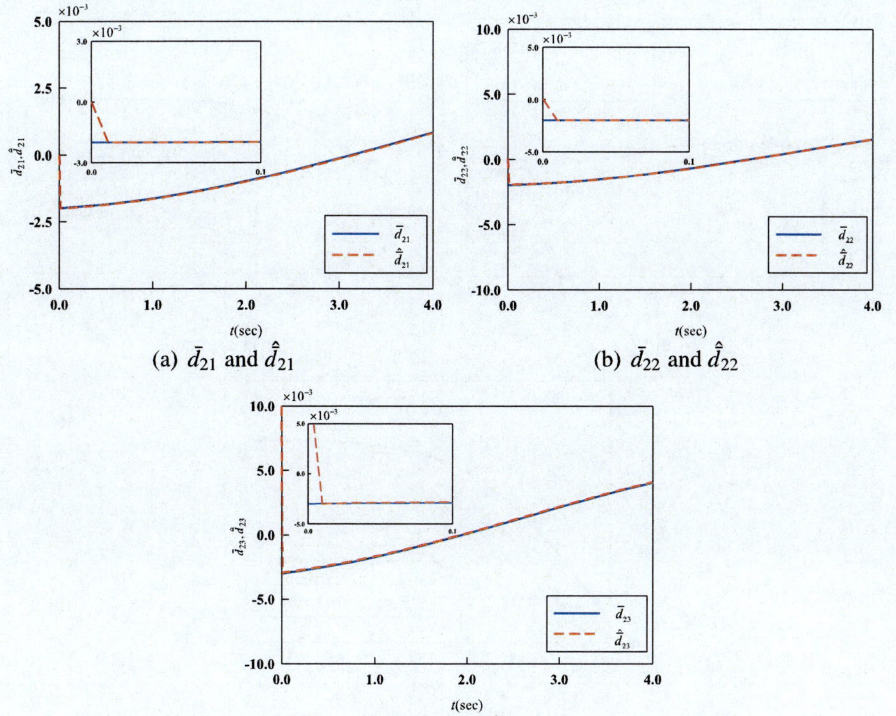

(a) $\bar{d}_{21}$ and $\hat{\bar{d}}_{21}$ (b) $\bar{d}_{22}$ and $\hat{\bar{d}}_{22}$

(c) $\bar{d}_{23}$ and $\hat{\bar{d}}_{23}$

Fig. 5.8 The estimation performance of the designed DTDO

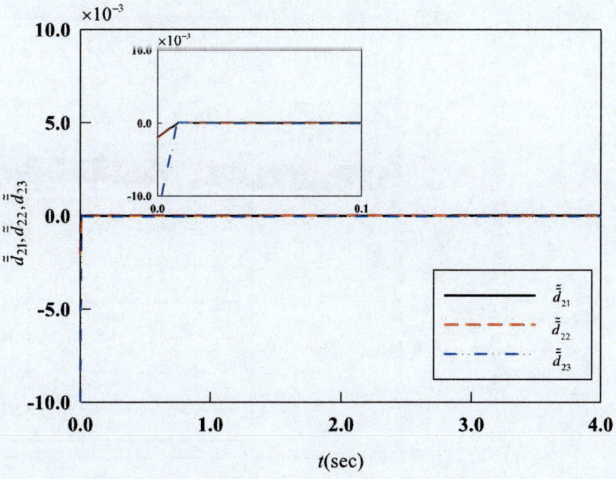

Fig. 5.9 Estimation errors of the disturbance observer $\tilde{\bar{d}}_{21}$, $\tilde{\bar{d}}_{22}$ and $\tilde{\bar{d}}_{23}$

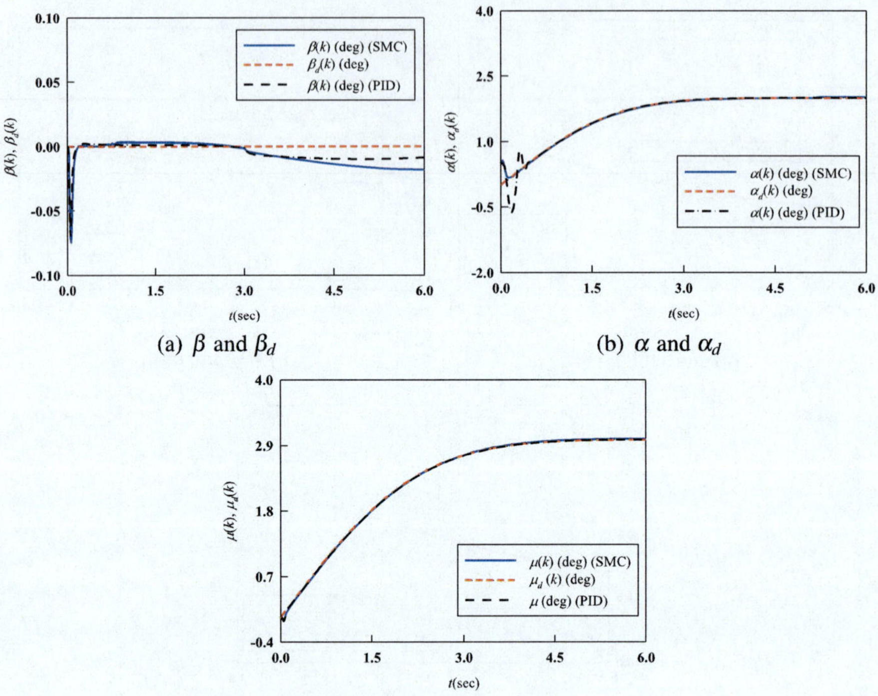

(a) β and β_d (b) α and α_d

(c) μ and μ_d

Fig. 5.10 The tracking results of attitude angles for Case 1 and Case 2

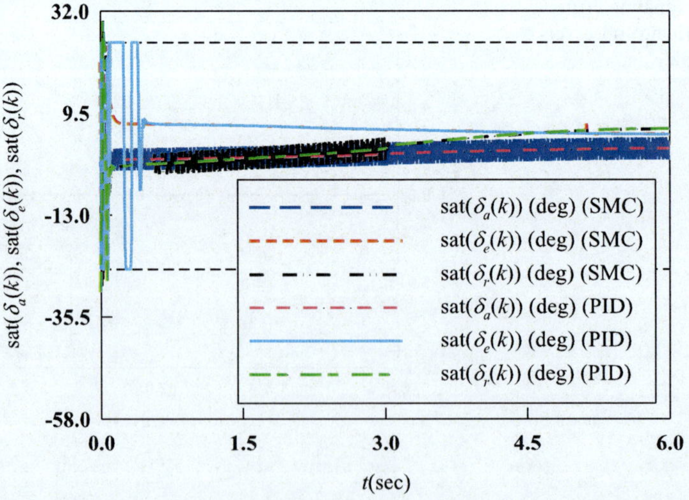

Fig. 5.11 The control inputs for Case 1 and Case 2 $\text{sat}(\delta_a(k))$, $\text{sat}(\delta_e(k))$ and $\text{sat}(\delta_r(k))$

Case2: The designed DO and the NN in the designed controller are replaced using the PID controller to tackle the uncertainties. By properly adjusting PID control parameters, the good tracking control performance can be obtained. Other conditions are identical to the Case 1. In Fig. 5.10, the simulations of attitude tracking are presented, and the control inputs are displayed in Fig. 5.11. According to the simulations in Fig. 5.10, the bounded control performance can also be acquired by properly adjusting the PID control parameters. However, the maximum overshoot of the output signals in Case 2 are larger than that of the proposed DTDO-based adaptive neural control approach in this paper. Furthermore, the saturated time of input signals is longer than the proposed control scheme based on Fig. 5.10.

According to Figs. 5.2, 5.5, 5.10 and 5.11, it can be obtained that the overall control performance using the developed adaptive neural control strategy is better than the SMC and PID control methods presented in Case 1 and Case 2 respectively for dealing with uncertainties in the UAV system with disturbances considered in this chapter.

5.5 Conclusions

For the discrete-time nonlinear dynamics of the UAV with system uncertainties, external disturbances and input saturation, the DTDO-based adaptive neural control scheme has been developed. The following advantages of the proposed adaptive neural control approach can be summarized: (1) system uncertainties, external disturbances and input saturation can be tackled together in tracking control of a class of uncertain nonlinear systems, with the application for the UAV; (2) a nonlinear DTDO has been exploited in the integrated control design strategy to eliminate the adverse effects of unknown disturbances; (3) the system uncertainties have been approximated by employing a radial basis function NN, and the adverse impacts of input saturation have been compensated. Finally, simulation results based on a nonlinear UAV model have been shown to demonstrate the availability of the developed adaptive neural control scheme.

References

1. Sonneveldt, L.: Nonlinear F-16 model description. Technical report. Delft University of Technology, Netherlands (2006)
2. Hu, Q.: Adaptive output feedback sliding-mode manoeuvring and vibration control of flexible spacecraft with input saturation. IET Control Theory Appl. **2**(6), 467–478 (2008)
3. Zhu, Z., Xia, Y., Fu, M., et al.: Adaptive sliding mode control for attitude stabilization with actuator saturation. IEEE Trans. Ind. Electron. **58**(10), 4898–4907 (2011)
4. Bošković, J.D., Li, S.-M., Mehra, R.K.: Robust adaptive variable structure control of spacecraft under control input saturation. J. Guid. Control Dyn. **24**(1), 14–22 (2001)

5. Hu, Q., Xiao, B., Friswell, M.: Robust fault-tolerant control for spacecraft attitude stabilisation subject to input saturation. IET Control Theory Appl. **5**(2), 271–282 (2011)
6. Bustan, D., Pariz, N., Sani, S.K.H.: Robust fault-tolerant tracking control design for spacecraft under control input saturation. ISA Trans. **53**(4), 1073–1080 (2014)
7. Boskovic, J.D., Li, S.-M., Mehra, R.K.: Robust tracking control design for spacecraft under control input saturation. J. Guid. Control Dyn. **27**(4), 627–633 (2004)
8. Chen, M., Yu, J.: Adaptive dynamic surface control of NSVs with input saturation using a disturbance observer. Chin. J. Aeronaut. **28**(3), 853–864 (2015)
9. Osa, Y., Mabuchi, T., Uchikado, S.: Synthesis of discrete time adaptive flight control system using nonlinear model matching. IEEE Int. Symp. Ind. Electron. **1**, 58–63 (2001)
10. Xiong, J.-J., Zhang, G.: Discrete-time sliding mode control for a quadrotor UAV. Optik **127**(8), 3718–3722 (2016)
11. Jiang, B., Chowdhury, F.N.: Fault estimation and accommodation for linear MIMO discrete-time systems. IEEE Trans. Control Syst. Technol. **13**(3), 493–499 (2005)
12. Xu, B., Sun, F., Yang, C., et al.: Adaptive discrete-time controller design with neural network for hypersonic flight vehicle via back-stepping. Int. J. Control **84**(9), 1543–1552 (2011)
13. Xu, B., Wang, D., Sun, F., et al.: Direct neural discrete control of hypersonic flight vehicle. Nonlinear Dyn. **70**(1), 269–278 (2012)
14. Xu, B., Zhang, Y.: Neural discrete back-stepping control of hypersonic flight vehicle with equivalent prediction model. Neurocomputing **154**, 337–346 (2015)
15. Shin, D.-H., Kim, Y.: Nonlinear discrete-time reconfigurable flight control law using neural networks. IEEE Trans. Control Syst. Technol. **14**(3), 408–422 (2006)
16. Mareels, I.M., Penfold, H., Evans, R.J.: Controlling nonlinear time-varying systems via euler approximations. Automatica **28**(4), 681–696 (1992)
17. Chen, M., Shao, S.-Y., Jiang, B.: Adaptive neural control of uncertain nonlinear systems using disturbance observer. IEEE Trans. Cybern. **47**(10), 3110–3123 (2017)
18. Yang, Q., Chen, M.: Adaptive neural prescribed performance tracking control for near space vehicles with input nonlinearity. Neurocomputing **174**, 780–789 (2016)

Chapter 6
Discrete-Time Control for Uncertain UAV System Based on SMDO and NN

6.1 Introduction

The design idea of SMC is to use high frequency switching control signal to drive the controlled signal to reach and stay near the SM surface in finite time [1]. In addition, the SMC method provides a systematic method for maintaining stability and consistency performance under the condition of low modeling accuracy, and the SMC has strong robustness against internal and external disturbances and parameter uncertainties. Up to now, researchers have applied the SMC technology to the design of flight control schemes [2–10], and improved the control performance of the systems by using the SMC method. The above research results put forward the SMC schemes for continuous time systems. However, so far, few researches have been conducted on the application of DTDO design method, SMC technology and NN approximation theory in discrete-time flight control design. Therefore, the above problems need further study in the design of discrete-time flight control scheme. According to the above description of the problems, and based on the research of the fourth chapter, for the longitudinal flight control system model with wind disturbances and system uncertainties and the attitude dynamic model of the uncertain UAV system with wind disturbances and input saturation in this chapter, an discrete-time tracking control scheme is proposed based on the SMDO and an adaptive discrete-time attitude control scheme is presented based on the SMDO, an auxiliary system and the NN.

6.2 Problem Formulation

6.2.1 Model of Longitudinal Flight Motion

In this section, the influence of thrust vector on UAV is ignored, and attitude angle commands β and μ are set to be zero, so only longitudinal trajectory motion is

S. Shao et al., *Robust Discrete-Time Flight Control of UAV with External Disturbances*, Studies in Systems, Decision and Control 317,
https://doi.org/10.1007/978-3-030-57957-9_6

considered. According to (2.30), (2.31) and (2.33), the expressions of height $\bar{z}_g$, speed $\bar{V}_A$ and the track angle γ are given as follows:

$$\dot{\bar{z}}_g = -\bar{V}_A \sin\gamma + d_z \tag{6.1}$$

$$\dot{\bar{V}}_A = \frac{1}{M}\left(-\bar{D} + \bar{T}\cos\alpha\right) - g\sin\gamma + \Delta F_v + d_v \tag{6.2}$$

$$\dot{\gamma} = \frac{1}{M\bar{V}_A}\left[\bar{L} - Mg\cos\gamma + \bar{T}\sin\alpha\right] + \Delta F_\gamma + d_\gamma \tag{6.3}$$

where $\bar{T}$ is the engine thrust, $d_z = w_{Wg}$, $d_v = -\dot{u}_{Wg}\cos\gamma\cos\chi - \dot{v}_{Wg}\cos\gamma\sin\chi + \dot{w}_{Wg}\sin\gamma$, and $d_\gamma = \frac{1}{\bar{V}_A}[\dot{u}_{Wg}(\cos\mu\sin\gamma\cos\chi + \sin\mu\sin\chi) + \dot{v}_{Wg}(\cos\mu\sin\gamma \sin\chi - \sin\mu\cos\chi) + \dot{w}_{Wg}\cos\mu\cos\gamma]$ are external disturbances, ΔF_v and ΔF_γ are the unknown nonlinear functions.

In order to design the NN control method based on the SMDO, the Euler approximation method [11] is used to convert (6.1), (6.2) and (6.3) into the approximate DT form, which can be written as

$$\bar{z}_g(k+1) = \bar{z}_g(k) - \Delta T\bar{V}_A(k)\sin\gamma(k) + \bar{d}_z(k) \tag{6.4}$$

$$\begin{aligned}\bar{V}_A(k+1) = \bar{V}_A(k) + \Delta T\left(\frac{1}{M}\left(-\bar{D}(k) + \bar{T}(k)\cos\alpha(k)\right) - g\sin\gamma(k)\right)\\ +\Delta\bar{F}_v(k) + \bar{d}_v(k)\end{aligned} \tag{6.5}$$

$$\begin{aligned}\gamma(k+1) = \gamma(k) + \Delta T\frac{1}{M\bar{V}_A(k)}\left(\bar{L} - Mg\cos\gamma(k) + \bar{T}(k)\sin\alpha(k)\right)\\ +\Delta\bar{F}_\gamma(k) + \bar{d}_\gamma(k)\end{aligned} \tag{6.6}$$

where $\bar{d}_z(k) = \Delta T d_z(k)$, ΔT is the sampling period, the external disturbances $\bar{d}_v(k) = -\bar{u}_{Wg}(k)\cos\gamma(k)\cos\chi(k) - \bar{v}_{Wg}(k)\cos\gamma(k)\sin\chi(k) + \bar{w}_{Wg}(k)\sin\gamma(k)$, and $\bar{d}_\gamma = \frac{1}{\bar{V}_A}[\bar{u}_{Wg}(k)(\cos\mu(k)\sin\gamma(k)\cos\chi(k) + \sin\mu(k)\sin\chi(k)) + \bar{w}_{Wg}(k)\cos\mu(k)\cos\gamma(k) + \bar{v}_{Wg}(k)(\cos\mu(k)\sin\gamma(k)\sin\chi(k) - \sin\mu(k)\cos\chi(k))]$, $\Delta\bar{F}_v(k) = \Delta T F_v$ and $\Delta\bar{F}_\gamma(k) = \Delta T F_\gamma$.

6.2.2 Model of Uncertain Attitude Dynamics System

According to (2.41) and (4.1), the dynamic model of uncertain UAV attitude with external disturbances and input saturation can be expressed as

$$\begin{cases} \dot{x}_1 = F_1(x_1) + \Delta F_1(x_1) + G_1(x_1)x_2 + d_1 \\ \dot{x}_2 = F_2(x) + \Delta F_2(x) + G_2(x)v + d_2 \\ v = \text{sat}(u)y = x_1 \end{cases} \tag{6.7}$$

where $F_1(x_1) \in \Re^3$ and $F_2(x) \in \Re^3$ are known nonlinear functions, $\Delta F_1(x_1) \in \Re^3$ and $\Delta F_2(x) \in \Re^3$ are unknown nonlinear functions, and $\Delta F_2(x)$ is the system uncertainty caused by the wind gradient, d_1 is the disturbance caused by wind, d_2 is the introduced external unknown disturbances, $v = \text{sat}(u)$ is the saturation function, and the saturation function $\text{sat}(u)$ satisfies that $\text{sat}(u) = [\text{sat}(u_1), \text{sat}(u_2), \text{sat}(u_3)]^{\mathrm{T}} \in \Re^3$, and one has

$$\text{sat}(u_{\bar{i}}) = \begin{cases} \bar{v}_{\bar{i}}\text{sign}(u_{\bar{i}}), & |u_{\bar{i}}| \geq \bar{v}_{\bar{i}}, \\ u_{\bar{i}}, & |u_{\bar{i}}| < \bar{v}_{\bar{i}}. \end{cases} \tag{6.8}$$

where $\bar{v}_{\bar{i}}$ is the bound of $v_{\bar{i}}(k)$, $\text{sign}(\cdot)$ denotes a sign function and $\bar{i} = 1, 2, 3$.

In order to design NN control method based on the SMDO and an auxiliary system, the Euler approximation method [11] is first used to convert the attitude dynamics model (6.7) of UAV with external disturbances and system uncertainties into an approximate discrete-time form, and its expression can be described as

$$\begin{cases} x_1(k+1) = \bar{F}_1(k) + \Delta\bar{F}_1(k) + \bar{G}_1(k)x_2(k) + \bar{d}_1(k) \\ x_2(k+1) = \bar{F}_2(k) + \Delta\bar{F}_2(k) + \bar{G}_2(k)\text{sat}(u(k)) + \bar{d}_2(k) \\ y(k) = x_1(k) \end{cases} \tag{6.9}$$

where $\bar{F}_1(k) = \Delta T F_1(x_1(k)) + x_1(k)$, $\bar{G}_1(k) = \Delta T G_1(x_1(k))$, $\bar{F}_2(k) = \Delta T F_2(x(k)) + x_2(k)$, $\bar{G}_2(k) = \Delta T G_2(x(k))$, $\Delta\bar{F}_1(k) = \Delta T \Delta F_1(x_1)$ and $\Delta\bar{F}_2(k) = \Delta T \Delta F_2(x)$. Furthermore, $\bar{d}_1(k) = [\bar{d}_{11}, \bar{d}_{12}, \bar{d}_{13}]^{\mathrm{T}}$ and $\bar{d}_2(k) = [\bar{d}_{21}, \bar{d}_{22}, \bar{d}_{23}]^{\mathrm{T}}$ are defined, and $\bar{d}_1(k)$ can be found in (4.3)–(4.3).

This chapter focuses on the trajectory control system model with wind disturbances and system uncertainties (6.4)–(6.6) and the uncertain UAV attitude dynamic model with wind disturbances and input saturation (6.9), a discrete-time adaptive NN control scheme is proposed based on the SMDO. The designed DT adaptive controller can ensure the errors between the output signals $\bar{z}_g(k)$, $\bar{V}_A(k)$, $\gamma(k)$ and $y(k)$ and the reference signals $\bar{z}_{gd}(k)$, $\bar{V}_{Ad}(k)$, $\gamma_d(k)$ and $y_d(k)$ are bounded, and all the signals in the closed loop system are also bounded. According to the design thread of the discrete-time control scheme, the developed control strategy is shown by a block diagram in Fig. 6.1.

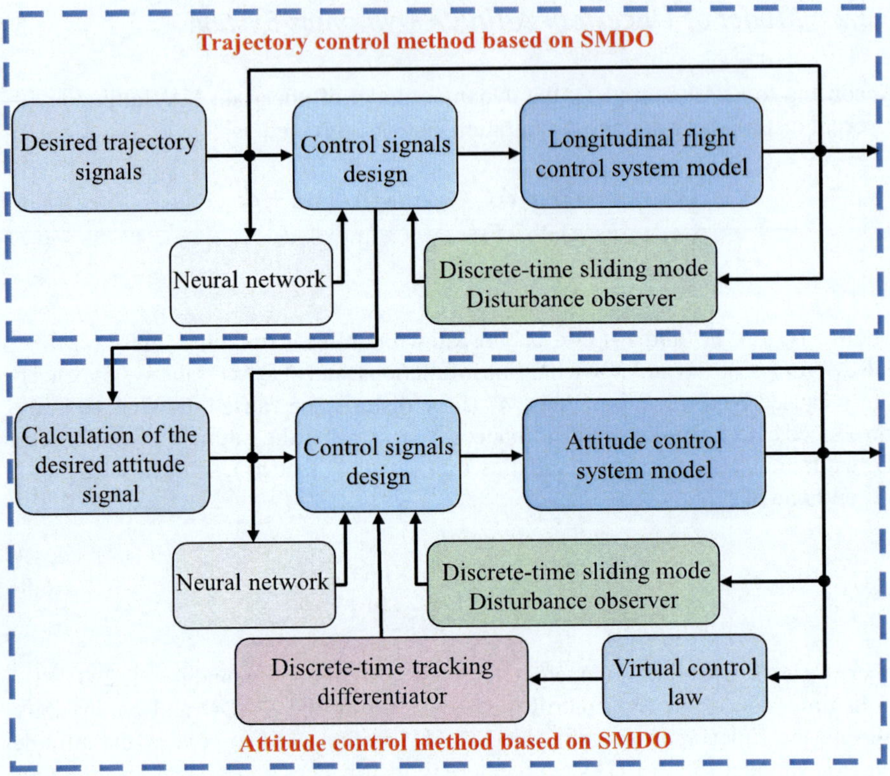

Fig. 6.1 The block diagram of the developed control strategy

In order to design the NN-based discrete-time adaptive flight control method, the following assumptions are given:

Assumption 6.1 The system control matrix $\bar{G}_i(k)(i = 1, 2)$ in the attitude dynamics model of UAV is invertible, and there are two positive constants $\underline{g}_i > 0$ and $\bar{g}_i > 0$ that can make $\underline{g}_i \leq \left\|\bar{G}_i(k)\right\| \leq \bar{g}_i$.

Assumption 6.2 Assume the wind is considered as gusty and the external wind speed is bounded, i.e. external disturbances $\bar{d}_z(k), \bar{d}_v(k), \bar{d}_\gamma(k)$ and $\bar{d}_i(k)$ are bounded, and $\left|\Delta\bar{d}_z(k)\right| \leq \Delta T\mu_z$, $\left|\Delta\bar{d}_v(k)\right| \leq \Delta T\mu_v$, $\left|\Delta\bar{d}_\gamma(k)\right| \leq \Delta T\mu_\gamma$ and $\left\|\Delta\bar{d}_i(k)\right\| \leq \Delta T\mu_i$, where $\Delta\bar{d}_z(k) = \bar{d}_z(k+1) - \bar{d}_z(k)$, $\Delta\bar{d}_v(k) = \bar{d}_v(k+1) - \bar{d}_v(k)$, $\Delta\bar{d}_\gamma(k) = \bar{d}_\gamma(k+1) - \bar{d}_\gamma(k)$, $\Delta\bar{d}_i(k) = \bar{d}_i(k+1) - \bar{d}_i(k)$, and μ_z, μ_v, μ_γ and μ_i are positive constants.

6.3 Discrete-Time Control Scheme Based on an Auxiliary System and SMDO

6.3.1 Trajectory Control Method Based on SMDO

This section is for a trajectory control system model with external disturbances (6.4)–(6.6), an adaptive NN control scheme is proposed based on a SMDO. Firstly, the height tracking error is defined as $\eta_z(k) = \bar{z}_g(k) - \bar{z}_{gd}(k)$. According to (6.4), one has

$$\eta_z(k+1) = \bar{z}_g(k) - \Delta T \bar{V}_A(k) u_z(k) + \bar{d}_z(k) - \bar{z}_{gd}(k+1) \tag{6.10}$$

where $u_z(k) = \sin\gamma(k)$. Therefore, the track angle signal $\gamma_d(k)$ can be calculated by $\gamma_d(k) = \arcsin\gamma(k)$.

In order to compensate the negative affects of external disturbance $\bar{d}_z(k)$ on the control performance of the system, a SMDO is designed. To facilitate the design of discrete-time SMDO, the following auxiliary system is first given:

$$\begin{cases} s_z(k) = \sigma_z(k) - \eta_z(k) \\ \sigma_z(k+1) = \bar{z}_g(k) + \hat{\bar{d}}_z(k) - \bar{z}_{gd}(k+1) - \Delta T \bar{V}_A(k) u_z(k) - \lambda_z \mathrm{sign}(s_z(k)) \end{cases} \tag{6.11}$$

where $\hat{\bar{d}}_z(k)$ is the estimation of $\bar{d}_z(k)$, and $\lambda_z > 0$ is a positive constant.

According to (6.10) and (6.11), one has

$$\begin{aligned} s_z(k+1) &= \sigma_z(k+1) - \eta_z(k+1) \\ &= -\lambda_1 \mathrm{sign}(s_z(k)) - \tilde{\bar{d}}_z(k) \end{aligned} \tag{6.12}$$

where $\tilde{\bar{d}}_z(k) = \bar{d}_z(k) - \hat{\bar{d}}_z(k)$.

Then, the discrete-time SMDO is designed as follows:

$$\begin{cases} \hat{\bar{d}}_z(k) = \delta_z(\vartheta_z(k) - s_z(k)) \\ \vartheta_z(k+1) = -\lambda_z \mathrm{sign}(s_z(k)) + \frac{1}{\delta_z}\hat{d}_z(k) \end{cases} \tag{6.13}$$

where $\delta_z > 0$ a positive constant.

Based on (6.13), it yields

$$\hat{\bar{d}}_z(k+1) = \hat{\bar{d}}_z(k) + \delta_z \tilde{\bar{d}}_z(k) \tag{6.14}$$

$$\tilde{\bar{d}}_z(k+1) = \Delta \bar{d}_z(k) + (1 - \delta_z)\tilde{\bar{d}}_z(k) \tag{6.15}$$

From Assumption 6.2 and (6.15), one can obtain

$$\tilde{\tilde{d}}_z^2(k+1) \leq \tau_z(1-\delta_z)^2\tilde{\tilde{d}}_z^2(k) + \frac{\Delta T^2\mu_z^2}{\tau_z} + (1-\delta_z)^2\tilde{\tilde{d}}_z^2(k) + \Delta T^2\mu_z^2 \quad (6.16)$$

where τ_z a positive constant.

Moreover, according to (6.12), one has

$$s_z^2(k+1) \leq 2\lambda_z^2 + 2\tilde{\tilde{d}}_z^2(k) \quad (6.17)$$

The height control law $u_z(k)$ is given as follows:

$$u_z(k) = (1/(\Delta T\bar{V}_A))(\bar{z}_g(k) + \hat{\hat{d}}_1(k) - \bar{z}_{gd}(k+1) + \rho_z\eta_z(k)) \quad (6.18)$$

where $\rho_z > 0$ a positive constant.

Combining (6.10) and (6.18), $\eta_1(k+1)$ can be written as

$$\eta_z(k+1) = \tilde{\tilde{d}}_z(k) + \rho_z\eta_z(k) \quad (6.19)$$

According to (6.19), it yields

$$\eta_z^2(k+1) \leq 2\tilde{\tilde{d}}_z^2(k) + 2\rho_z^2\eta_z^2(k) \quad (6.20)$$

The following Lyapunov function is chosen as

$$V_z(k) = \delta_z^2\eta_z^2(k) + \delta_z^2 s_z^2(k) + \tilde{\tilde{d}}_z^2(k) \quad (6.21)$$

From (6.21), the first order difference of Lyapunov function $V_z(k)$ is described as

$$\begin{aligned}\Delta V_z(k) = {} & \delta_z^2\eta_z^2(k+1) + \delta_z^2 s_z^2(k+1) + \tilde{\tilde{d}}_z^2(k+1) - \delta_z^2\eta_z^2(k) \\ & -\delta_z^2 s_z^2(k) - \tilde{\tilde{d}}_z^2(k)\end{aligned} \quad (6.22)$$

According to (6.16), (6.17) and (6.20), one has

$$\begin{aligned}\Delta V_z(k) \leq {} & 2\delta_z^2\lambda_z^2 + 4\delta_z^2\tilde{\tilde{d}}_z^2(k) + 2\delta_z^2\rho_z^2\eta_z^2(k) + (1-\delta_z)^2\tilde{\tilde{d}}_z^2(k)\tau_z + \frac{\Delta T^2\mu_z^2}{\tau_z} \\ & -\delta_z^2\eta_z^2(k) - \delta_z^2 s_z^2(k) - \tilde{\tilde{d}}_z^2(k) + (1-\delta_z)^2\tilde{\tilde{d}}_z^2(k) + \Delta T^2\mu_z^2\end{aligned} \quad (6.23)$$

The following control analysis is conducted for the velocity $\bar{V}_A$ and the track angle γ. On the basis of (6.5) and (6.6), one can obtain

$$H(k+1) = F_H(k) + \Delta\bar{F}_H(k) + G_H(k)u_H(k) + \bar{d}_H(k) \tag{6.24}$$

where $\bar{d}_H(k) = [\bar{d}_v(k), \bar{d}_\gamma(k)]^{\mathrm{T}}$, $\Delta\bar{F}_H(k) = [\Delta\bar{F}_v(k), \Delta\bar{F}_\gamma(k)]^{\mathrm{T}}$, $F_H(k) = [-\Delta T g \sin\gamma(k) + \bar{V}_A(k), -\Delta T g \cos\gamma(k)/\bar{V}_A(k) + \gamma(k)]^{\mathrm{T}}$, $H(k+1) = [\bar{V}_A(k+1), \gamma(k+1)]^{\mathrm{T}}$, $u_H(k) = [-\bar{D}(k) + \bar{T}(k)\cos\alpha(k), \bar{L} + \bar{T}(k)\sin\alpha(k)]^{\mathrm{T}}$, $H(k+1) = [\bar{V}_A(k+1), \gamma(k+1)]^{\mathrm{T}}$, $G_H(k) = \mathrm{diag}[\Delta T/M, \Delta T/M\bar{V}_A(k)]$. Furthermore, the tracking signal of attack angle α_d and the thrust $\bar{T}(k)$ can be obtained by u_H.

Define $\eta_H(k) = H(k) - H_d(k)$ and $H_d(k) = [\bar{V}_{Ad}(k), \gamma_d(k)]^{\mathrm{T}}$. Therefore, based on (6.24), one has

$$\begin{aligned}\eta_H(k+1) &= H(k+1) - H_d(k+1)\\ &= F_H(k) + \Delta\bar{F}_H(k) + G_H(k)u_H(k) + \bar{d}_H(k) - H_d(k+1)\end{aligned} \tag{6.25}$$

According to Lemma 2.2, and using NN to approximate the unknown function $\tau_H\Delta\bar{F}_H(k)$, $\eta_H(k+1)$ can be written as

$$\eta_H(k+1) = F_H(k) + \Theta_H(k) + \Upsilon_H(k) + G_H(k)u_H(k) + \bar{d}_H(k) - H_d(k+1) \tag{6.26}$$

where $\Theta_H(k) = [\frac{1}{\tau_H}\zeta_{H1}^{*\mathrm{T}}(k)\varphi_{H1}(H(k)), \frac{1}{\tau_H}\zeta_{H2}^{*\mathrm{T}}(k)\varphi_{H2}(H(k))]^{\mathrm{T}}$, τ_H is a positive constant, $\zeta_{H\bar{i}}^{*}(k)$ is the optimal weight vector, $\Upsilon_H(k) = [\frac{1}{\tau_H}\varepsilon_{H1}(k), \frac{1}{\tau_H}\varepsilon_{H2}(k)]^{\mathrm{T}}$, $\varepsilon_{H\bar{i}}(k)$ is the minimum approximation error, $\varphi_{H\bar{i}}(\cdot)$ is the basis function vector, and $\bar{i} = 1, 2$. To facilitate the design of control scheme, define $\varphi_{H\bar{i}}(H(k)) = \varphi_{H\bar{i}}(k)$.

In order to deal with the disturbance $\bar{d}_H(k)$, a NN-SMDO is designed. To facilitate the design of discrete-time SMDO, the following auxiliary system is first given:

$$\begin{cases} s_H(k) = \sigma_H(k) - \eta_H(k)\\ \sigma_H(k+1) = F_H(k) + \hat{\Theta}_H(k) + G_H(k)u_H(k) + \hat{\bar{d}}_H(k)\\ \qquad\qquad - H_d(k+1) - \lambda_H \mathrm{sign}(s_H(k)) \end{cases} \tag{6.27}$$

where $\hat{\Theta}_H(k) = [\frac{1}{\tau_H}\hat{\zeta}_{H1}^{\mathrm{T}}(k)\varphi_{H1}(k), \frac{1}{\tau_H}\hat{\zeta}_{H2}^{\mathrm{T}}(k)\varphi_{H2}(k)]^{\mathrm{T}}$, $\hat{\Theta}_H(k)$ is the estimation of $\Theta_H(k)$, $\hat{\bar{d}}_H(k)$ is the estimation of $\bar{d}_H(k)$, and $\lambda_H > 0$ is design constant.

According to (6.26) and (6.27), one has

$$\begin{aligned}s_H(k+1) &= \sigma_H(k+1) - \eta_H(k+1)\\ &= -\lambda_H \mathrm{sign}(s_H(k)) - \tilde{\bar{d}}_H(k) + \tilde{\Theta}_H(k) - \Upsilon_H(k)\end{aligned} \tag{6.28}$$

where $\tilde{\Theta}_H(k) = \hat{\Theta}_H(k) - \Theta_H(k)$, $\tilde{\Theta}_H(k) = [\frac{1}{\tau_H}\tilde{\zeta}_{H1}^{\mathrm{T}}(k)\varphi_{H1}(k), \frac{1}{\tau_H}\tilde{\zeta}_{H2}^{\mathrm{T}}(k)\varphi_{H2}(k)]^{\mathrm{T}}$, and $\tilde{\bar{d}}_H(k) = \bar{d}_H(k) - \hat{\bar{d}}_H(k)$.

Then, the discrete-time SMDO is designed as follows:

$$\begin{cases} \hat{\bar{d}}_H(k) = \delta_H(\vartheta_H(k) - s_H(k)) \\ \vartheta_H(k+1) = -\lambda_H \text{sign}(s_H(k)) + \frac{1}{\delta_H}\hat{d}_H(k) \end{cases} \tag{6.29}$$

where $\delta_H > 0$ is design constant

Based on (6.29), it yields

$$\hat{\bar{d}}_H(k+1) = \hat{\bar{d}}_H(k) + \delta_H \tilde{\bar{d}}_H(k) - \delta_H \tilde{\Theta}_H(k) + \delta_H \Upsilon_H(k) \tag{6.30}$$

$$\tilde{\bar{d}}_H(k+1) = \Delta\bar{d}_H(k) + (1-\delta_H)\tilde{\bar{d}}_H(k) + \delta_H \tilde{\Theta}_H(k) - \delta_H \Upsilon_H(k) \tag{6.31}$$

where $\Delta\bar{d}_H(k) = \bar{d}_H(k+1) - \bar{d}_H(k)$. Furthermore, according to Assumption 6.2, one can obtain that $\left\|\Delta\bar{d}_H(k)\right\| \leq \Delta T \mu_H$, and μ_H is a positive constant.

From (6.31), one has

$$\begin{aligned} \left\|\tilde{\bar{d}}_H(k+1)\right\|^2 \leq & (1-\delta_H)^2 \left\|\tilde{\bar{d}}_H(k)\right\|^2 + 2\delta_H^2 \left\|\tilde{\bar{d}}_H(k)\right\|^2 + 3\delta_H^2 \left\|\tilde{\Theta}_H(k)\right\|^2 \\ & + (1-\delta_H)^2 \left\|\tilde{\Theta}_H(k)\right\|^2 + \varpi_{H1} \left\|\tilde{\bar{d}}_H(k)\right\|^2 + 2\delta_H^2 \bar{\Upsilon}_H^2 + 2\Delta T^2 \mu_H^2 \\ & + 2\Delta T \mu_H \delta_H \bar{\Upsilon}_H + \frac{(1-\delta_H)^2 \Delta T^2 \mu_H^2}{\varpi_{H1}} + (1-\delta_H)^2 \bar{\Upsilon}_H^2 \end{aligned} \tag{6.32}$$

where ϖ_{H1} is a positive constant, $\|\Upsilon_H(k)\| \leq \bar{\Upsilon}_H$, and $\bar{\Upsilon}_H > 0$ is a positive constant.

In addition, based on the (6.28), the following expression can be obtained

$$\|s_H(k+1)\|^2 \leq 8\lambda_H^2 + 4\left\|\tilde{\bar{d}}_H(k)\right\|^2 + 4\left\|\tilde{\Theta}_H(k)\right\|^2 + 4\bar{\Upsilon}_H^2 \tag{6.33}$$

The control law $u_H(k)$ is designed by

$$u_H(k) = -\bar{G}_H^{-1}(k)(\bar{F}_H(k) + \hat{\Theta}_H(k) + \rho_H \eta_H(k) + \hat{\bar{d}}_H(k) - H_d(k+1)) \tag{6.34}$$

where $\rho_H > 0$ is design constant.

Combining (6.26) and (6.34), $\eta_H(k+1)$ can be written as

$$\eta_H(k+1) = -\tilde{\Theta}_H(k) + \tilde{\bar{d}}_H(k) + \Upsilon_H(k) - \rho_H \eta_H(k) \tag{6.35}$$

According to Lemma 2.2 and (6.35), one has

$$\begin{aligned} \|\eta_H(k+1)\|^2 \leq & 3\left\|\tilde{\Theta}_H(k)\right\|^2 + \frac{1}{\varpi_{H2}}\left\|\tilde{\Theta}_H(k)\right\|^2 + 3\bar{\Upsilon}_H^2 + \frac{1}{\varpi_{H2}}\bar{\Upsilon}_H^2 \\ & + 4\left\|\tilde{\bar{d}}_H(k)\right\|^2 + 2\rho_H^2\|\eta_H(k)\|^2 + 2\varpi_{H2}\rho_H^2\|\eta_H(k)\|^2 \end{aligned} \tag{6.36}$$

Moreover, the adaptive law $\hat{\zeta}_{H\bar{i}}(k)$ is designed by

$$\hat{\zeta}_{H\bar{i}}(k+1) = \kappa_{H\bar{i}}\varphi_{H\bar{i}}(k)\eta_{H\bar{i}}(k) - (\omega_{H\bar{i}} - 1)\hat{\zeta}_{H\bar{i}}(k) \tag{6.37}$$

where $\kappa_{H\bar{i}} > 0$ and $\omega_{H\bar{i}} > 0$ are design constants, $\eta_{H\bar{i}}(k)$ is the $\bar{i}$th variables of $\eta_H(k)$, and $\bar{i} = 1, 2$.

According to (6.37), one has

$$\tilde{\zeta}_{H\bar{i}}(k+1) = \tilde{\zeta}_{H\bar{i}}(k) + \kappa_{H\bar{i}}\varphi_{H\bar{i}}(k)\eta_{H\bar{i}}(k) - \omega_{H\bar{i}}\hat{\zeta}_{H\bar{i}}(k) \tag{6.38}$$

where $\tilde{\zeta}_{H\bar{i}}(k+1) = \hat{\zeta}_{H\bar{i}}(k+1) - \zeta^*_{H\bar{i}}(k+1)$, and $\tilde{\zeta}_{H\bar{i}}(k) = \hat{\zeta}_{H\bar{i}}(k) - \zeta^*_{H\bar{i}}(k)$.

Based on (6.38), it yields

$$\begin{aligned}\left\|\tilde{\zeta}_{H\bar{i}}(k+1)\right\|^2 - \left\|\tilde{\zeta}_{H\bar{i}}(k)\right\|^2 &= \kappa^2_{H\bar{i}}\left\|\varphi_{H\bar{i}}(k)\right\|^2\eta^2_{H\bar{i}}(k) \\ &+\omega^2_{H\bar{i}}\left\|\hat{\zeta}_{H\bar{i}}(k)\right\|^2 - 2\kappa_{H\bar{i}}\omega_{H\bar{i}}\varphi^{\mathrm{T}}_{H\bar{i}}(k)\hat{\zeta}_{H\bar{i}}(k)\eta_{H\bar{i}}(k) \\ &+2\kappa_{H\bar{i}}\tilde{\zeta}^{\mathrm{T}}_{H\bar{i}}(k)\varphi_{H\bar{i}}(k)\eta_{H\bar{i}}(k) - 2\omega_{H\bar{i}}\tilde{\zeta}^{\mathrm{T}}_{H\bar{i}}(k)\hat{\zeta}_{H\bar{i}}(k)\end{aligned} \tag{6.39}$$

Furthermore, one has

$$-2\omega^2_{H\bar{i}}\tilde{\zeta}^{\mathrm{T}}_{H\bar{i}}(k)\hat{\zeta}_{H\bar{i}}(k) = -\omega^2_{H\bar{i}}\left\|\tilde{\zeta}_{H\bar{i}}(k)\right\|^2 - \omega^2_{H\bar{i}}\left\|\hat{\zeta}_{H\bar{i}}(k)\right\|^2 + \omega^2_{H\bar{i}}\left\|\zeta^*_{H\bar{i}}(k)\right\|^2 \tag{6.40}$$

Combining (6.39) and (6.40), one has

$$\begin{aligned}\omega_{H\bar{i}}\left(\left\|\tilde{\zeta}_{H\bar{i}}(k+1)\right\|^2 - \left\|\tilde{\zeta}_{H\bar{i}}(k)\right\|^2\right) &\le (\omega^2_{H\bar{i}} - \omega^3_{H\bar{i}} - \kappa^2_{H\bar{i}}\omega^2_{H\bar{i}})\left\|\hat{\zeta}_{H\bar{i}}(k)\right\|^2 \\ &+(2\omega^2_{H\bar{i}}\bar{\varphi}_H + \omega_{H\bar{i}}\kappa^2_{H\bar{i}}\bar{\varphi}_H)\eta^2_{H\bar{i}}(k) - (\omega^2_{H\bar{i}} - \kappa^2_{H\bar{i}})\left\|\tilde{\zeta}_{H\bar{i}}(k)\right\|^2 + \omega^2_{H\bar{i}}\bar{\zeta}^*_H\end{aligned} \tag{6.41}$$

where $\left\|\varphi_{H\bar{i}}(k)\right\|^2 \le \bar{\varphi}_H$, $\left\|\zeta^*_{H\bar{i}}(k)\right\|^2 \le \bar{\zeta}^*_H$, $\bar{\varphi}_H > 0$ and $\bar{\zeta}^*_H > 0$ are positive constants.

The following Lyapunov function is chosen as:

$$\begin{aligned}V_H(k) = \omega_{H\min}\|\eta_H(k)\|^2 + \left\|\tilde{\bar{d}}_H(k)\right\|^2 \\ +\omega_{H\min}\|s_H(k)\|^2 + \sum_{\bar{i}=1}^{2}\omega_{H\bar{i}}\left\|\tilde{\zeta}_{H\bar{i}}(k)\right\|^2\end{aligned} \tag{6.42}$$

where $\omega_{H\min}$ is the minimum value of $\omega_{H\bar{i}}$, and $\bar{i} = 1, 2$.

According to (6.42), the first order difference of Lyapunov function $V_H(k)$ can be written as

$$
\begin{aligned}
\Delta V_H(k) = \omega_{H\min}(\|\eta_H(k+1)\|^2 + \|s_H(k+1)\|^2) + \left\|\tilde{\bar{d}}_H(k+1)\right\|^2 - \left\|\tilde{\bar{d}}_H(k)\right\|^2 \\
+ \sum_{\bar{i}=1}^{2} \omega_{H\bar{i}} \left\|\tilde{\zeta}_{H\bar{i}}(k+1)\right\|^2 - \omega_{H\min}(\|\eta_H(k)\|^2 + \|s_H(k)\|^2) \\
- \sum_{\bar{i}=1}^{m} \omega_{H\bar{i}} \left\|\tilde{\zeta}_{H\bar{i}}(k)\right\|^2
\end{aligned} \tag{6.43}
$$

On the basis of (6.32), (6.33), (6.36), (6.41) and (6.43), $\Delta V_H(k)$ satisfies the following expression

$$
\begin{aligned}
\Delta V_H(k) \le -\omega_{H\min}(1 - 2\rho_H^2 - 2\varpi_{H2}\rho_H^2)\|\eta_H(k)\|^2 - \omega_{H\min}\|s_H(k)\|^2 \\
-(1 - 8\omega_{H\min} - (1-\delta_H)^2 - 2\delta_H{}^2 - \varpi_{H1})\left\|\tilde{\bar{d}}_H(k)\right\|^2 \\
-(\min(\omega_{H\bar{i}}^2 - \kappa_{H\bar{i}}^2) - 7\omega_{H\min}\tfrac{\bar{\varphi}_H}{\tau_H^2} - 3\delta_H^2\tfrac{\bar{\varphi}_H}{\tau_H^2})\sum_{\bar{i}=1}^{2}\left\|\tilde{\zeta}_{H\bar{i}}(k)\right\|^2 + 2\Delta T\mu_H\delta_H\bar{\Upsilon}_H \\
+(\tfrac{\bar{\varphi}_H(1-\delta_H)^2}{\tau_H^2} + \tfrac{\bar{\varphi}_H\omega_{H\min}}{\tau_H^2\varpi_{H2}})\sum_{\bar{i}=1}^{2}\left\|\tilde{\zeta}_{H\bar{i}}(k)\right\|^2 + \tfrac{\omega_{H\min}}{\varpi_{H2}}\bar{\Upsilon}_H^2 + 8\omega_{H\min}\lambda_H^2 + \sum_{\bar{i}=1}^{2}\omega_{H\bar{i}}^2\bar{\zeta}_H^* \\
-\min(\omega_{H\bar{i}}^2 - \omega_{H\bar{i}}^2 - \kappa_{H\bar{i}}^2\omega_{H\bar{i}}^2)\sum_{\bar{i}=1}^{2}\left\|\hat{\zeta}_{H\bar{i}}(k)\right\|^2 + 7\omega_{H\min}\bar{\Upsilon}_H^2 + 2\Delta T^2\mu_H^2 \\
+2\delta_H^2\bar{\Upsilon}_H^2 + \tfrac{(1-\delta_H)^2\Delta T^2\mu_H^2}{\varpi_{H1}} + (1-\delta_H)^2\bar{\Upsilon}_H^2 \\
+\max(2\omega_{H\bar{i}}^2\bar{\varphi}_H + \omega_{H\bar{i}}\kappa_{H\bar{i}}^2\bar{\varphi}_H)\|\eta_H(k)\|^2
\end{aligned} \tag{6.44}
$$

where $\min(\cdot)$ denotes the minimum value and $\max(\cdot)$ denotes the maximum value.

6.3.2 Attitude Control Scheme Based on SMDO

In this section, an adaptive neural control scheme based on the SMDO and an auxiliary system is proposed for the uncertain attitude dynamics model of UAV in the present of external disturbances and input saturation (6.9). To compensate for the adverse effects of input saturation, the following auxiliary system is designed by [12]

$$
\begin{cases}
\chi_1(k+1) = -C_1\chi_1(k) + \bar{G}_1(k)\chi_2(k) \\
\chi_2(k+1) = -C_2\chi_2(k) + \bar{G}_2(k)\Delta v(k)
\end{cases} \tag{6.45}
$$

where $\chi_1(k) \in \Re^3$ and $\chi_2(k) \in \Re^3$ are the state variables of auxiliary system, $C_1 = C_1^{\mathrm{T}} > 0$ and $C_2 = C_2^{\mathrm{T}} > 0$ are the designed constant matrices, and $\Delta v(k) = \mathrm{sat}(u(k)) - u(k)$.

Remark 6.1 In actual control systems, the actuators of the systems usually provide only limited control power. Therefore, it is assumed that $\Delta v(k)$ is bounded for the physical properties of the UAV system. $\Delta v(k)$ satisfies that $\|\Delta v(k)\| \leq \nu$, and $\nu > 0$ is a positive constant [12–14].

Considering $\Delta v(k) = \mathrm{sat}(u(k)) - u(k)$, (6.9) can be written as

$$\begin{cases} x_1(k+1) = \bar{F}_1(k) + \bar{G}_1(k)x_2(k) + \Delta\bar{F}_1(k) + \bar{d}_1(k) \\ x_2(k+1) = \bar{F}_2(k) + \bar{G}_2(k)(\Delta v(k) + u(k)) + \Delta\bar{F}_2(k) + \bar{d}_2(k) \\ y(k) = x_1(k) \end{cases} \tag{6.46}$$

For the uncertain discrete-time UAV attitude dynamic system (6.46), the following adaptive controller design process of discrete-time attitude based on the NN and the SMDO is presented by the BC technology:

Step 1: Defining error variables as $\eta_1(k) = x_1(k) - y_d(k) - \chi_1(k)$ and $\eta_2(k) = x_2(k) - \mathrm{sat}(\alpha_1(k)) - \chi_2(k)$, and the variable $\alpha_1(k)$ denotes the designed control signal. According to (6.46), one has

$$\begin{aligned} \eta_1(k+1) &= x_1(k+1) - y_d(k+1) - \chi_1(k+1) \\ &= \bar{F}_1(k) + \Delta\bar{F}_1(k) + \bar{G}_1(k)(\eta_2(k) + \mathrm{sat}(\alpha_1(k))) \\ &\quad + \bar{G}_1(k)\chi_2(k) + \bar{d}_1(k) - y_d(k+1) - \chi_1(k+1) \end{aligned} \tag{6.47}$$

where $y_d(k)$ is tracking signal which can be obtained based on the analysis in Sect. 6.3.1.

According to Lemma 2.2, by using the NN to approximate $\tau_1\Delta\bar{F}_1(k)$, it yields

$$\begin{aligned} \eta_1(k+1) = \bar{F}_1(k) + \Theta_1(k) + \Upsilon_1(k) + \bar{G}_1(k)(\eta_2(k) + \mathrm{sat}(\alpha_1(k))) \\ + \bar{G}_1(k)\chi_2(k) + \bar{d}_1(k) - y_d(k+1) - \chi_1(k+1) \end{aligned} \tag{6.48}$$

where $\Theta_1(k) = [\frac{1}{\tau_1}\zeta_{11}^{*\mathrm{T}}(k)\varphi_{11}(z_1), \frac{1}{\tau_1}\zeta_{12}^{*\mathrm{T}}(k)\varphi_{12}(z_1), \frac{1}{\tau_1}\zeta_{13}^{*\mathrm{T}}(k)\varphi_{13}(z_1)]^{\mathrm{T}}$, $\tau_1 > 0$ is a design constant, $\Upsilon_1(k) = [\frac{1}{\tau_1}\varepsilon_{11}(k), \frac{1}{\tau_1}\varepsilon_{12}(k), \frac{1}{\tau_1}\varepsilon_{13}(k)]^{\mathrm{T}}$, $\zeta_{1\bar{j}}^*(k)$ is the optimal weight vector of the NN, $\varepsilon_{1\bar{j}}(k)$ is the minimum approximation error, $\varphi_{1\bar{j}}(\cdot)$ is the basis function vector, $\bar{j} = 1, 2, 3$, and $z_1 = x_1(k)$. To facilitate the design of control scheme, define $\varphi_{1\bar{j}}(z_1) = \varphi_{1\bar{j}}(k)$.

For the variable $y_d(k+1)$, $y_d(k+1)$ is predicted by the discrete-time tracking differentiator given in Lemma 2.3. The discrete-time tracking differentiator can be written as

$$\begin{cases} \bar{\mathscr{B}}_{11\bar{j}}(k+1) = \bar{\mathscr{B}}_{11\bar{j}}(k) + q_{01\bar{j}}\bar{\mathscr{B}}_{12\bar{j}}(k) \\ \bar{\mathscr{B}}_{12\bar{j}}(k+1) = \bar{\mathscr{B}}_{12\bar{j}}(k) - q_{01\bar{j}}p_{01\bar{j}}\bar{\mathscr{B}}_{13\bar{j}}(\bar{\mathscr{B}}_{14\bar{j}}(k), \delta_{01\bar{j}}) \end{cases} \tag{6.49}$$

where $\bar{j} = 1, 2, 3$, $q_{01\bar{j}}$ and $p_{01\bar{j}}$ are designed constants, $\delta_{01\bar{j}} = q_{01\bar{j}}p_{01\bar{j}}$, $\bar{\mathscr{B}}_{11\bar{j}}(k)$ and $\bar{\mathscr{B}}_{12\bar{j}}(k)$ are state variables of the discrete-time tracking differentiator.

According to (6.49) and Lemma 2.3, one has

$$y_d(k+1) = \bar{\mathscr{B}}_{11}(k) + q_{01}\bar{\mathscr{B}}_{12}(k) - \tilde{\mathscr{B}}_1(k) \tag{6.50}$$

where $\bar{\mathscr{B}}_{11}(k) = [\bar{\mathscr{B}}_{111}(k), \bar{\mathscr{B}}_{112}(k), \bar{\mathscr{B}}_{113}(k)]^{\mathrm{T}}$, $\bar{\mathscr{B}}_{12}(k) = [\bar{\mathscr{B}}_{121}(k), \bar{\mathscr{B}}_{122}(k), \bar{\mathscr{B}}_{123}(k)]^{\mathrm{T}}$, $q_{01} = \mathrm{diag}[q_{011}, q_{012}, q_{013}]$, $p_{01} = \mathrm{diag}[p_{011}, p_{012}, p_{013}]$, the estimated error $\tilde{\mathscr{B}}_2(k) = [\tilde{\mathscr{B}}_{11}(k), \tilde{\mathscr{B}}_{12}(k), \tilde{\mathscr{B}}_{13}(k)]^{\mathrm{T}}$, and $\tilde{\mathscr{B}}_1(k)$ is bounded.

Substituting (6.50) into (6.48), one has

$$\begin{aligned}\eta_1(k+1) = {} & \bar{F}_1(k) + \Theta_1(k) + \Upsilon_1(k) + \bar{G}_1(k)(\eta_2(k) + \mathrm{sat}(\alpha_1(k))) \\ & + \bar{G}_1(k)\chi_2(k) + \bar{d}_1(k) - \bar{\mathscr{B}}_{11}(k) - q_{01}\bar{\mathscr{B}}_{12}(k) \\ & + \tilde{\mathscr{B}}_1(k) - \chi_1(k+1)\end{aligned} \tag{6.51}$$

To compensate for the influence of external disturbance $\bar{d}_1(k)$ on the control performance of UAV system, a SMDO based on the NN is designed. In order to design the discrete-time SMDO, the following auxiliary system is firstly given:

$$\begin{cases} s_1(k) = \sigma_1(k) - \eta_1(k) \\ \sigma_1(k+1) = \bar{F}_1(k) + \hat{\Theta}_1(k) + \bar{G}_1(k)x_2(k) + \hat{\bar{d}}_1(k) - q_{01}\bar{\mathscr{B}}_{12}(k) \\ \qquad + \tilde{\mathscr{B}}_1(k) + C_1\chi_1(k) - \bar{G}_1\chi_2(k) - \lambda_1 \mathrm{sign}(s_1(k)) \end{cases} \tag{6.52}$$

where $\hat{\Theta}_1(k) = [\frac{1}{\tau_1}\hat{\zeta}_{11}^{\mathrm{T}}(k)\varphi_{11}(k), \frac{1}{\tau_1}\hat{\zeta}_{12}^{\mathrm{T}}(k)\varphi_{12}(k), \frac{1}{\tau_1}\hat{\zeta}_{13}^{\mathrm{T}}(k)\varphi_{13}(k)]^{\mathrm{T}}$, $\hat{\Theta}_1(k)$ is the estimation of $\Theta_1(k)$, $\hat{\bar{d}}_1(k)$ is the estimation of $\bar{d}_1(k)$, and $\lambda_1 > 0$ is a design constant.

According to (6.51) and (6.52), one has

$$\begin{aligned} s_1(k+1) &= \sigma_1(k+1) - \eta_1(k+1) \\ &= -\lambda_1 \mathrm{sign}(s_1(k)) - \tilde{\bar{d}}_1(k) + \tilde{\Theta}_1(k) - \Upsilon_1^*(k)\end{aligned} \tag{6.53}$$

where $\tilde{\Theta}_1(k) = \hat{\Theta}_1(k) - \Theta_1(k)$, $\Upsilon_1^*(k) = \Upsilon_1(k) + \tilde{\mathscr{B}}_1(k)$, $\tilde{\bar{d}}_1(k) = \bar{d}_1(k) - \hat{\bar{d}}_1(k)$, and $\tilde{\Theta}_1(k) = [\frac{1}{\tau_1}\tilde{\zeta}_{11}^{\mathrm{T}}(k)\varphi_{11}(k), \frac{1}{\tau_1}\tilde{\zeta}_{12}^{\mathrm{T}}(k)\varphi_{12}(k), \frac{1}{\tau_1}\tilde{\zeta}_{13}^{\mathrm{T}}(k)\varphi_{13}(k)]^{\mathrm{T}}$.

Then, the following discrete-time SMDO is designed by:

$$\begin{cases} \hat{\bar{d}}_1(k) = \delta_1(\vartheta_1(k) - s_1(k)) \\ \vartheta_1(k+1) = -\lambda_1 \mathrm{sign}(s_1(k)) + \frac{1}{\delta_1}\hat{d}_1(k) \end{cases} \tag{6.54}$$

where $\delta_1 > 0$ is a design constant.

Based on (6.54), it yields

$$\hat{\bar{d}}_1(k+1) = \hat{\bar{d}}_1(k) + \delta_1\tilde{\bar{d}}_1(k) - \delta_1\tilde{\Theta}_1(k) + \delta_1\Upsilon_1^*(k) \tag{6.55}$$

$$\tilde{\bar{d}}_1(k+1) = \Delta\bar{d}_1(k) + (1-\delta_1)\tilde{\bar{d}}_1(k) + \delta_1\tilde{\Theta}_1(k) - \delta_1\Upsilon_1^*(k) \tag{6.56}$$

According to Assumption 6.2 and (6.56), one has

$$\begin{aligned}\left\|\tilde{\bar{d}}_1(k+1)\right\|^2 \le\ & (1-\delta_1)^2\left\|\tilde{\bar{d}}_1(k)\right\|^2 + 2\delta_1^2\left\|\tilde{\bar{d}}_1(k)\right\|^2 + 3\delta_1^2\left\|\tilde{\Theta}_1(k)\right\|^2 \\ & +(1-\delta_1)^2\left\|\tilde{\Theta}_1(k)\right\|^2 + \varpi_{11}\left\|\tilde{d}_1(k)\right\|^2 + 2\delta_1^2\bar{\Upsilon}_1^2 + 2\Delta T^2\mu_1^2 \\ & +2\Delta T\mu_1\delta_1\bar{\Upsilon}_1 + \frac{(1-\delta_1)^2\Delta T^2\mu_1^2}{\varpi_{11}} + (1-\delta_1)^2\bar{\Upsilon}_1^2\end{aligned} \tag{6.57}$$

where ϖ_{11} is a positive constant, $\|\Upsilon_1^*(k)\| \le \bar{\Upsilon}_1$ and $\bar{\Upsilon}_1 > 0$ is a constant.

From (6.53), the following expression is given by:

$$\|s_1(k+1)\|^2 \le 12\lambda_1^2 + 4\left\|\tilde{\bar{d}}_1(k)\right\|^2 + 4\left\|\tilde{\Theta}_1(k)\right\|^2 + 4\bar{\Upsilon}_1^2 \tag{6.58}$$

For the saturation function $\text{sat}(\alpha_1(k))$, the handling method in Chap. 4 is used. Then, (6.51) can be written as

$$\begin{aligned}\eta_1(k+1) =\ & \bar{F}_1(k) + \Theta_1(k) + \Upsilon_1(k) + \bar{G}_1(k)(\eta_2(k) + \mathscr{Y}(k) + \bar{\lambda}(k)\alpha_1(k)) \\ & +\bar{G}_1(k)\chi_2(k) + \bar{d}_1(k) - \bar{\mathscr{B}}_{11}(k) - q_{01}\bar{\mathscr{B}}_{12}(k) \\ & +\tilde{\bar{\mathscr{B}}}_1(k) - \chi_1(k+1)\end{aligned} \tag{6.59}$$

where $\mathscr{Y}(k)$ is the approximation error, $\bar{\lambda}(k)$ is the unknown and bounded diagonal matrix with $\|\bar{\lambda}(k)\| \le \bar{\ell}_0$, and $\bar{\ell}_0$ is a positive constant.

According to the designed method of control signal $x_{vd}(k)$ in Chap. 4, Lemma 2.2 and (6.59), the NN is used to approximate the uncertainty $(1/\tau_0)\Delta\bar{\bar{\mathscr{F}}}_{1\bar{j}}(k)(\bar{j} = 1, 2, 3)$ with $\Delta\bar{\mathscr{F}}_{1\bar{j}}(k)$ being the $\bar{j}$th variable of $\Delta\bar{\mathscr{F}}_1(k) = -(\bar{\lambda}(k)\bar{G}_1(k))^{-1}(\bar{F}_1(k) + \hat{\Theta}_1(k) + C_1\chi_1(k) + \rho_1\eta_1(k) + \hat{\bar{d}}_1(k) - \bar{\mathscr{B}}_{11}(k) - q_{01}\bar{\mathscr{B}}_{12}(k)) + \bar{G}_1^{-1}(k)(\bar{F}_1(k) + \hat{\Theta}_1(k) + C_1\chi_1(k) + \rho_1\eta_1(k) + \hat{\bar{d}}_1(k) - \bar{\mathscr{B}}_{11}(k) - q_{01}\bar{\mathscr{B}}_{12}(k))$, the control signal $\alpha_1(k)$ can be written as

$$\begin{aligned}\alpha_1(k) =\ & -\bar{G}_1^{-1}(k)(\bar{F}_1(k) + \hat{\Theta}_1(k) + C_1\chi_1(k) + \rho_1\eta_1(k)) \\ & -\bar{G}_1^{-1}(k)(\hat{\bar{d}}_1(k) - \bar{\mathscr{B}}_{11}(k) - q_{01}\bar{\mathscr{B}}_{12}(k)) + \Theta_0(k) + \Upsilon_0(k)\end{aligned} \tag{6.60}$$

where $\Theta_0(k) = [\frac{1}{\tau_0}\zeta_{01}^{*\text{T}}(k)\varphi_{01}(z_0), \frac{1}{\tau_0}\zeta_{02}^{*\text{T}}(k)\varphi_{02}(z_0), \frac{1}{\tau_0}\zeta_{03}^{*\text{T}}(k)\varphi_{03}(z_1)]^\text{T}$, $\tau_0 > 0$ is a design constant, $\Upsilon_0(k) = [\frac{1}{\tau_0}\varepsilon_{01}(k), \frac{1}{\tau_0}\varepsilon_{02}(k), \frac{1}{\tau_0}\varepsilon_{03}(k)]^\text{T}$, $\zeta_{0\bar{j}}^*(k)$ is the optimal weight vector of the NN, $\varepsilon_{0\bar{j}}(k)$ is the minimum approximation error, $\varphi_{0\bar{j}}(\cdot)$ is the basis function vector, $z_0 = [x_1^\text{T}(k), \eta_1^\text{T}(k), \chi_1{}^\text{T}(k), \bar{\mathscr{B}}_{11}^\text{T}(k), \bar{\mathscr{B}}_{12}^\text{T}(k), \hat{\Theta}_1^\text{T}(k), \hat{\bar{d}}_1^\text{T}(k)]^\text{T}$, $\bar{j} = 1, 2, 3$, and $\rho_1 > 0$ is a design constant. To facilitate the design of control scheme, define $\varphi_{0\bar{j}}(z_0) = \varphi_{0\bar{j}}(k)$.

Then, the control signal $\alpha_1(k)$ is designed by

$$\begin{aligned}\alpha_1(k) = &-\bar{G}_1^{-1}(k)(\bar{F}_1(k) + \hat{\Theta}_1(k) + C_1\chi_1(k) + \rho_1\eta_1(k)) \\ &-\bar{G}_1^{-1}(k)(\hat{\bar{d}}_1(k) - \mathscr{B}_{11}(k) - q_{01}\mathscr{B}_{12}(k)) + \hat{\Theta}_0(k)\end{aligned} \tag{6.61}$$

where $\hat{\Theta}_0(k) = [\frac{1}{\tau_0}\hat{\zeta}_{01}^{\mathrm{T}}(k)\varphi_{01}(k), \frac{1}{\tau_0}\hat{\zeta}_{02}^{\mathrm{T}}(k)\varphi_{02}(k), \frac{1}{\tau_0}\hat{\zeta}_{03}^{\mathrm{T}}(k)\varphi_{03}(k)]^{\mathrm{T}}$, and $\hat{\Theta}_0(k)$ is the estimation of $\Theta_0(k)$.

Combining (6.45), (6.59) and (6.61), $\eta_1(k+1)$ can be written as

$$\begin{aligned}\eta_1(k+1) = &-\tilde{\Theta}_1(k) + \bar{G}_1(k)\eta_2(k) + \tilde{\bar{d}}_1(k) + \Upsilon_1^*(k) - \rho_1\eta_1(k) \\ &+\bar{\lambda}(k)\bar{G}_1(k)\tilde{\Theta}_0(k) - \bar{\lambda}(k)\bar{G}_1(k)\Upsilon_0(k) + \bar{G}_1(k)\mathscr{Y}(k)\end{aligned} \tag{6.62}$$

where $\tilde{\Theta}_0(k) = [\frac{1}{\tau_0}\tilde{\zeta}_{01}^{\mathrm{T}}(k)\varphi_{01}(k), \frac{1}{\tau_0}\tilde{\zeta}_{02}^{\mathrm{T}}(k)\varphi_{02}(k), \frac{1}{\tau_0}\tilde{\zeta}_{03}^{\mathrm{T}}(k)\varphi_{03}(k)]^{\mathrm{T}}$ and $\tilde{\Theta}_0(k) = \hat{\Theta}_0(k) - \Theta_0(k)$.

From (6.62), one has

$$\begin{aligned}\eta_1(k+1) = &-\tilde{\Theta}_1(k) + \bar{G}_1(k)\eta_2(k) + \tilde{\bar{d}}_1(k) \\ &+\bar{\Upsilon}_1^*(k) - \rho_1\eta_1(k) + \bar{\lambda}(k)\bar{G}_1(k)\tilde{\Theta}_0(k)\end{aligned} \tag{6.63}$$

where $\bar{\Upsilon}_1^*(k) = \Upsilon_1^*(k) - \bar{\lambda}(k)\bar{G}_1(k)\Upsilon_0(k) + \bar{G}_1(k)\mathscr{Y}(k)$ with $\left\|\bar{\Upsilon}_1^*(k)\right\| \le \bar{\Upsilon}_1$, and $\bar{\Upsilon}_1$ is a positive constant.

According to Lemma 2.2 and (6.63), one has

$$\begin{aligned}\|\eta_1(k+1)\|^2 \le\ &4\left\|\tilde{\Theta}_1(k)\right\|^2 + \frac{2}{\varpi_{12}}\left\|\tilde{\Theta}_1(k)\right\|^2 + 4\bar{\Upsilon}^2 + 3\varpi_{12}\bar{g}_1^2\bar{\ell}_0^2\left\|\tilde{\Theta}_0(k)\right\|^2 \\ &+\frac{2}{\varpi_{12}}\bar{g}_1^2\|\eta_2(k)\|^2 + 6\left\|\tilde{\bar{d}}_1(k)\right\|^2 + 3\rho_1^2\|\eta_1(k)\|^2 \\ &+3\varpi_{12}\rho_1^2\|\eta_1(k)\|^2 + \frac{2}{\varpi_{12}}\bar{\Upsilon}^2 + 4\bar{g}_1^2\|\eta_2(k)\|^2 \\ &+3\bar{g}_1^2\bar{\ell}_0^2\left\|\tilde{\Theta}_0(k)\right\|^2\end{aligned} \tag{6.64}$$

where ϖ_{12} is a positive constant.

Moreover, the adaptive law $\hat{\zeta}_{1\bar{j}}(k)$ is designed by

$$\hat{\zeta}_{1\bar{j}}(k+1) = \kappa_{1\bar{j}}\varphi_{1\bar{j}}(k)\eta_{1\bar{j}}(k) - (\omega_{1\bar{j}} - 1)\hat{\zeta}_{1\bar{j}}(k) \tag{6.65}$$

where $\kappa_{1\bar{j}} > 0$ and $\omega_{1\bar{j}} > 0$ are design constants, $\eta_{1\bar{j}}(k)$ is the $\bar{j}$th variables of $\eta_1(k)$, and $\bar{j} = 1, 2, 3$.

According to (6.65), one has

$$\tilde{\zeta}_{1\bar{j}}(k+1) = \tilde{\zeta}_{1\bar{j}}(k) + \kappa_{1\bar{j}}\varphi_{1\bar{j}}(k)\eta_{1\bar{j}}(k) - \omega_{1\bar{j}}\hat{\zeta}_{1\bar{j}}(k) \tag{6.66}$$

where $\tilde{\zeta}_{1\bar{j}}(k+1) = \hat{\zeta}_{1\bar{j}}(k+1) - \zeta^*_{1\bar{j}}(k+1)$, and $\tilde{\zeta}_{1\bar{j}}(k) = \hat{\zeta}_{1\bar{j}}(k) - \zeta^*_{1\bar{j}}(k)$.

From (6.66), it yields

$$\begin{aligned}\left\|\tilde{\zeta}_{1\bar{j}}(k+1)\right\|^2 - \left\|\tilde{\zeta}_{1\bar{j}}(k)\right\|^2 &= \kappa^2_{1\bar{j}}\|\varphi_{1\bar{j}}(k)\|^2\eta^2_{1\bar{j}}(k)\\ &+\omega^2_{1\bar{j}}\left\|\hat{\zeta}_{1\bar{j}}(k)\right\|^2 - 2\kappa_{1\bar{j}}\omega_{1\bar{j}}\varphi^{\mathrm{T}}_{1\bar{j}}(k)\hat{\zeta}_{1\bar{j}}(k)\eta_{1\bar{j}}(k)\\ &+2\kappa_{1\bar{j}}\tilde{\zeta}^{\mathrm{T}}_{1\bar{j}}(k)\varphi_{1\bar{j}}(k)\eta_{1\bar{j}}(k) - 2\omega_{1\bar{j}}\tilde{\zeta}^{\mathrm{T}}_{1\bar{j}}(k)\hat{\zeta}_{1\bar{j}}(k)\end{aligned} \tag{6.67}$$

Furthermore, one can obtain

$$-2\omega^2_{1\bar{j}}\tilde{\zeta}^{\mathrm{T}}_{1\bar{j}}(k)\hat{\zeta}_{1\bar{j}}(k) = -\omega^2_{1\bar{j}}\left\|\tilde{\zeta}_{1\bar{j}}(k)\right\|^2 - \omega^2_{1\bar{j}}\left\|\hat{\zeta}_{1\bar{j}}(k)\right\|^2 + \omega^2_{1\bar{j}}\left\|\zeta^*_{1\bar{j}}(k)\right\|^2 \tag{6.68}$$

Combining (6.67) and (6.68), one has

$$\begin{aligned}\omega_{1\bar{j}}(\left\|\tilde{\zeta}_{1\bar{j}}(k+1)\right\|^2 - \left\|\tilde{\zeta}_{1\bar{j}}(k)\right\|^2) &\le -(\omega^2_{1\bar{j}} - \omega^3_{1\bar{j}} - \kappa^2_{1\bar{j}}\omega^2_{1\bar{j}})\left\|\hat{\zeta}_{1\bar{j}}(k)\right\|^2\\ +(2\omega^2_{1\bar{j}}\bar{\varphi}_1 + \omega_{1\bar{j}}\kappa^2_{1\bar{j}}\bar{\varphi}_1)\eta^2_{1\bar{j}}(k) &- (\omega^2_{1\bar{j}} - \kappa^2_{1\bar{j}})\left\|\tilde{\zeta}_{1\bar{j}}(k)\right\|^2 + \omega^2_{1\bar{j}}\bar{\zeta}^*_1\end{aligned} \tag{6.69}$$

where $\|\varphi_{1\bar{j}}(k)\|^2 \le \bar{\varphi}_1$, $\left\|\zeta^*_{1\bar{j}}(k)\right\|^2 \le \bar{\zeta}^*_1$, $\bar{\varphi}_1 > 0$ and $\bar{\zeta}^*_1 > 0$ are constants.

Moreover, the adaptive law $\hat{\zeta}_{0\bar{j}}(k)$ is designed by

$$\hat{\zeta}_{0\bar{j}}(k+1) = \kappa_{0\bar{j}}\varphi_{0\bar{j}}(k)\eta_{1\bar{j}}(k) - (\omega_{0\bar{j}} - 1)\hat{\zeta}_{0\bar{j}}(k) \tag{6.70}$$

where $\kappa_{0\bar{j}} > 0$ and $\omega_{0\bar{j}} > 0$ are design constants.

According to (6.70), one has

$$\tilde{\zeta}_{0\bar{j}}(k+1) = \tilde{\zeta}_{0\bar{j}}(k) + \kappa_{0\bar{j}}\varphi_{0\bar{j}}(k)\eta_{1\bar{j}}(k) - \omega_{0\bar{j}}\hat{\zeta}_{0\bar{j}}(k) \tag{6.71}$$

where $\tilde{\zeta}_{0\bar{j}}(k+1) = \hat{\zeta}_{0\bar{j}}(k+1) - \zeta^*_{0\bar{j}}(k+1)$, and $\tilde{\zeta}_{0\bar{j}}(k) = \hat{\zeta}_{0\bar{j}}(k) - \zeta^*_{0\bar{j}}(k)$.

From (6.71), it yields

$$\begin{aligned}\left\|\tilde{\zeta}_{0\bar{j}}(k+1)\right\|^2 - \left\|\tilde{\zeta}_{0\bar{j}}(k)\right\|^2 &= \kappa^2_{0\bar{j}}\|\varphi_{0\bar{j}}(k)\|^2\eta^2_{1\bar{j}}(k)\\ &+\omega^2_{0\bar{j}}\left\|\hat{\zeta}_{0\bar{j}}(k)\right\|^2 - 2\kappa_{0\bar{j}}\omega_{0\bar{j}}\varphi^{\mathrm{T}}_{0\bar{j}}(k)\hat{\zeta}_{0\bar{j}}(k)\eta_{1\bar{j}}(k)\\ &+2\kappa_{0\bar{j}}\tilde{\zeta}^{\mathrm{T}}_{0\bar{j}}(k)\varphi_{0\bar{j}}(k)\eta_{1\bar{j}}(k) - 2\omega_{0\bar{j}}\tilde{\zeta}^{\mathrm{T}}_{0\bar{j}}(k)\hat{\zeta}_{0\bar{j}}(k)\end{aligned} \tag{6.72}$$

Furthermore, one can obtain

$$-2\omega^2_{0\bar{j}}\tilde{\zeta}^{\mathrm{T}}_{0\bar{j}}(k)\hat{\zeta}_{0\bar{j}}(k) = -\omega^2_{0\bar{j}}\left\|\tilde{\zeta}_{0\bar{j}}(k)\right\|^2 - \omega^2_{0\bar{j}}\left\|\hat{\zeta}_{0\bar{j}}(k)\right\|^2 + \omega^2_{0\bar{j}}\left\|\zeta^*_{0\bar{j}}(k)\right\|^2 \tag{6.73}$$

Combining (6.72) and (6.73), one has

$$\begin{aligned}\omega_{0\bar{j}}(\left\|\tilde{\zeta}_{0\bar{j}}(k+1)\right\|^2 - \left\|\tilde{\zeta}_{0\bar{j}}(k)\right\|^2) \leq -(\omega_{0\bar{j}}^2 - \omega_{0\bar{j}}^3 - \kappa_{0\bar{j}}^2\omega_{0\bar{j}}^2)\left\|\hat{\zeta}_{0\bar{j}}(k)\right\|^2 \\ +(2\omega_{0\bar{j}}^2\bar{\varphi}_0 + \omega_{0\bar{j}}\kappa_{0\bar{j}}^2\bar{\varphi}_1)\eta_{1\bar{j}}^2(k) - (\omega_{0\bar{j}}^2 - \kappa_{0\bar{j}}^2)\left\|\tilde{\zeta}_{0\bar{j}}(k)\right\|^2 + \omega_{0\bar{j}}^2\bar{\zeta}_0^* \end{aligned} \tag{6.74}$$

where $\|\varphi_{0\bar{j}}(k)\|^2 \leq \bar{\varphi}_0$, $\left\|\zeta_{0\bar{j}}^*(k)\right\|^2 \leq \bar{\zeta}_0^*$, $\bar{\varphi}_0 > 0$ and $\bar{\zeta}_0^* > 0$ are constants.

According to (6.45), one has

$$\chi_1^{\mathrm{T}}(k+1)\chi_1(k+1) \leq 2\|C_1\|^2\|\chi_1(k)\|^2 + 2\bar{g}_1^2\|\chi_2(k)\|^2 \tag{6.75}$$

The following Lyapunov function is chosen as

$$\begin{aligned}V_1(k) = \omega_{1\min}\|\eta_1(k)\|^2 + \left\|\tilde{\bar{d}}_1(k)\right\|^2 + \omega_{1\min}\|s_1(k)\|^2 + \omega_{1\min}\|\chi_1(k)\|^2 \\ +\sum_{\bar{j}=1}^{3}\omega_{1\bar{j}}\left\|\tilde{\zeta}_{1\bar{j}}(k)\right\|^2 + \sum_{\bar{j}=1}^{3}\omega_{0\bar{j}}\left\|\tilde{\zeta}_{0\bar{j}}(k)\right\|^2\end{aligned} \tag{6.76}$$

where $\omega_{1\min}$ is the minimum value of $\omega_{1\bar{j}}$, and $\bar{j} = 1, 2, 3$.

On the basis of (6.76), the first order difference of the Lyapunov function $V_1(k)$ can be written as

$$\begin{aligned}\Delta V_1(k) = \omega_{1\min}\|\eta_1(k+1)\|^2 + \left\|\tilde{\bar{d}}_1(k+1)\right\|^2 - \left\|\tilde{\bar{d}}_1(k)\right\|^2 \\ +\sum_{\bar{j}=1}^{3}\omega_{1\bar{j}}\left\|\tilde{\zeta}_{1\bar{j}}(k+1)\right\|^2 - \omega_{1\min}\|\eta_1(k)\|^2 + \omega_{1\min}\|\chi_1(k+1)\|^2 \\ -\sum_{\bar{j}=1}^{3}\omega_{1\bar{j}}\left\|\tilde{\zeta}_{1\bar{j}}(k)\right\|^2 - \omega_{1\min}(\|\chi_1(k)\|^2 - \|s_1(k+1)\|^2 + \|s_1(k)\|^2) \\ +\sum_{\bar{j}=1}^{3}\omega_{0\bar{j}}\left\|\tilde{\zeta}_{0\bar{j}}(k+1)\right\|^2 - \sum_{\bar{j}=1}^{3}\omega_{0\bar{j}}\left\|\tilde{\zeta}_{0\bar{j}}(k)\right\|^2\end{aligned} \tag{6.77}$$

According to (6.57), (6.58), (6.64)–(6.75), $\Delta V_1(k)$ satisfies that

$$\begin{aligned}\Delta V_1(k) \leq -\omega_{1\min}(1 - 3\rho_1^2 - 3\varpi_{12}\rho_1^2)\|\eta_1(k)\|^2 - \omega_{1\min}\|s_1(k)\|^2 \\ -(1 - 10\omega_{1\min} - (1-\delta_1)^2 - 2{\delta_1}^2 - \varpi_{11})\left\|\tilde{\bar{d}}_1(k)\right\|^2 \\ +2\delta_1^2\bar{\Upsilon}_1^2 - (\omega_{1\min} - 2\omega_{1\min}\|C_1\|^2)\|\chi_1(k)\|^2 + \tfrac{(1-\delta_1)^2\Delta T^2\mu_1^2}{\varpi_{11}} + (1-\delta_1)^2\bar{\Upsilon}_1^2 \\ -(\min(\omega_{1\bar{j}}^2 - \kappa_{1\bar{j}}^2) - 8\omega_{1\min}\tfrac{\bar{\varphi}_1}{\tau_1^2} - 3\delta_1^2\tfrac{\bar{\varphi}_1}{\tau_1^2})\sum_{\bar{j}=1}^{3}\left\|\tilde{\zeta}_{1\bar{j}}(k)\right\|^2 + 2\Delta T\mu_1\delta_1\bar{\Upsilon}_1\end{aligned}$$

$$+(\frac{\bar{\varphi}_1(1-\delta_1)^2}{\tau_1^2}+\frac{2\bar{\varphi}_1\omega_{1\min}}{\tau_1^2\varpi_{12}})\sum_{\bar{j}=1}^{3}\left\|\tilde{\zeta}_{1\bar{j}}(k)\right\|^2+\frac{2\omega_{1\min}}{\varpi_{12}}\bar{\Upsilon}_1^2+12\omega_{1\min}\lambda_1^2+\sum_{\bar{j}=1}^{3}\omega_{1\bar{j}}^2\bar{\zeta}_1^*$$

$$-\min(\omega_{1\bar{j}}^2-\omega_{1\bar{j}}^3-\kappa_{1\bar{j}}^2\omega_{1\bar{j}}^2)\sum_{\bar{j}=1}^{3}\left\|\hat{\zeta}_{1\bar{j}}(k)\right\|^2+4\omega_{1\min}(\bar{\Upsilon}_1^2+\bar{\Upsilon}^2)+2\Delta T^2\mu_1^2$$

$$+(4+\frac{2}{\varpi_{12}})\omega_{1\min}\bar{g}_1^2\|\eta_2(k)\|^2+\max(2\omega_{1\bar{j}}^2\bar{\varphi}_1+\omega_{1\bar{j}}\kappa_{1\bar{j}}^2\bar{\varphi}_1)\|\eta_1(k)\|^2+\sum_{\bar{j}=1}^{3}\omega_{0\bar{j}}^2\bar{\zeta}_0^*$$

$$-(\min(\omega_{0\bar{j}}^2-\kappa_{0\bar{j}}^2)-\frac{3\varpi_{12}\bar{g}_1^2\bar{\ell}_0^2\bar{\varphi}_0\omega_{1\min}}{\tau_0^2}-\frac{3\bar{\varphi}_0\bar{g}_1^2\bar{\ell}_0^2}{\tau_0^2})\sum_{\bar{j}=1}^{3}\left\|\tilde{\zeta}_{0\bar{j}}(k)\right\|^2$$

$$-\min(\omega_{0\bar{j}}^2-\omega_{0\bar{j}}^3-\kappa_{0\bar{j}}^2\omega_{0\bar{j}}^2)\sum_{\bar{j}=1}^{3}\left\|\hat{\zeta}_{0\bar{j}}(k)\right\|^2+2\omega_{1\min}\bar{g}_1^2\|\chi_2(k)\|^2$$

$$+\max(2\omega_{0\bar{j}}^2\bar{\varphi}_0+\omega_{0\bar{j}}\kappa_{0\bar{j}}^2\bar{\varphi}_0)\|\eta_1(k)\|^2 \tag{6.78}$$

Step 2: Considering (6.45), (6.46) and $\eta_2(k)=x_2(k)-\mathrm{sat}(\alpha_1(k))-\chi_2(k)$, one has

$$\begin{aligned}\eta_2(k+1)&=x_2(k+1)-v_d(k+1)-\chi_2(k+1)\\&=\bar{F}_2(k)+\Delta\bar{F}_2(k)+\bar{G}_2(k)u(k)+\bar{d}_2(k)\\&\quad+C_2(k)\chi_2(k)-v_d(k+1)\end{aligned} \tag{6.79}$$

On the basis of Lemma 2.2, the NN is used to approximate $\tau_2\Delta F_2(k)$, (6.79) can be written as

$$\begin{aligned}\eta_2(k+1)=&\bar{F}_2(k)+\Theta_2(k)+\bar{G}_2(k)u(k)+C_2(k)\chi_2(k)\\&+\bar{d}_2(k)-v_d(k+1)+\Upsilon_2(k)\end{aligned} \tag{6.80}$$

where $\Theta_2(k)=[\frac{1}{\tau_2}\zeta_{21}^{*\mathrm{T}}(k)\varphi_{21}(z_2),\frac{1}{\tau_2}\zeta_{22}^{*\mathrm{T}}(k)\varphi_{22}(z_2),\frac{1}{\tau_2}\zeta_{23}^{*\mathrm{T}}(k)\varphi_{2m}(z_2)]^{\mathrm{T}}$, $\tau_2>0$ is a design constant, τ_2 is a constant, $\zeta_{2\bar{j}}^*(k)$ is the optimal weight vector, $\Upsilon_2(k)=[\frac{1}{\tau_2}\varepsilon_{21}(k),\frac{1}{\tau_2}\varepsilon_{22}(k),\frac{1}{\tau_2}\varepsilon_{23}(k)]^{\mathrm{T}}$, $\varepsilon_{2\bar{j}}(k)$ is the minimum approximation error, $\varphi_{2\bar{j}}(\cdot)$ is the basis function vector, $\bar{j}=1,2,3$ and $z_2=[x_1^{\mathrm{T}}(k),x_2^{\mathrm{T}}(k)]^{\mathrm{T}}$. For convenience, $\varphi_{2\bar{j}}(z_2)=\varphi_{2\bar{j}}(k)$ is defined.

To forecast $v_d(k+1)$, the following discrete-time tracking differentiator based on Lemma 2.3 is employed to estimate $v_d(k+1)$:

$$\begin{cases}\bar{\mathscr{B}}_{21\bar{j}}(k+1)=\bar{\mathscr{B}}_{21\bar{j}}(k)+q_{02\bar{j}}\bar{\mathscr{B}}_{22\bar{j}}(k)\\\bar{\mathscr{B}}_{22\bar{j}}(k+1)=\bar{\mathscr{B}}_{22\bar{j}}(k)-q_{02\bar{j}}p_{02\bar{j}}\bar{\mathscr{B}}_{23\bar{j}}(\bar{\mathscr{B}}_{24\bar{j}}(k),\delta_{02\bar{j}})\end{cases} \tag{6.81}$$

where $\bar{j}=1,2,3$, $q_{02\bar{j}}$ and $p_{02\bar{j}}$ are design constants, $\delta_{02\bar{j}}=q_{02\bar{j}}p_{02\bar{j}}$, $\bar{\mathscr{B}}_{21\bar{j}}(k)$ and $\bar{\mathscr{B}}_{22\bar{j}}(k)$ are the state variables of the discrete-time tracking differentiator.

According to (6.81) and Lemma 2.3, one has

$$v_d(k+1) = \bar{\mathscr{B}}_{21}(k) + q_{02}\bar{\mathscr{B}}_{22}(k) - \tilde{\mathscr{B}}_2(k) \tag{6.82}$$

where $p_{02} = \mathrm{diag}[p_{021}, p_{022}, p_{023}]$, $\bar{\mathscr{B}}_{21}(k) = [\bar{\mathscr{B}}_{211}(k), \bar{\mathscr{B}}_{212}(k), \bar{\mathscr{B}}_{213}(k)]^{\mathrm{T}}$, $\bar{\mathscr{B}}_{22}(k) = [\bar{\mathscr{B}}_{221}(k), \bar{\mathscr{B}}_{222}(k), \bar{\mathscr{B}}_{223}(k)]^{\mathrm{T}}$, $q_{02} = \mathrm{diag}[q_{021}, q_{022}, q_{023}]$, the estimated error $\tilde{\mathscr{B}}_2(k) = [\tilde{\mathscr{B}}_{21}(k), \tilde{\mathscr{B}}_{22}(k), \tilde{\mathscr{B}}_{23}(k)]^{\mathrm{T}}$, $\tilde{\mathscr{B}}_2(k)$ is bounded, $\left\|\tilde{\mathscr{B}}_2(k)\right\| \le \bar{\mathscr{B}}_2$, and $\bar{\mathscr{B}}_2$ is a positive constant.

Substituting (6.82) into (6.80), one can obtain

$$\begin{aligned}\eta_2(k+1) = &\bar{F}_2(k) + \Theta_2(k) + \bar{G}_2(k)u(k) + C_2(k)\chi_2(k) \\ &+\bar{d}_2(k) - q_{02}\bar{\mathscr{B}}_{22}(k) - \bar{\mathscr{B}}_{21}(k) + \tilde{\mathscr{B}}_2(k) + \Upsilon_2(k)\end{aligned} \tag{6.83}$$

In order to deal with the disturbance $\bar{d}_2(k)$, the NN-based SMDO is designed. In order to design the discrete-time SMDO, the following auxiliary system is firstly given:

$$\begin{cases} s_2(k) = \sigma_2(k) - \eta_2(k) \\ \sigma_2(k+1) = \bar{F}_2(k) + \hat{\Theta}_2(k) + \bar{G}_2(k)u(k) + \hat{\bar{d}}_2(k) \\ \qquad\qquad + C_2\chi_2(k) - \lambda_2\mathrm{sign}(s_2(k)) - q_{02}\bar{\mathscr{B}}_{22}(k) - \bar{\mathscr{B}}_{21}(k) \end{cases} \tag{6.84}$$

where $\hat{\Theta}_2(k) = [\frac{1}{\tau_2}\hat{\zeta}_{21}^{\mathrm{T}}(k)\varphi_{21}(k), \frac{1}{\tau_2}\hat{\zeta}_{22}^{\mathrm{T}}(k)\varphi_{22}(k), \frac{1}{\tau_2}\hat{\zeta}_{23}^{\mathrm{T}}(k)\varphi_{23}(k)]^{\mathrm{T}}$, $\hat{\Theta}_2(k)$ is the estimation of $\Theta_2(k)$, $\hat{\bar{d}}_2(k)$ is the estimation of $\bar{d}_2(k)$, and $\lambda_2 > 0$ is a design constant.

According to (6.80) and (6.84), one has

$$\begin{aligned} s_2(k+1) &= \sigma_2(k+1) - \eta_2(k+1) \\ &= -\lambda_2\mathrm{sign}(s_2(k)) - \tilde{\bar{d}}_2(k) + \tilde{\Theta}_2(k) - \tilde{\mathscr{B}}_2(k) - \Upsilon_2(k) \end{aligned} \tag{6.85}$$

where $\tilde{\Theta}_2(k) = [\frac{1}{\tau_2}\tilde{\zeta}_{21}^{\mathrm{T}}(k)\varphi_{21}(k), \frac{1}{\tau_2}\tilde{\zeta}_{22}^{\mathrm{T}}(k)\varphi_{22}(k), \frac{1}{\tau_2}\tilde{\zeta}_{23}^{\mathrm{T}}(k)\varphi_{23}(k)]^{\mathrm{T}}$, $\tilde{\Theta}_2(k) = \hat{\Theta}_2(k) - \Theta_2(k)$, and $\tilde{\bar{d}}_2(k) = \bar{d}_2(k) - \hat{\bar{d}}_2(k)$.

Then, the discrete-time SMDO is designed as follows:

$$\begin{cases} \hat{\bar{d}}_2(k) = \delta_2(\vartheta_2(k) - s_2(k)) \\ \vartheta_2(k+1) = -\lambda_2\mathrm{sign}(s_2(k)) + \frac{1}{\delta_2}\hat{\bar{d}}_2(k) \end{cases} \tag{6.86}$$

where $\delta_2 > 0$ is a design constant.

Based on (6.86), one has

$$\hat{\bar{d}}_2(k+1) = \hat{\bar{d}}_2(k) + \delta_2\tilde{\bar{d}}_2(k) - \delta_2\tilde{\Theta}_2(k) + \delta_2\tilde{\mathscr{B}}_2(k) + \delta_2\Upsilon_2(k) \tag{6.87}$$

$$\tilde{\bar{d}}_2(k+1) = \Delta\bar{d}_2(k) + (1-\delta_2)\tilde{\bar{d}}_2(k) + \delta_2\tilde{\Theta}_2(k) - \delta_2\tilde{\mathscr{B}}_2(k) - \delta_2\Upsilon_2(k) \quad (6.88)$$

According to Assumption 6.2 and (6.88), one can obtain

$$\begin{aligned}\left\|\tilde{\bar{d}}_2(k+1)\right\|^2 &\le (1-\delta_2)^2\left\|\tilde{\bar{d}}_2(k)\right\|^2 + 2\delta_2^2\left\|\tilde{\bar{d}}_2(k)\right\|^2 + 4\delta_2^2\left\|\tilde{\Theta}_2(k)\right\|^2 \\ &+(1-\delta_2)^2\left\|\tilde{\Theta}_2(k)\right\|^2 + 2\varpi_{21}\left\|\tilde{\bar{d}}_2(k)\right\|^2 + 2\delta_2^2\bar{\Upsilon}_2^2 + 2\Delta T^2\mu_2^2 \\ &+2\Delta T\mu_2\delta_2\bar{\Upsilon}_2 + \tfrac{(1-\delta_2)^2\Delta T^2\mu_2^2}{\varpi_{21}} + \delta_2^2\bar{\mathscr{B}}_2^2 + (1-\delta_2)^2\bar{\Upsilon}_2^2 \\ &+2\Delta T\mu_2\delta_2\bar{\mathscr{B}}_2 + 2\delta_2^2\bar{\Upsilon}_2\bar{\mathscr{B}}_2 + \delta_2^2\bar{\mathscr{B}}_2^2 + \tfrac{(1-\delta_2)^2\delta_2^2}{\varpi_{21}}\bar{\mathscr{B}}_2^2 \end{aligned} \quad (6.89)$$

where ϖ_{21} is a positive constant, $\|\Upsilon_2(k)\| \le \bar{\Upsilon}_2$, and $\bar{\Upsilon}_2 > 0$ is a positive constant.

From (6.85), it yields

$$\|s_2(k+1)\|^2 \le 15\lambda_2^2 + 5\left\|\tilde{\bar{d}}_2(k)\right\|^2 + 5\left\|\tilde{\Theta}_2(k)\right\|^2 + 5\bar{\Upsilon}_2^2 + 5\bar{\mathscr{B}}_2^2 \quad (6.90)$$

The following controller $u(k)$ is given by

$$\begin{aligned} u(k) &= -\bar{G}_2^{-1}(k)(\bar{F}_2(k) + \hat{\Theta}_2(k) + C_2\chi_2(k)) \\ &\quad -\bar{G}_2^{-1}(k)(\rho_2\eta_2(k) + \hat{\bar{d}}_2(k) - q_{02}\bar{\mathscr{B}}_{22}(k)) + \bar{G}_2^{-1}(k)\bar{\mathscr{B}}_{21}(k) \end{aligned} \quad (6.91)$$

where $\rho_2 > 0$ is a design constant.

Combining (6.45), (6.83) and (6.91), $\eta_2(k+1)$ can be written as

$$\eta_2(k+1) = -\tilde{\Theta}_2(k) + \tilde{\bar{d}}_2(k) + \Upsilon_2(k) - \rho_2\eta_2(k) + \tilde{B}_2(k) \quad (6.92)$$

According to Lemma 2.2 and (6.92), one has

$$\begin{aligned}\|\eta_2(k+1)\|^2 &\le 4\left\|\tilde{\Theta}_2(k)\right\|^2 + \frac{1}{\varpi_{22}}\left\|\tilde{\Theta}_2(k)\right\|^2 + \frac{1}{\varpi_{22}}\bar{\mathscr{B}}_2^2 + 3\varpi_{22}\rho_2^2\|\eta_2(k)\|^2 \\ &\quad +4\bar{\Upsilon}_2^2 + \frac{1}{\varpi_{22}}\bar{\Upsilon}_2^2 + 2\rho_2^2\|\eta_2(k)\|^2 + 4\bar{\mathscr{B}}_i^2 + 5\left\|\tilde{\bar{d}}_2(k)\right\|^2 \end{aligned} \quad (6.93)$$

where ϖ_{22} is a positive constant.

In addition, the adaptive law $\hat{\zeta}_{2\bar{j}}(k)$ is designed by

$$\hat{\zeta}_{2\bar{j}}(k+1) = \kappa_{2\bar{j}}\varphi_{2\bar{j}}(k)\eta_{2\bar{j}}(k) - (\omega_{2\bar{j}} - 1)\hat{\zeta}_{2\bar{j}}(k) \quad (6.94)$$

where $\kappa_{2\bar{j}} > 0$ and $\omega_{2\bar{j}} > 0$ are design constants, $\eta_{2\bar{j}}(k)$ is the $\bar{j}$th variable of $\eta_2(k)$, and $\bar{j} = 1, 2, 3$.

According to (6.94), one has

$$\tilde{\zeta}_{2\bar{j}}(k+1) = \tilde{\zeta}_{2\bar{j}}(k) + \kappa_{2\bar{j}}\varphi_{2\bar{j}}(k)\eta_{2\bar{j}}(k) - \omega_{2\bar{j}}\hat{\zeta}_{2\bar{j}}(k) \tag{6.95}$$

where $\tilde{\zeta}_{2\bar{j}}(k+1) = \hat{\zeta}_{2\bar{j}}(k+1) - \zeta^*_{2\bar{j}}(k+1)$ and $\tilde{\zeta}_{2\bar{j}}(k) = \hat{\zeta}_{2\bar{j}}(k) - \zeta^*_{2\bar{j}}(k)$.

Based on (6.95), it yields

$$\begin{aligned}\left\|\tilde{\zeta}_{2\bar{j}}(k+1)\right\|^2 - \left\|\tilde{\zeta}_{2\bar{j}}(k)\right\|^2 &= \kappa^2_{2\bar{j}}\left\|\varphi_{2\bar{j}}(k)\right\|^2\eta^2_{2\bar{j}}(k) - 2\omega_{2\bar{j}}\tilde{\zeta}^{\mathrm{T}}_{2\bar{j}}(k)\hat{\zeta}_{2\bar{j}}(k)\\ +\omega^2_{2\bar{j}}\left\|\hat{\zeta}_{2\bar{j}}(k)\right\|^2 &- 2\kappa_{2\bar{j}}\omega_{2\bar{j}}\varphi^{\mathrm{T}}_{2\bar{j}}(k)\hat{\zeta}_{2\bar{j}}(k)\eta_{2\bar{j}}(k) + 2\kappa_{2\bar{j}}\tilde{\zeta}^{\mathrm{T}}_{2\bar{j}}(k)\varphi_{2\bar{j}}(k)\eta_{2\bar{j}}(k)\end{aligned} \tag{6.96}$$

Furthermore, one can obtain

$$-2\omega^2_{2\bar{j}}\tilde{\zeta}^{\mathrm{T}}_{2\bar{j}}(k)\hat{\zeta}_{2\bar{j}}(k) = -\omega^2_{2\bar{j}}\left\|\tilde{\zeta}_{2\bar{j}}(k)\right\|^2 - \omega^2_{2\bar{j}}\left\|\hat{\zeta}_{2\bar{j}}(k)\right\|^2 + \omega^2_{2\bar{j}}\left\|\zeta^*_{2\bar{j}}(k)\right\|^2 \tag{6.97}$$

Combining (6.96) and (6.97), one has

$$\begin{aligned}\omega_{2\bar{j}}(\left\|\tilde{\zeta}_{2\bar{j}}(k+1)\right\|^2 - \left\|\tilde{\zeta}_{2\bar{j}}(k)\right\|^2) &\le (2\omega^2_{2\bar{j}}\bar{\varphi}_2 + \omega_{2\bar{j}}\kappa^2_{2\bar{j}}\bar{\varphi}_2)\eta^2_{2\bar{j}}(k)\\ -(\omega^2_{2\bar{j}} - \kappa^2_{2\bar{j}})\left\|\tilde{\zeta}_{2\bar{j}}(k)\right\|^2 &- (\omega^2_{2\bar{j}} - \omega^3_{2\bar{j}} - \kappa^2_{2\bar{j}}\omega^2_{2\bar{j}})\left\|\hat{\zeta}_{2\bar{j}}(k)\right\|^2 + \omega^2_{2\bar{j}}\bar{\zeta}^*_2\end{aligned} \tag{6.98}$$

where $\left\|\varphi_{2\bar{j}}(k)\right\|^2 \le \bar{\varphi}_2$, $\left\|\zeta^*_{2\bar{j}}(k)\right\|^2 \le \bar{\zeta}^*_2$, $\bar{\varphi}_2 > 0$ and $\bar{\zeta}^*_2 > 0$ are constants.

According to (6.45), one has

$$\|\chi_2(k+1)\|^2 \le 2\|C_2\|^2\|\chi_2(k)\|^2 + 2\bar{g}^2_2\nu^2 \tag{6.99}$$

The following Lyapunov function is chosen as

$$\begin{aligned}V_2(k) = \omega_{2\min}\|\eta_2(k)\|^2 &+ \left\|\tilde{\bar{d}}_2(k)\right\|^2 + \omega_{2\min}\|s_2(k)\|^2\\ &+\sum_{\bar{j}=1}^{3}\omega_{2\bar{j}}\left\|\tilde{\zeta}_{2\bar{j}}(k)\right\|^2 + \omega_{2\min}\|\chi_2(k)\|^2\end{aligned} \tag{6.100}$$

where $\omega_{2\min}$ is the minimum value of $\omega_{2\bar{j}}$, and $\bar{j} = 1, 2, 3$.

Based on (6.100), the first order difference of the Lyapunov function $V_2(k)$ can be written as

$$\begin{aligned}\Delta V_2(k) &= \omega_{2\min}\|\eta_2(k+1)\|^2 + \left\|\tilde{\bar{d}}_2(k+1)\right\|^2 + \sum_{\bar{j}=1}^{3}\omega_{2\bar{j}}\left\|\tilde{\zeta}_{2\bar{j}}(k+1)\right\|^2\\ &-\left\|\tilde{\bar{d}}_2(k)\right\|^2 - \omega_{2\min}\|\eta_2(k)\|^2 + \omega_{2\min}\|\chi_2(k+1)\|^2 - \sum_{\bar{j}=1}^{3}\omega_{2\bar{j}}\left\|\tilde{\zeta}_{2\bar{j}}(k)\right\|^2\\ &-\omega_{2\min}(\|\chi_2(k)\|^2 - \|s_2(k+1)\|^2 + \|s_2(k)\|^2)\end{aligned} \tag{6.101}$$

According to (6.89), (6.90), (6.93) and (6.99), one has

$$
\begin{aligned}
\Delta V_2(k) \leq & -\omega_{2\min}(1-2\rho_2^2-3\varpi_{22}\rho_2^2)\|\eta_2(k)\|^2 \\
& -(1-10\omega_{2\min}-(1-\delta_2)^2-2\delta_2{}^2-2\varpi_{21})\left\|\tilde{\bar{d}}_2(k)\right\|^2+2\delta_2^2\bar{\Upsilon}_2^2-\omega_{2\min}\|s_2(k)\|^2 \\
& -(\omega_{2\min}-2\omega_{2\min}\|C_2\|^2)\|\chi_2(k)\|^2+4\delta_2^2\tfrac{\bar{\varphi}_2}{\tau_2^2}\sum_{\bar{j}=1}^{3}\left\|\tilde{\zeta}_{2\bar{j}}(k)\right\|^2+2\delta_2^2\bar{\Upsilon}_2\bar{\mathscr{B}}_2+\delta_2^2\bar{\mathscr{B}}_2^2 \\
& -(\min(\omega_{2\bar{j}}^2-\kappa_{2\bar{j}}^2)-9\omega_{2\min}\tfrac{\bar{\varphi}_2}{\tau_2^2})\sum_{\bar{j}=1}^{3}\left\|\tilde{\zeta}_{2\bar{j}}(k)\right\|^2+\tfrac{(1-\delta_2)^2\delta_2^2}{\varpi_{21}}\bar{\mathscr{B}}_2^2+2\omega_{2\min}\bar{g}_2^2\nu^2 \\
& +(\tfrac{\bar{\varphi}_2(1-\delta_2)^2}{\tau_2^2}+\tfrac{\bar{\varphi}_2\omega_{2\min}}{\tau_2^2\varpi_{22}})\sum_{\bar{j}=1}^{3}\left\|\tilde{\zeta}_{2\bar{j}}(k)\right\|^2+\tfrac{\omega_{2\min}}{\varpi_{22}}\bar{\Upsilon}_2^2+\delta_2^2\bar{\mathscr{B}}_2^2+2\Delta T\mu_2\delta_2\bar{\mathscr{B}}_2 \\
& -\min(\omega_{2\bar{j}}^2-\omega_{2\bar{j}}^3-\kappa_{2\bar{j}}^2\omega_{2\bar{j}}^2)\sum_{\bar{j}=1}^{3}\left\|\hat{\zeta}_{2\bar{j}}(k)\right\|^2+9\omega_{2\min}\bar{\mathscr{B}}_2^2+\tfrac{\omega_{2\min}}{\varpi_{22}}\bar{\mathscr{B}}_2^2 \\
& +9\omega_{2\min}\bar{\Upsilon}_2^2+2\Delta T^2\mu_2^2+15\omega_{2\min}\lambda_2^2+\sum_{\bar{j}=1}^{3}\omega_{2\bar{j}}^2\bar{\zeta}_2^*+2\Delta T\mu_2\delta_2\bar{\Upsilon}_2 \\
& +\tfrac{(1-\delta_2)^2\Delta T^2\mu_2^2}{\varpi_{21}}+(1-\delta_2)^2\bar{\Upsilon}_2^2+\max(2\omega_{2\bar{j}}^2\bar{\varphi}_2+\omega_{2\bar{j}}\kappa_{2\bar{j}}^2\bar{\varphi}_2)\|\eta_2(k)\|^2
\end{aligned} \tag{6.102}
$$

Aiming at the trajectory control system model with external disturbances (6.4)–(6.6) and the attitude dynamics model of uncertain UAV in discrete-time form with external disturbances and input saturation (6.9), the above designed discrete-time tracking control scheme based on the SMDO with discrete-time form can be summarized as the following theorem

Theorem 6.1 *For the trajectory control system model with external disturbances (6.4)–(6.6) and the attitude dynamics model of uncertain discrete-time UAV system with external disturbances and input saturation (6.9), and the state information of the system can be measured. The designed SMDOs with discrete-time form are given by (6.11), (6.13), (6.27), (6.29), (6.52), (6.54), (6.84) and (6.86), the height control law is chosen as (6.18), the speed and track angle control laws are designed by (6.35), the control signal $\alpha_1(k)$ is selected as (6.61), the adaptive laws of weight estimation are designed by (6.37), (6.65) and (6.94), and the discrete-time adaptive neural attitude controller is designed by (6.91). On the basis of the proposed adaptive neural control scheme, all signals in the closed-loop system are bounded.*

Proof In order to prove the stability of the entire closed-loop system, the following Lyapunov function is given by:

$$
V(k)=\sum_{\iota=1}^{2}V_\iota(k)+V_z(k)+V_H(k) \tag{6.103}
$$

According to (6.78) and (6.102), $\Delta V(k)=\sum_{\iota=1}^{2}V_\iota(k+1)-\sum_{\iota=1}^{2}V_\iota(k)+\Delta V_z(k)+\Delta V_H(k)$ satisfies that

$$\begin{aligned}
\Delta V(k)\le{}& -R_{11}\|\eta_1(k)\|^2-R_{12}\left\|\tilde{\bar{d}}_1(k)\right\|^2-R_{13}\|s_1(k)\|^2-R_{14}\|\chi_1(k)\|^2\\
&-R_{15}\sum_{\bar{j}=1}^{3}\left\|\tilde{\zeta}_{1\bar{j}}(k)\right\|^2-R_{16}\sum_{\bar{j}=1}^{3}\left\|\hat{\zeta}_{1\bar{j}}(k)\right\|^2+R_{17}-R_{18}\sum_{\bar{j}=1}^{3}\left\|\tilde{\zeta}_{0\bar{j}}(k)\right\|^2\\
&-R_{19}\sum_{\bar{j}=1}^{3}\left\|\hat{\zeta}_{0\bar{j}}(k)\right\|^2-R_{21}\|\eta_2(k)\|^2-R_{22}\left\|\tilde{\bar{d}}_2(k)\right\|^2-R_{23}\|\chi_2(k)\|^2\\
&-R_{24}\sum_{\bar{j}=1}^{3}\left\|\tilde{\zeta}_{2\bar{j}}(k)\right\|^2-R_{25}\|s_2(k)\|^2-R_{26}\sum_{\bar{j}=1}^{3}\left\|\hat{\zeta}_{2\bar{j}}(k)\right\|^2+R_{27}-R_{31}\eta_z^2(k)\\
&-R_{32}\tilde{\bar{d}}_z^2(k)-R_{33}s_z^2(k)+R_{34}-R_{41}\|\eta_H(k)\|^2-R_{42}\left\|\tilde{\bar{d}}_H(k)\right\|^2\\
&-R_{43}\sum_{\bar{i}=1}^{2}\left\|\tilde{\zeta}_{H\bar{i}}(k)\right\|^2-R_{44}\|s_H(k)\|^2-R_{45}\sum_{\bar{i}=1}^{2}\left\|\hat{\zeta}_{H\bar{i}}(k)\right\|^2+R_{46}
\end{aligned}\tag{6.104}$$

where $R_{11}=\omega_{1\min}(1-3\rho_1^2-3\varpi_{12}\rho_1^2)-\max(2\omega_{1\bar{j}}^2\bar{\varphi}_1+\omega_{1\bar{j}}\kappa_{1\bar{j}}^2\bar{\varphi}_1)-\max(2\omega_{0\bar{j}}^2\bar{\varphi}_0+\omega_{0\bar{j}}\kappa_{0\bar{j}}^2\bar{\varphi}_0)$, $R_{12}=1-10\omega_{1\min}-(1-\delta_1)^2-2{\delta_1}^2-\varpi_{11}$, $R_{13}=\omega_{1\min}$, $R_{14}=\omega_{1\min}-2\omega_{1\min}\|C_1\|^2$, $R_{15}=\min(\omega_{1\bar{j}}^2-\kappa_{1\bar{j}}^2)-8\omega_{1\min}\frac{\bar{\varphi}_1}{\tau_1^2}-3\delta_1^2\frac{\bar{\varphi}_1}{\tau_1^2}-\frac{\bar{\varphi}_1(1-\delta_1)^2}{\tau_1^2}-\frac{2\bar{\varphi}_1\omega_{1\min}}{\tau_1^2\varpi_{12}}$, $R_{16}=\min(\omega_{1\bar{j}}^2-\omega_{1\bar{j}}^3-\kappa_{1\bar{j}}^2\omega_{1\bar{j}}^2)$, $R_{17}=2\delta_1^2\bar{\Upsilon}_1^2+\frac{\omega_{1\min}}{\varpi_{12}}\bar{\Upsilon}_1^2+8\omega_{1\min}\bar{\Upsilon}_1^2+2\Delta T^2\mu_1^2+12\omega_{1\min}\lambda_1^2+\sum_{\bar{j}=1}^{3}\omega_{1\bar{j}}^2\bar{\zeta}_1^*+2\Delta T\mu_1\delta_1\bar{\Upsilon}_1+\frac{(1-\delta_1)^2\Delta T^2\mu_1^2}{\varpi_{11}}+(1-\delta_1)^2\bar{\Upsilon}_1^2+\sum_{\bar{j}=1}^{3}\omega_{0\bar{j}}^2\bar{\zeta}_0^*$, $R_{18}=-(\min(\omega_{0\bar{j}}^2-\kappa_{0\bar{j}}^2)-\frac{3\varpi_{12}\bar{g}_1^2\bar{\ell}_0^2\bar{\varphi}_0\omega_{1\min}}{\tau_0^2}-\frac{3\bar{\varphi}_0\bar{g}_1^2\bar{\ell}_0^2}{\tau_0^2})$, $R_{19}=-\min(\omega_{0\bar{j}}^2-\omega_{0\bar{j}}^3-\kappa_{0\bar{j}}^2\omega_{0\bar{j}}^2)$, $R_{22}=1-10\omega_{2\min}-(1-\delta_2)^2-2{\delta_2}^2-2\varpi_{21}$, $R_{23}=\omega_{2\min}-2\omega_{2\min}\|C_2\|^2-2\omega_{1\min}\bar{g}_1^2$, $R_{24}=\min(\omega_{2\bar{j}}^2-\kappa_{2\bar{j}}^2)-9\omega_{2\min}\frac{\bar{\varphi}_2}{\tau_2^2}-4\delta_2^2\frac{\bar{\varphi}_2}{\tau_2^2}-\frac{\bar{\varphi}_2(1-\delta_2)^2}{\tau_2^2}-\frac{\bar{\varphi}_2\omega_{2\min}}{\tau_2^2\varpi_{22}}$, $R_{25}=\omega_{2\min}$, $R_{26}=\min(\omega_{2\bar{j}}^2-\omega_{2\bar{j}}^3-\kappa_{2\bar{j}}^2\omega_{2\bar{j}}^2)$, $R_{21}=\omega_{2\min}(1-2\rho_2^2-3\varpi_{22}\rho_2^2)-\max(2\omega_{2\bar{j}}^2\bar{\varphi}_2+\omega_{2\bar{j}}\kappa_{2\bar{j}}^2\bar{\varphi}_2)-(4+\frac{2}{\varpi_{12}})\omega_{1\min}\bar{g}_1^2$, $R_{32}=1-4\delta_z^2-(1-\delta_z)^2\tau_z-(1-\delta_z)^2$, $R_{27}=2\delta_2^2\bar{\Upsilon}_2^2+9\omega_{2\min}\bar{\mathscr{B}}_2^2+\frac{\omega_{2\min}}{\varpi_{22}}\bar{\mathscr{B}}_2^2+\delta_2^2\bar{\mathscr{B}}_2^2+2\Delta T\mu_2\delta_2\bar{\mathscr{B}}_2+2\delta_2^2\bar{\Upsilon}_2\bar{\mathscr{B}}_2+\delta_2^2\bar{\mathscr{B}}_2^2+\frac{(1-\delta_2)^2\delta_2^2}{\varpi_{21}}\bar{\mathscr{B}}_2^2+2\omega_{2\min}\bar{g}_2^2\nu^2+\frac{\omega_{2\min}}{\varpi_{22}}\bar{\Upsilon}_2^2+9\omega_{2\min}\bar{\Upsilon}_2^2+2\Delta T^2\mu_2^2+15\omega_{2\min}\lambda_2^2+\sum_{\bar{j}=1}^{3}\omega_{2\bar{j}}^2\bar{\zeta}_2^*+2\Delta T\mu_2\delta_2\bar{\Upsilon}_2+(1-\delta_2)^2\bar{\Upsilon}_2^2+\frac{(1-\delta_2)^2\Delta T^2\mu_2^2}{\varpi_{21}}$, $R_{31}=\delta_z^2-2\delta_z^2\rho_z^2$, $R_{33}=\delta_z^2$, $R_{34}=2\delta_z^2\lambda_z^2+\Delta T^2\mu_z^2+\frac{\Delta T^2\mu_z^2}{\tau_z}$, $R_{41}=\omega_{H\min}(1-2\rho_H^2-2\varpi_{H2}\rho_H^2)-\max(2\omega_{H\bar{i}}^2\bar{\varphi}_H+\omega_{H\bar{i}}\kappa_{H\bar{i}}^2\bar{\varphi}_H)$, $R_{42}=1-8\omega_{H\min}-(1-\delta_H)^2-2{\delta_H}^2-\varpi_{H1}$, $R_{43}=\min(\omega_{H\bar{i}}^2-\kappa_{H\bar{i}}^2)-7\omega_{H\min}\frac{\bar{\varphi}_H}{\tau_H^2}-3\delta_H^2\frac{\bar{\varphi}_H}{\tau_H^2}-\frac{\bar{\varphi}_H(1-\delta_H)^2}{\tau_H^2}-\frac{\bar{\varphi}_H\omega_{H\min}}{\tau_H^2\varpi_{H2}}$,

$R_{44} = \omega_{H\min}$, $R_{45} = \min(\omega^2_{H\bar{i}} - \omega^2_{H\bar{i}} - \kappa^2_{H\bar{i}}\omega^2_{H\bar{i}})$ and $R_{46} = 2\Delta T \mu_H \delta_H \bar{\Upsilon}_H + \frac{\omega_{H\min}}{\varpi_{H2}} \bar{\Upsilon}^2_H + 8\omega_{H\min}\lambda^2_H + \sum_{\bar{i}=1}^{2} \omega^2_{H\bar{i}}\bar{\zeta}^*_H + 7\omega_{H\min}\bar{\Upsilon}^2_H + 2\Delta T^2\mu^2_H + 2\delta^2_H\bar{\Upsilon}^2_H + \frac{(1-\delta_H)^2\Delta T^2\mu^2_H}{\varpi_{H1}} + (1-\delta_H)^2\bar{\Upsilon}^2_H$.

According to (6.104), if the selected control parameters can make that $R_{11} > 0$, $R_{12} > 0$, $R_{13} > 0$, $R_{14} > 0$, $R_{15} > 0$, $R_{16} > 0$, $R_{18} > 0$, $R_{19} > 0$, $R_{21} > 0$, $R_{22} > 0$, $R_{23} > 0$, $R_{24} > 0$, $R_{25} > 0$, $R_{26} > 0$, $R_{31} > 0$, $R_{32} > 0$, $R_{33} > 0$, $R_{41} > 0$, $R_{42} > 0$, $R_{43} > 0$, $R_{44} > 0$, $R_{45} > 0$ and $R_{46} > 0$, then all signals in the closed loop system are bounded. Furthermore, the errors between the output signals $\bar{z}_g(k)$, $\bar{V}_A(k)$, $\gamma(k)$ and $x_1(k)$ with the reference signals $\bar{z}_{gd}(k)$, $\bar{V}_{Ad}(k)$, $\gamma_d(k)$ and $x_d(k)$ are bounded, and the disturbance estimation errors are also bounded. ◇

6.4 Simulation Study

In this section, the simulation results are given for the trajectory control system model (6.4), (6.5) and (6.6) with wind disturbances and system uncertainties, and the attitude dynamic model of the uncertain UAV system with wind disturbances and input saturation (6.9) to illustrate the effectiveness of SMDO-based discrete-time tracking control scheme.

In the simulation, the effects of gust on the control of UAV system are considered, and the specific form of gust is shown in (2.40). In this numerical simulation, the maximum external wind speed is assumed as $\bar{V}_{Wm} = [5, 4, 3]^{\mathrm{T}}\mathrm{m/s}$, and $t_m = 3\mathrm{s}$. The external harmonic disturbances are set as $\bar{d}_{21} = \Delta T(0.2\sin(0.1t) - 0.2\cos(0.4t))$, $\bar{d}_{22} = \Delta T(0.1\sin(0.2t) - 0.2\cos(0.5t))$ and $\bar{d}_{23} = \Delta T(0.3\sin(0.3t) - 0.3\cos(0.5t))$. The sampling period is selected as $\Delta T = 0.005$. The saturation levels of control inputs are $\bar{v}_{m1} = 21.5\mathrm{deg}$ $\bar{v}_{m2} = 25\mathrm{deg}$ and $\bar{v}_{m3} = 30\mathrm{deg}$. The initial values are chosen as $\beta_0 = 2\mathrm{deg}$, $\alpha_0 = 2\mathrm{deg}$, $\mu_0 = 2\mathrm{deg}$, and $p_0 = q_0 = r_0 = 0\mathrm{deg/s}$. The flight altitude is 1005m and the flight velocity is 100m/s. The desired tracking signals are given as $\beta_d = \mu_d = 0\mathrm{deg}$, and the desired signals α_d and γ_d can be obtained by calculating $u_H(k)$ and $u_z(k)$ respectively. The tracking signals of velocity and altitude are selected as $100\mathrm{erf}(0.2t) + 100$ and $200\mathrm{erf}(0.2t) + 1000$. The control parameters in the SMDO are chosen as $\delta_1 = 0.1$, $\lambda_1 = 0.1$, $\delta_2 = 0.1$, $\lambda_2 = 0.1$, $\delta_z = 0.1$, $\lambda_z = 0.1$, $\delta_H = 0.1$ and $\lambda_H = 0.1$. Set the control parameters of the adaptive laws as $\kappa_{H1} = \kappa_{H2} = \kappa_{H3} = 0.0005$, $\omega_{H1} = \omega_{H2} = \omega_{H3} = 0.01$, $\kappa_{11} = \kappa_{12} = \kappa_{13} = 0.0005$, $\omega_{11} = \omega_{12} = \omega_{13} = 0.01$, $\kappa_{21} = \kappa_{22} = \kappa_{23} = 0.00005$ and $\omega_{21} = \omega_{22} = \omega_{23} = 0.01$. The control parameters of the discrete-time tracking differentiator are selected as $q_{02} = \mathrm{diag}[10, 10, 10]$ and $p_{02} = \mathrm{diag}[20, 20, 20]$. The Control parameters of auxiliary system are selected as $C_1 = C_2 = \mathrm{diag}[0.002, 0.002, 0.002]$. Furthermore, the following definitions are that $\hat{\bar{d}}_v$, $\hat{\bar{d}}_\gamma$, $\hat{\bar{d}}_{11}$, $\hat{\bar{d}}_{12}$, $\hat{\bar{d}}_{13}$, $\hat{\bar{d}}_{21}$, $\hat{\bar{d}}_{22}$ and $\hat{\bar{d}}_{23}$ are the estimations of $\bar{d}_v$, $\bar{d}_\gamma$, $\bar{d}_{11}$, $\bar{d}_{12}$, $\bar{d}_{13}$, $\bar{d}_{21}$, $\bar{d}_{22}$ and $\bar{d}_{23}$ respectively. Moreover, the disturbance estimation errors are defined as $\tilde{\bar{d}}_v = \bar{d}_v - \hat{\bar{d}}_v$, $\tilde{\bar{d}}_\gamma = \bar{d}_\gamma - \hat{\bar{d}}_\gamma$,

$\tilde{\bar{d}}_{11} = \bar{d}_{11} - \hat{\bar{d}}_{11}$, $\tilde{\bar{d}}_{12} = \bar{d}_{12} - \hat{\bar{d}}_{12}$, $\tilde{\bar{d}}_{13} = \bar{d}_{13} - \hat{\bar{d}}_{13}$, $\tilde{\bar{d}}_{21} = \bar{d}_{21} - \hat{\bar{d}}_{21}$, $\tilde{\bar{d}}_{22} = \bar{d}_{22} - \hat{\bar{d}}_{22}$, $\tilde{\bar{d}}_{23} = \bar{d}_{23} - \hat{\bar{d}}_{23}$, and the tracking errors are defined as $e_z = \bar{z}_g - \bar{z}_{gd}$, $e_v = \bar{V}_A - \bar{V}_{Ad}$, $e_\gamma = \gamma - \gamma_d$, $e_\beta = \beta - \beta_d$, $e_\alpha = \alpha - \alpha_d$ and $e_\mu = \mu - \mu_d$.

On the basis of the above simulation parameters, the simulations are presented in Figs. 6.2, 6.3, 6.4, 6.5, 6.6, 6.7, 6.8, 6.9, 6.10, 6.11, 6.12, 6.13 and 6.14. According to the tracking results in Fig. 6.3, the UAV can fly at the desired altitude. The flight velocity control performance of the UAV is presented in Fig. 6.3. It can also show that the velocity of UAV under discrete time control can fly according to the desired velocity. According to Figs. 6.4 and 6.5, the track angle γ of the UAV can keep up with the desired signal, and the thrust $\bar{T}$ changes over time. Furthermore, the tracking control results of attitude angles are shown in Fig. 6.6. From Fig. 6.6, the desired signal $x_d(k)$ can be followed by the output signal $x_1(k)$. Under the action of NN-based discrete-time control method, the tracking error responses of attitude angles are given in Fig. 6.7, and the tracking errors e_β, e_α and e_μ can converge bounded. Then, the control inputs are shown in Fig. 6.8. It is shown that the negative impacts of input saturation can be restrained using the proposed control scheme. The estimation performance of the designed SMDO is shown in Figs. 6.9, 6.10, 6.11, 6.12 and 6.13. According to Figs. 6.9, 6.11 and 6.13, the designed SMDO can estimate the wind disturbances. Therefore, on the basis of the above numerical simulation results, the adaptive discrete-time control method based on SMDO can be used to obtain satisfactory control performance of the UAV system.

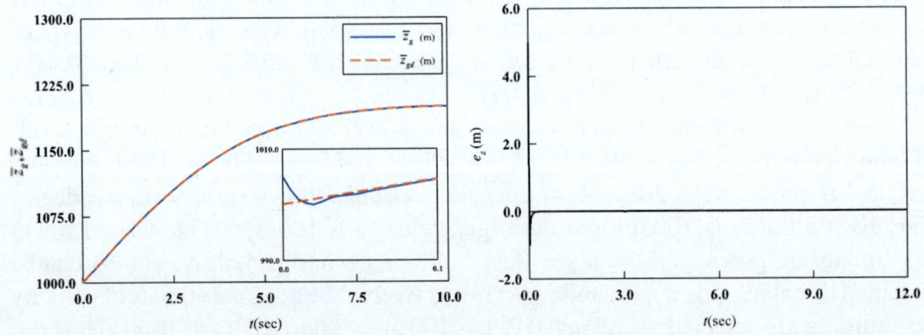

(a) Height tracking signals $\bar{z}_g$ and $\bar{z}_{gd}$ (b) Height tracking error e_z

Fig. 6.2 Tracking control results of UAV flight height

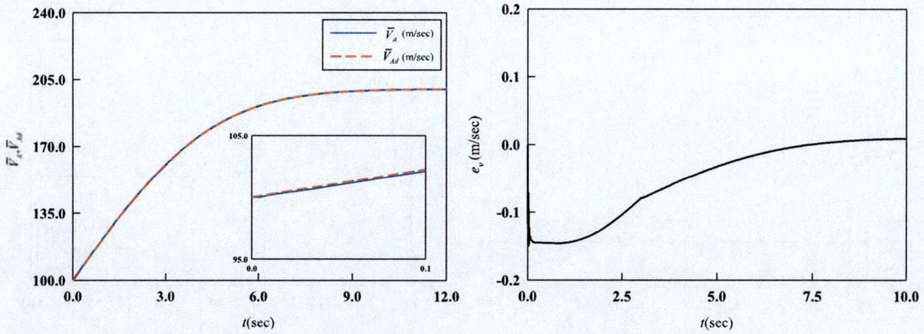

(a) Velocity tracking response $\bar{V}_A$ and $\bar{V}_{Ad}$ (b) Velocity tracking error e_v

Fig. 6.3 Tracking control results of flight velocity

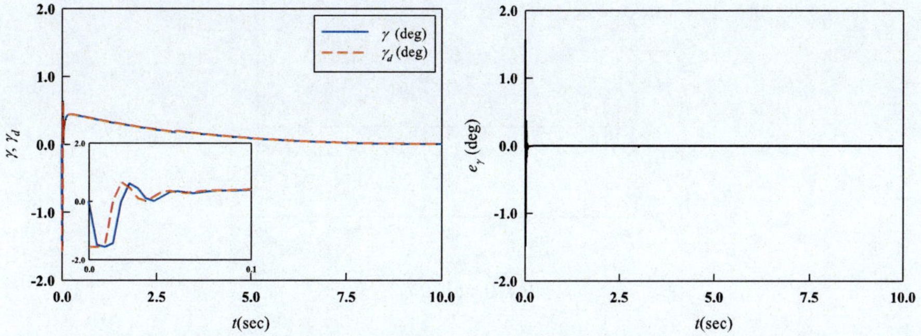

(a) Track angle response γ and γ_d (b) Track angle tracking error e_γ

Fig. 6.4 Tracking control results of track angle

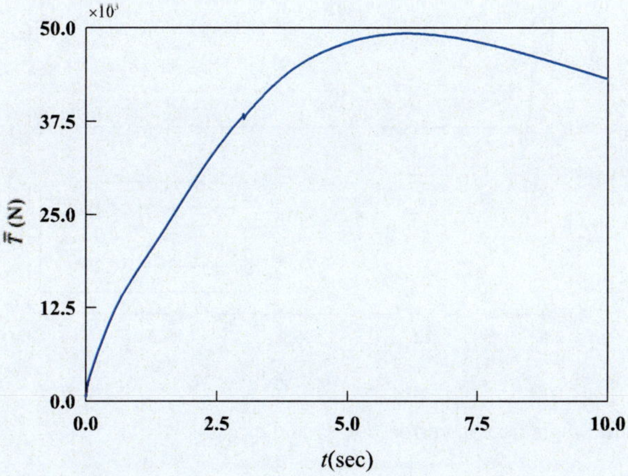

Fig. 6.5 Engine thrust response $\bar{T}$

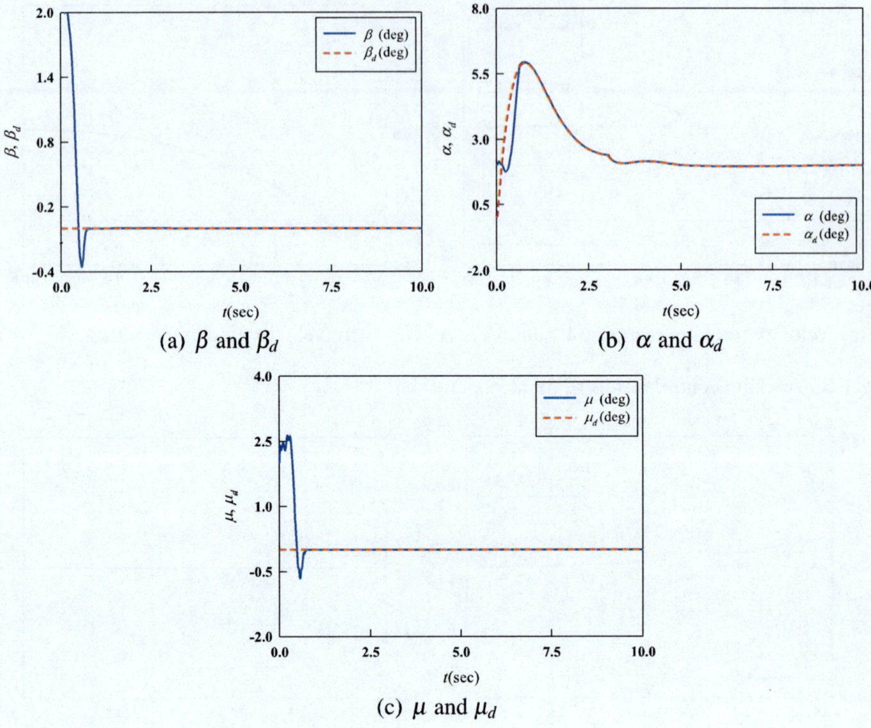

Fig. 6.6 The tracking results of attitude angles

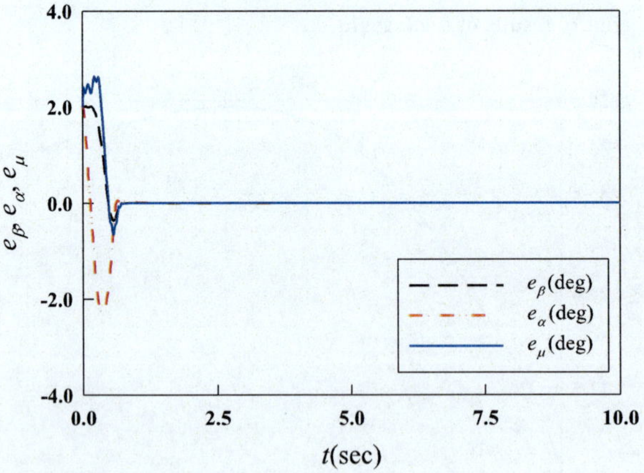

Fig. 6.7 The attitude angle tracking errors e_β, e_α and e_μ

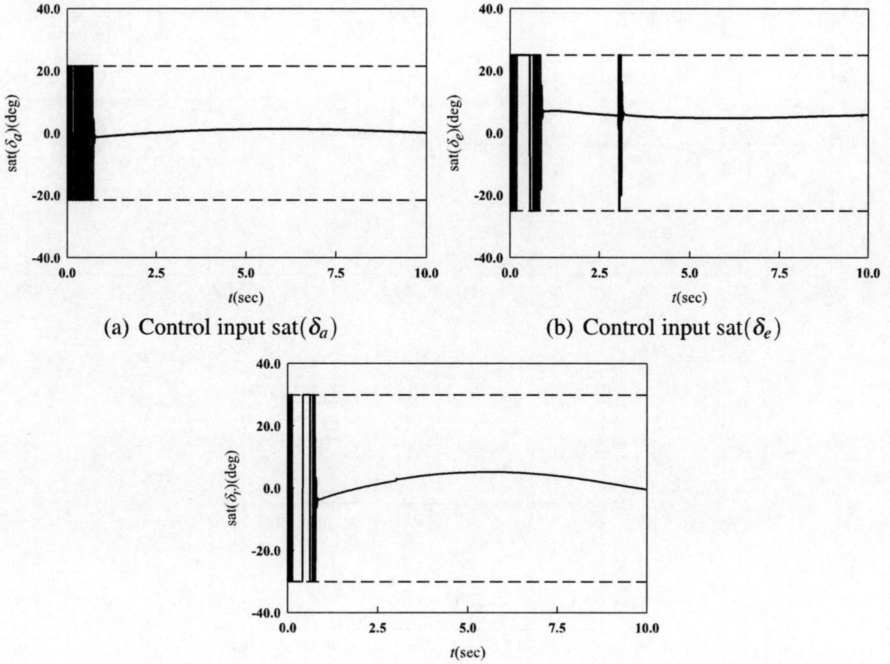

(a) Control input sat(δ_a)

(b) Control input sat(δ_e)

(c) Control input sat(δ_r)

Fig. 6.8 Control input signals

6.5 Conclusions

In this chapter, firstly, the discrete-time control method of the flight height, the flight velocity and the track angle has been studied for the trajectory control system model with wind disturbances and system uncertainties. Secondly, an adaptive discrete-time attitude control scheme based on the designed SMDO, the auxiliary system and the NN has been proposed for the uncertain attitude dynamics model of UAV with wind disturbances and input saturation. The control scheme studied in this chapter, the uncertainties in the system have been processed by the NN approximation, and the negative affects of external disturbances in the system have been suppressed by the designed discrete-time SMDO. In addition, the boundedness of closed-loop system signals has been proven by discrete-time Lyapunov stability theory. Finally, numerical simulation results have been given to show the effectiveness of the discrete-time control scheme based on SMDO.

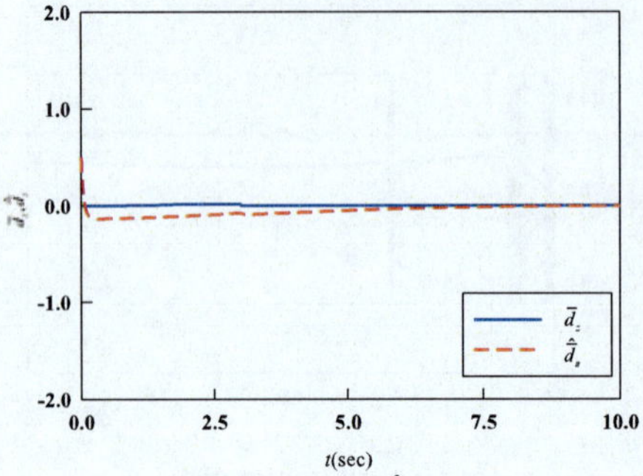

(a) $\bar{d}_z$ and $\hat{\bar{d}}_z$

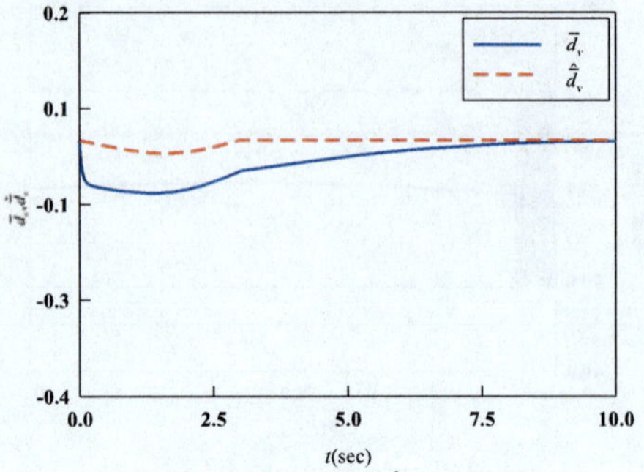

(b) $\bar{d}_v$ and $\hat{\bar{d}}_v$

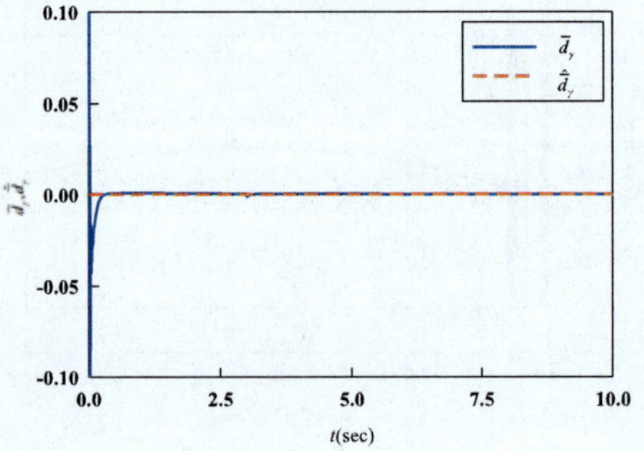

(c) $\bar{d}_\gamma$ and $\hat{\bar{d}}_\gamma$

Fig. 6.9 The estimation performance of the designed SMDO

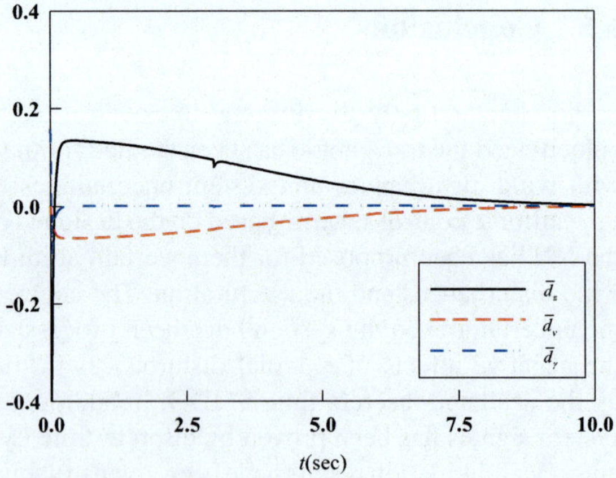

Fig. 6.10 Estimation errors of the disturbance observer $\tilde{\bar{d}}_z$, $\tilde{\bar{d}}_v$ and $\tilde{\bar{d}}_\gamma$

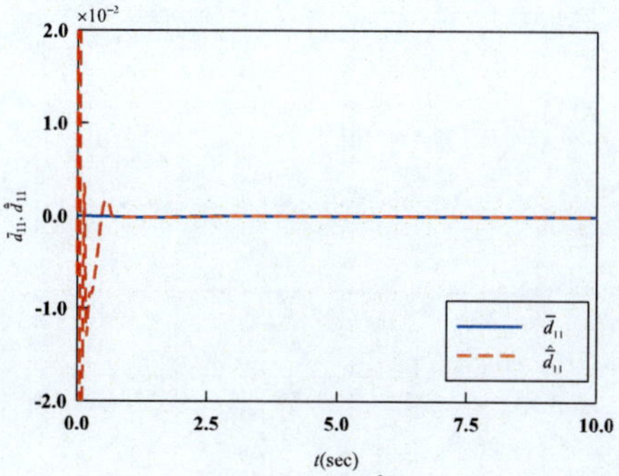

(a) $\bar{d}_{11}$ and $\hat{\bar{d}}_{11}$

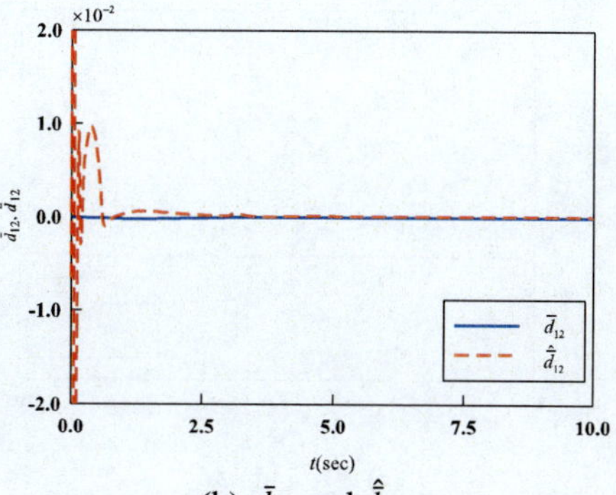

(b) $\bar{d}_{12}$ and $\hat{\bar{d}}_{12}$

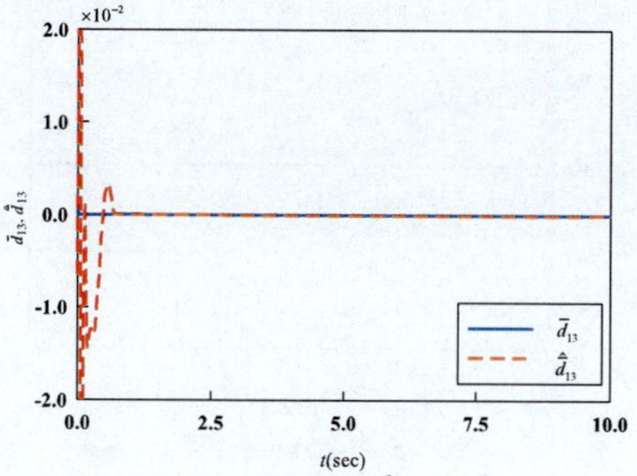

(c) $\bar{d}_{13}$ and $\hat{\bar{d}}_{13}$

Fig. 6.11 The estimation performance of the designed SMDO

Fig. 6.12 Estimation errors of the disturbance observer $\tilde{\bar{d}}_{11}$, $\tilde{\bar{d}}_{12}$ and $\tilde{\bar{d}}_{13}$

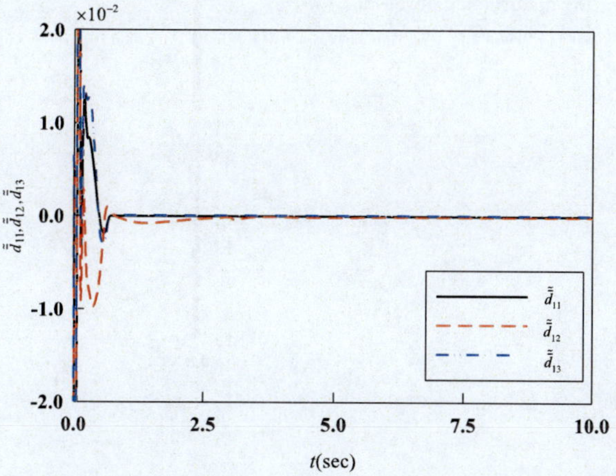

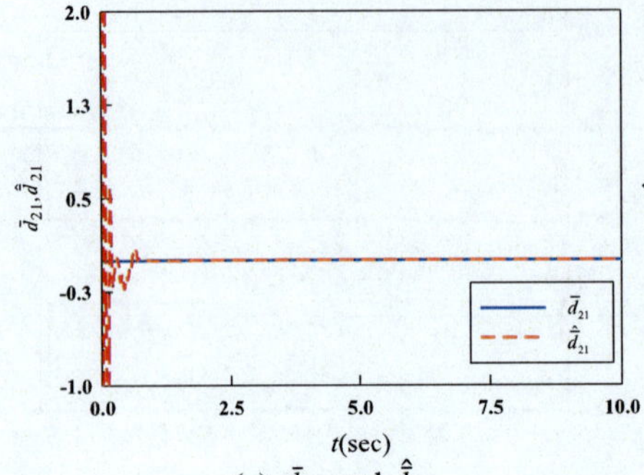

(a) $\bar{d}_{21}$ and $\hat{\bar{d}}_{21}$

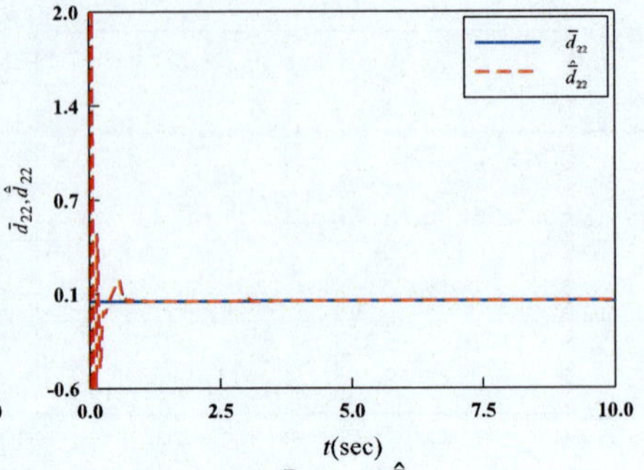

(b) $\bar{d}_{22}$ and $\hat{\bar{d}}_{22}$

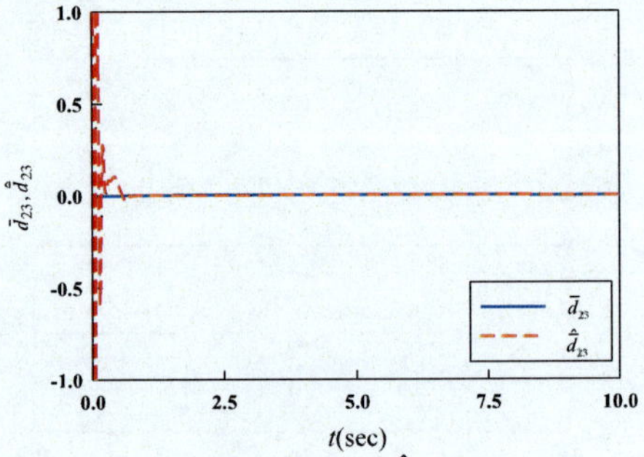

(c) $\bar{d}_{23}$ and $\hat{\bar{d}}_{23}$

Fig. 6.13 The estimation performance of the designed SMDO

Fig. 6.14 Estimation errors of the disturbance observer $\tilde{\bar{d}}_{21}$, $\tilde{\bar{d}}_{22}$ and $\tilde{\bar{d}}_{23}$

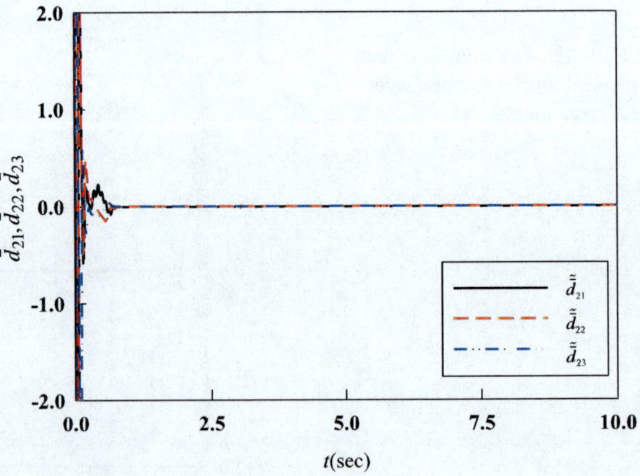

References

1. Chen, F., Jiang, R., Zhang, K., et al.: Robust backstepping sliding-mode control and observer-based fault estimation for a quadrotor UAV. IEEE Trans. Ind. Electron. **63**(8), 5044–5056 (2016)
2. Ma, D., Xia, Y., Shen, G., et al.: Flatness-based adaptive sliding mode tracking control for a quadrotor with disturbances. J. Frankl. Inst. **355**(14), 6300–6322 (2018)
3. Zheng, Z., Jin, Z., Sun, L., et al.: Adaptive sliding mode relative motion control for autonomous carrier landing of fixed-wing unmanned aerial vehicles. IEEE Access **5**, 5556–5565 (2017)
4. Espinoza, T., Dzul, A., Lozano, R., et al.: Backstepping-sliding mode controllers applied to a fixed-wing UAV. J. Intell. Robot. Syst. **73**(1–4), 67–79 (2014)
5. Oh, H., Kim, S., Tsourdos, A., et al.: Decentralised standoff tracking of moving targets using adaptive sliding mode control for UAVs. J. Intell. Robot. Syst. **76**(1), 169–183 (2014)
6. Mu, B., Zhang, K., Shi, Y.: Integral sliding mode flight controller design for a quadrotor and the application in a heterogeneous multi-agent system. IEEE Trans. Ind. Electron. **64**(12), 9389–9398 (2017)
7. Castañeda, H., Salas-Peña, O.S., de León-Morales, J.: Extended observer based on adaptive second order sliding mode control for a fixed wing UAV. ISA Trans. **66**, 226–232 (2017)
8. Xiong, J.-J., Zhang, G.-B.: Global fast dynamic terminal sliding mode control for a quadrotor UAV. ISA Trans. **66**, 233–240 (2017)
9. Wu, K., Cai, Z., Zhao, J., et al.: Target tracking based on a nonsingular fast terminal sliding mode guidance law by fixed-wing UAV. Appl. Sci. **7**(4), 333–1–18 (2017)
10. Hua, C.-C., Wang, K., Chen, J.-N., et al.: Tracking differentiator and extended state observer-based nonsingular fast terminal sliding mode attitude control for a quadrotor. Nonlinear Dyn. **94**(1), 343–354 (2018)
11. Mareels, I.M., Penfold, H., Evans, R.J.: Controlling nonlinear time-varying systems via Euler approximations. Automatica **28**(4), 681–696 (1992)
12. Chen, M., Shao, S.-Y., Jiang, B.: Adaptive neural control of uncertain nonlinear systems using disturbance observer. IEEE Trans. Cybern. **47**(10), 3110–3123 (2017)
13. Sonneveldt, L.: Nonlinear F-16 model description, Tech. rep., Delft University of Technology, Netherlands (2006)
14. Yang, Q., Chen, M.: Adaptive neural prescribed performance tracking control for near space vehicles with input nonlinearity. Neurocomputing **174**, 780–789 (2016)

Chapter 7
DTFO Control for Uncertain UAV Attitude System Based on NN and Prescribed Performance Method

7.1 Introduction

Among the existing tracking control methods for fixed-wing UAV system, most of them focus on how to ensure that the tracking error converts to the bounded region or asymptotically converts to the origin, which belongs to steady-state performance studies. Up to now, studies on transient performance such as overshooting and convergence rate of fixed-wing UAV system have been relatively few. In fact, the study of transient performance plays an important role in improving the performance of UAV flight control system. For example, the excessive overshoot may cause the UAV actuator to exceed the physical limit, resulting in instability of the closed-loop UAV system. Although the prescribed performance control method has been applied in the control of aircraft [1, 2], the DTFC method based on the prescribed performance control method has not been studied much. Therefore, the transient and steady-state performance of UAV systems need to be further considered when studying the design of robust discrete-time flight control scheme. On the other hand, the FOC was a powerful way to describe data memory and inheritance [3]. Over the past few decades, some researchers have found that FOC can describe some non-classical phenomena in natural science and engineering applications, and the FOC theory has been successfully applied to many practical fields [4–18]. The most important advantage of using FOC was its non-local nature, which means that FOC has historical dependency and long-range correlation [19, 20]. In addition, the introduction of FOC in the design of the control method can increase the additional adjustment freedom, thus improving the control performance of the controller. Although FOC of continuous time has been used in flight control design [21, 21–26], the flight control scheme for DTFO has not been studied much. Therefore, the problem of combining the design of FOC and flight control in the form of discrete-time needs further study. For the problems mentioned above, this chapter proposes an adaptive DTFO control scheme based on the NN, DTDO and the prescribed performance method for the attitude dynamics model of UAV with external disturbances and system uncertainties.

S. Shao et al., *Robust Discrete-Time Flight Control of UAV with External Disturbances*, Studies in Systems, Decision and Control 317,
https://doi.org/10.1007/978-3-030-57957-9_7

7.2 Problem Formulation

In this chapter, system uncertainties and external disturbances are considered in the attitude dynamics model of UAV in the form of continuous time (2.41), the dynamic model can be described as

$$\begin{cases} \dot{x}_1 = F_1(x_1) + \Delta F_1(x_1) + G_1(x_1)x_2 + d_1 \\ \dot{x}_2 = F_2(x) + \Delta F_2(x) + G_2(x)u + d_2 \\ y = x_1 \end{cases} \tag{7.1}$$

where $F_1(x_1)$ and $F_2(x)$ are the known nonlinear function vectors, $\Delta F_1(x_1)$ and $\Delta F_2(x)$ are the unknown nonlinear function vectors, $\Delta F_1(x_1)$ is the system uncertainty or modeling error, $\Delta F_2(x)$ is the system uncertainty caused by the wind gradient, d_1 is the disturbance caused by the wind, and d_2 is the external disturbance.

In this chapter, a NN-based DTFO controller is designed for the tracking control of the uncertain UAV system with prescribed performance using the DTDO and the backstepping technology. This chapter aims at developing an adaptive discrete-time neural flight control scheme, so that the system output y can track the given desired signal $x_{1d} = [\beta_d, \alpha_d, \mu_d]^T$ in the presence of system uncertainties and external disturbances, and the tracking errors are in a bounded compact set which satisfies the prescribed performance. Moreover, the proposed control scheme will guarantee that all closed-loop system signals are ultimately bounded. According to the design thread of the discrete-time control scheme, the developed control strategy is shown by a block diagram in Fig. 7.1.

To develop the design of the DTFO tracking control scheme for the system (7.1) with prescribed performance, the following definitions, lemmas and assumptions are given:

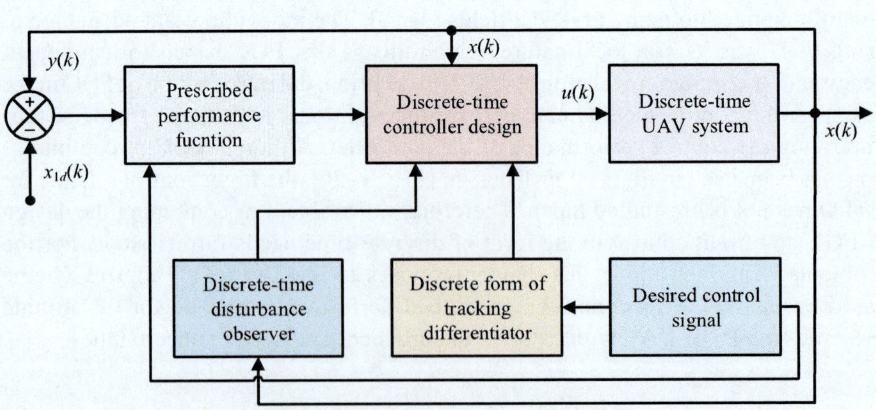

Fig. 7.1 The block diagram of the developed control strategy

Assumption 7.1 According to the dynamic characteristic of the UAV system (7.1), we can assume that the known control coefficient matrices $G_1(x_1)$ and $G_2(x)$ are invertible. Furthermore, there are two positive constants $\underline{g}_1 > 0$ and $\bar{g}_1 > 0$ that can make $\underline{g}_1 \leq \|\bar{G}_1(k)\| \leq \bar{g}_1$.

Assumption 7.2 Assume that the external wind speed is bounded, thus $\bar{d}_1(k)$, $\bar{d}_2(k)$, $\bar{d}_1(k+1)$ and $\bar{d}_2(k+1)$ are bounded disturbances.

Remark 7.1 According to the FO Definitions 2.1 and 2.2, for the function vector $\xi(k) = [\xi_1(k), \xi_2(k), \xi_3(k)]^T$, we define that

$$\nabla^{-\gamma}\xi(k) = \left[\sum_{j=0}^{k}\binom{\gamma}{k-j}\xi_1(j), \sum_{j=0}^{k}\binom{\gamma}{k-j}\xi_2(j), \sum_{j=0}^{k}\binom{\gamma}{k-j}\xi_3(j)\right]^T$$

and

$$\nabla^{\gamma}\xi(k) = \left[\sum_{j=0}^{k}(-1)^j\binom{\gamma}{j}\bar{\xi}_1, \sum_{j=0}^{k}(-1)^j\binom{\gamma}{j}\bar{\xi}_2, \sum_{j=0}^{k}(-1)^j\binom{\gamma}{j}\bar{\xi}_3\right]^T,$$

where γ is the fractional order and $\bar{\xi}_i = \xi_i(k-j)$.

7.3 System Transformation

In this section, in order to study the DTFO control of the uncertain UAV system based on the prescribed performance method, the attitude dynamics model of the constrained uncertain UAV (7.1) is converted into an unconstrained form. The specific conversion steps are as follows:

The tracking error for the system output x_1 and the desired signal x_{1d} is defined as

$$e_1 = x_1 - x_{1d} \tag{7.2}$$

where $e_1 = [e_{11}, e_{12}, e_{13}]^{\mathrm{T}}$.

According to Definition 2.3, the tracking error $e_{1i}(i = 1, 2, 3)$ should satisfy that [27]

$$\begin{cases} -\tau_{1i}\varpi_{1i}(t) < e_{1i}(t) < \varpi_{1i}(t), & e_{1i}(0) \geq 0. \\ -\varpi_{1i}(t) < e_{1i}(t) < \tau_{1i}\varpi_{1i}(t), & e_{1i}(0) \leq 0. \end{cases} \tag{7.3}$$

where $0 \leq \tau_{1i} \leq 1$ is a design constant and $e_{1i}(0)$ is the initial condition of $e_{1i}(t)$ with $|e_{1i}(0)| < \varpi_{1i}(0)$.

On the basis of the error transformation method [27, 28], an unconstrained form (7.1) can be obtained by transforming the originally constrained tracking error. The relationship of $e_{1i}(t)$ and $\varpi_{1i}(t)$ is described as

$$s_{1i}(t) = \Psi_{1i}^{-1}\left[\frac{e_{1i}(t)}{\varpi_{1i}(t)}\right] \tag{7.4}$$

where s_{1i} is the transformed tracking error variable, $\Psi_{1i}^{-1}[\cdot]$ is the inverse function of $\Psi_{1i}[\cdot]$, $\Psi_{1i}[\cdot]$ is the user-defined smooth, invertible function and strictly increasing which satisfies that [27]

$$\begin{cases} \lim\limits_{s_{1i}\to-\infty} \Psi_{1i}[s_{1i}] = -\tau_{1i} \\ \lim\limits_{s_{1i}\to\infty} \Psi_{1i}[s_{1i}] = 1, \quad e_{1i}(0) \geq 0. \end{cases} \tag{7.5}$$

$$\begin{cases} \lim\limits_{s_{1i}\to-\infty} \Psi_{1i}[s_{1i}] = -1 \\ \lim\limits_{s_{1i}\to\infty} \Psi_{1i}[s_{1i}] = \tau_{1i}, \quad e_{1i}(0) \leq 0. \end{cases} \tag{7.6}$$

According to (7.5) and (7.6), $\Psi_{1i}[\cdot]$ can be chosen as [28]

$$\begin{cases} \Psi_{1i}[s_{1i}] = \frac{\nu^{s_{1i}}-\tau_{1i}\nu^{-s_{1i}}}{\nu^{s_{1i}}+\nu^{-s_{1i}}}, \quad e_{1i}(0) \geq 0. \\ \Psi_{1i}[s_{1i}] = \frac{\tau_{1i}\nu^{s_{1i}}-\nu^{-s_{1i}}}{\nu^{s_{1i}}+\nu^{-s_{1i}}}, \quad e_{1i}(0) \leq 0. \end{cases} \tag{7.7}$$

where is a positive constant, and ν is chosen as $\nu = 2$.

Invoking (7.4) and (7.7), one has

$$\begin{cases} s_{1i}(t) = \frac{1}{2}\log_{\nu}\frac{e_{1i}(t)+\tau_{1i}\varpi_{1i}(t)}{\varpi_{1i}(t)-e_{1i}(t)}, \quad e_{1i}(0) \geq 0. \\ s_{1i}(t) = \frac{1}{2}\log_{\nu}\frac{e_{1i}(t)+\varpi_{1i}(t)}{\tau_{1i}\varpi_{1i}(t)-e_{1i}(t)}, \quad e_{1i}(0) \leq 0. \end{cases} \tag{7.8}$$

Hence, $s_{1i}(t)$ is bounded based on (7.8), and the following form is further satisfied:

$$\begin{cases} -\tau_{1i} < \Psi_{1i}[s_{1i}] < 1, \quad e_{1i}(0) \geq 0. \\ -1 < \Psi_{1i}[s_{1i}] < \tau_{1i}, \quad e_{1i}(0) \leq 0. \end{cases} \tag{7.9}$$

According to Definition 2.3, the initial condition $s_{1i}(0)$ is existing and bounded. Therefore, the objective of the tracking control problem with prescribed performance can be transformed to develop a DTFO controller to guarantee that s_{1i} is ultimately bounded.

From (7.4), it yields

$$\dot{s}_{1i}(t) = \frac{\partial \Psi_{1i}^{-1}}{\partial\left[\frac{e_{1i}(t)}{\varpi_{1i}(t)}\right]}\frac{1}{\varpi_{1i}(t)}\left[\dot{e}_{1i}(t) - \frac{e_{1i}(t)\dot{\varpi}_{1i}(t)}{\varpi_{1i}(t)}\right] \tag{7.10}$$

By combining (7.1) with (7.10), one has

$$\dot{s}_1 = \mathscr{N} + \mathscr{M}[F_1(x_1) + G_1(x_1)x_2 + \Delta F_1(x_1) + d_1 - \dot{x}_{1d}] \tag{7.11}$$

where $\dot{s}_1 = [\dot{s}_{11}, \dot{s}_{12}, \dot{s}_{13}]^T$, $\mathscr{N} = [\mathscr{N}_1, \mathscr{N}_2, \mathscr{N}_3]^T$, $\mathscr{M} = \text{diag}[\mathscr{M}_1, \mathscr{M}_2, \mathscr{M}_3]$, $\mathscr{N}_i = -\frac{\partial \Psi_{1i}^{-1}}{\partial [e_{1i}(t)/\varpi_{1i}(t)]} \frac{1}{\varpi_{1i}(t)} \frac{e_{1i}(t)\dot{\varpi}_{1i}(t)}{\varpi_{1i}(t)}$, $\mathscr{M}_i = \frac{\partial \Psi_{1i}^{-1}}{\partial [e_{1i}(t)/\varpi_{1i}(t)]} \frac{1}{\varpi_{1i}(t)}$, and diag[$\cdot$] denotes the diagonal matrix.

Hence, the transformed system dynamics can be written as

$$\begin{cases} \dot{s}_1 = \mathscr{N} + \mathscr{M}(\dot{x}_1 - \dot{x}_{1d}) \\ \dot{x}_2 = F_2(x) + \Delta F_2(x) + G_2(x)u + d_2 \end{cases} \tag{7.12}$$

Remark 7.2 According to the transformed system dynamics (7.12), the problem of tracking control has been transformed into a stability analysis problem of the transformed tracking error s_1 by designing a DTFO controller based on the NN and the DTDO. Furthermore, we can note that e_{1i}, $\Psi_{1i}^{-1}[\cdot]$, ϖ_{1i} and $\dot{\varpi}_{1i}$ are known [28]. $\mathscr{M}$ and $\mathscr{N}$ can be explicitly calculated and used in the controller design [28]. Moreover, according to Definition 2.3, we can obtain that ϖ_{1i} is known function. Hence, $\varpi_{1i}(k+1)$ can be applied to the design of discrete-time tracking control scheme.

7.4 NN-Based DTFO Control

The approximated DT dynamics of the UAV system (7.1) is given by the Euler approximation [29] in this section, and the NN-based DTDO is designed to compensate for the adverse effects of disturbances. The detailed design procedures of the NN-based adaptive DTFO controller are presented based on the backstepping technology and the designed DTDO.

7.4.1 DTDO Design Based on NN

In this subsection, a DTDO based on the NN will be designed to compensate for the negative effects of bounded disturbances. To design the DTDO, we first give the DT system model of the transformed system dynamics (7.12) by the Euler approximation [29] as follows:

$$\begin{cases} s_1(k+1) = s_1(k) + \mathscr{N}' + \mathscr{M}'[e_1(k+1) - e_1(k)] \\ x_2(k+1) = \bar{F}_2(k) + \Delta\bar{F}(k) + \bar{G}_2(k)u(k) + \bar{d}_2(k) \end{cases} \tag{7.13}$$

where ΔT is the sampling period, $\mathscr{N}' = [\mathscr{N}_1', \mathscr{N}_2', \mathscr{N}_3']^{\mathrm{T}}$, $\mathscr{M}' = \mathrm{diag}[\mathscr{M}_1', \mathscr{M}_2', \mathscr{M}_3']$, $\mathscr{N}_i' = -\frac{\partial \Psi_{1i}^{-1}(k)}{\partial[e_{1i}(k)/\varpi_{1i}(k)]} \frac{e_{1i}(k)[\varpi_{1i}(k+1)-\varpi_{1i}(k)]}{\varpi_{1i}^2(k)}$, $\bar{d}_2 = \Delta T d_2(k)$, $\mathscr{M}_i' = \frac{\partial \Psi_{1i}^{-1}(k)}{\partial[e_{1i}(k)/\varpi_{1i}(k)]} \frac{1}{\varpi_{1i}(k)}$, $\Psi_{1i}^{-1}(k) = \Psi_{1i}^{-1}[e_{1i}(k)/\varpi_{1i}(k)]$, $\bar{F}_2(k) = \Delta T F_2(x(k)) + x_2(k)$, $\Delta \bar{F}_2(k) = \Delta T \Delta F_2(x(k))$, $\bar{G}_2(k) = \Delta T G_2(x(k))$, $\bar{d}_2(k) = [\bar{d}_{21}(k), \bar{d}_{22}(k), \bar{d}_{23}(k)]^{\mathrm{T}}$ is the unknown disturbance vector, $i = 1, 2, 3$, and $e(k+1)$ can be described as

$$e_1(k+1) = \bar{F}_1(k) + \bar{G}_1(k)x_2(k) + \Delta \bar{F}_1(k) + \bar{d}_1(k) - x_{1d}(k+1) \tag{7.14}$$

where $\bar{F}_1(k) = \Delta T F_1(x_1(k)) + x_1(k)$, $\bar{G}_1(k) = \Delta T G_1(x_1(k))$, $\Delta \bar{F}_1(k) = \Delta T \Delta F_1(k)$, and $\bar{d}_1$ is given in Chap. 3. Furthermore, we define that $\bar{d}_1(k) = [\bar{d}_{11}, \bar{d}_{12}, \bar{d}_{13}]^{\mathrm{T}}$.

According to (7.14), (7.13) can be written as

$$\begin{cases} s_1(k+1) = s_1(k) + \mathscr{N}' + \mathscr{M}'[\bar{F}_1(k) + \bar{G}_1(k)x_2(k) \\ \qquad\qquad - x_{1d}(k+1) - e_1(k) + \Delta \bar{F}_1(k) + \bar{d}_1(k)] \\ x_2(k+1) = \bar{F}_2(k) + \Delta \bar{F}_2(k) + \bar{G}_2(k)u(k) + \bar{d}_2(k) \end{cases} \tag{7.15}$$

From (7.14) and (7.15), one has

$$\bar{x}(k+1) = \mathscr{F}(k) + \Delta\mathscr{F}(k) + \bar{u}(k) + D(k) \tag{7.16}$$

where $\bar{x}(k+1) = [x_1^{\mathrm{T}}(k+1), x_2^{\mathrm{T}}(k+1)]^{\mathrm{T}}$ is the state vector, $\mathscr{F}(k) = [\bar{F}_1^{\mathrm{T}}(k), \bar{F}_2^{\mathrm{T}}(k)]^{T}$ is the known nonlinear function vector, $D(k) = [\bar{d}_1^{\mathrm{T}}(k), \bar{d}_2^{\mathrm{T}}(k)]^{\mathrm{T}}$ is the unknown disturbance vector, $\Delta\mathscr{F}(k) = [\Delta F_1^{\mathrm{T}}(x_1(k)), \Delta F_2^{\mathrm{T}}(x(k))]^{\mathrm{T}}$ is the unknown nonlinear function vector and $\bar{u}(k) = [(\bar{G}_1(k)x_2(k))^{\mathrm{T}}, (\bar{G}_2(k)u(k))^{\mathrm{T}}]^{\mathrm{T}}$. Moreover, we define $\bar{x}(k+1) = [\bar{x}_1(k+1), \bar{x}_2(k+1), \ldots, \bar{x}_6(k+1)]^{\mathrm{T}}$, $\mathscr{F}(k) = [\mathscr{F}_1(k), \mathscr{F}_2(k), \ldots, \mathscr{F}_6(k)]^{\mathrm{T}}$, $\Delta\mathscr{F}(k) = [\Delta\mathscr{F}_1(k), \Delta\mathscr{F}_2(k), \ldots, \Delta\mathscr{F}_6(k)]^{\mathrm{T}}$, $\bar{u}(k) = [\bar{u}_1(k), \bar{u}_2(k), \ldots, \bar{u}_6(k)]^{\mathrm{T}}$, and $D(k) = [D_1(k), D_2(k), \ldots, D_6(k)]^{\mathrm{T}}$.

For the i_0th element of $\bar{x}(k+1)$ in (7.16), we have

$$\bar{x}_{i_0}(k+1) = \mathscr{F}_{i_0}(k) + \Delta\mathscr{F}_{i_0}(k) + \bar{u}_{i_0}(k) + D_{i_0}(k) \tag{7.17}$$

where $i_0 = 1, 2, \ldots, 6$.

On the basis of Lemma 2.2, the NN is used to approximate $\frac{1}{\bar{\rho}_{i_0}}\Delta\mathscr{F}_{i_0}(k)$ with $\bar{\rho}_{i_0} > 0$, we obtain

$$\bar{x}_{i_0}(k+1) = \mathscr{F}_{i_0}(k) + \bar{\rho}_{i_0}\theta_{i_0}^*(k)\phi_{i_0}(Z_{i_0}(k)) + \bar{\rho}_{i_0}\varepsilon_{i_0}(k) + \bar{u}_{i_0}(k) + D_{i_0}(k) \tag{7.18}$$

where $\theta_{i_0}^*(k)$ is the optimal weight vector of the NN, $\varepsilon_{i_0}(k)$ is the minimum approximation error, $\phi_{i_0}(\cdot)$ is the basis function vector, and $Z_{i_0}(k) = x(k)$. To facilitate the design of DTDO, define $\phi_{i_0}(Z_{i_0}(k)) = \phi_{i_0}(k)$.

To design the NN-based DTDO, an auxiliary variable $S_{i_0}(k)$ is employed as follows:

$$S_{i_0}(k) = \sigma_{i_0} D_{i_0}(k) - \sigma_{i_0} \bar{x}_{i_0}(k) \tag{7.19}$$

where σ_{i_0} is a design positive constant.

From (7.18) and (7.19), $S_{i_0}(k+1)$ is described as

$$\begin{aligned} S_{i_0}(k+1) = \sigma_{i_0}[D_{i_0}(k+1) - \bar{\rho}_{i_0}\theta^*_{i_0}(k)\phi_{i_0}(k) - \bar{u}_{i_0}(k)] \\ -\sigma_{i_0}[\mathscr{F}_{i_0}(k) + \bar{\rho}_{i_0}\varepsilon_{i_0}(k) + D_{i_0}(k)] \end{aligned} \tag{7.20}$$

Furthermore, the estimation of auxiliary variable $S_{i_0}(k)$ is designed by

$$\hat{S}_{i_0}(k+1) = -\sigma_{i_0}[\mathscr{F}_{i_0}(k) + \bar{\rho}_{i_0}\hat{\theta}_{i_0}(k)\phi_{i_0}(k) + \bar{u}_{i_0}(k)] \tag{7.21}$$

where $\hat{S}_{i_0}(k)$ is the estimation of $S_{i_0}(k)$, $\hat{\theta}_{i_0}(k)$ is the estimation of $\theta^*_{i_0}(k)$, and the adaptive law of $\hat{\theta}_{i_0}(k)$ will be designed in the following subsection.

According to (7.20) and (7.21), one has

$$\tilde{S}_{i_0}(k+1) = \sigma_{i_0}\Delta D_{i_0} + \sigma_{i_0}\bar{\rho}_{i_0}[\tilde{\theta}_{i_0}(k)\phi_{i_0}(k) - \varepsilon_{i_0}(k)] \tag{7.22}$$

where $\tilde{S}_{i_0}(k) = S_{i_0}(k) - \hat{S}_{i_0}(k)$, $\tilde{\theta}_{i_0}(k) = \hat{\theta}_{i_0}(k) - \theta^*_{i_0}(k)$ and $\Delta D_{i_0} = D_{i_0}(k+1) - D_{i_0}(k)$.

To estimate the disturbance $D_{i_0}(k)$, the disturbance estimation $\hat{D}_{i_0}(k)$ is designed by

$$\hat{D}_{i_0}(k) = \frac{1}{\sigma_{i_0}}\hat{S}_{i_0}(k) + \bar{x}_{i_0}(k) \tag{7.23}$$

Define $\tilde{D}_{i_0}(k) = D_{i_0}(k) - \hat{D}_{i_0}(k)$. Considering (7.19) and (7.23), it yields

$$\tilde{S}_{i_0}(k) = \sigma_{i_0}\tilde{D}_{i_0}(k) \tag{7.24}$$

Based on Assumption 7.2, one can obtain

$$\left|\Delta D_{i_0}(k)\right| \le \bar{D}_{i_0} \tag{7.25}$$

where $\bar{D}_{i_0}$ is a positive constant.

According to (7.22), (7.24) and (7.25), one has

$$\sigma_{i_0}\tilde{D}^2_{i_0}(k+1) \le 3\sigma_{i_0}\bar{D}^2_{i_0} + 3\bar{\phi}^2_{i_0}\sigma_{i_0}\bar{\rho}^2_{i_0}\left\|\tilde{\theta}_{i_0}(k)\right\|^2 + 3\sigma_{i_0}\bar{\rho}^2_{i_0}\bar{\varepsilon}^2_{i_0} \tag{7.26}$$

where $\left\|\phi_{i_0}(k)\right\| \le \bar{\phi}_{i_0}$ $\left|\varepsilon_{i_0}(k)\right| \le \bar{\varepsilon}_{i_0}$, $\bar{\phi}_{i_0}$ and $\bar{\varepsilon}_{i_0}$ are positive constants.

7.4.2 DTFO Controller Design and Stability Analysis

In this subsection, an adaptive DTFO tracking control scheme based on the NN and the designed DTDO is developed to guarantee that the tracking errors can satisfy the prescribed performance.

Remark 7.3 To guarantee that the tracking error $e_1(t)$ satisfies the prescribed performance, we must first ensure that $s_1(t)$ is bounded. In this chapter, a discrete-time controller with prescribed performance is designed to achieve the tracking control, and the transformed system signal $s_1(t)$ is discretized using a forward difference approach. On the basis of the literature [30], we know that the continuous-time signal is in the stability domain once the corresponding discrete-time signal is stable for the case of forward difference. Thence, we can guarantee the boundness of the continuous-time signal $s_1(t)$ by demonstrating that the discrete-time signal $s_1(k)$ is bounded and stable.

On the basis of the discrete-time system (7.15), the detailed design steps of the NN-based adaptive DTFO tracking controller is designed by backstepping technology as follows:

Step 1: Defining the error variable as $e_2(k) = x_2(k) - \text{sat}(x_{vd}(k))$, where $x_{vd}(k)$ is the control signal to be designed. From (7.15), one has

$$\begin{aligned} s_1(k+1) = s_1(k) + \mathscr{N}' + \mathscr{M}'[\bar{F}_1(k) + \bar{G}_1(k)(e_2(k) \\ +\text{sat}(x_{vd}(k))) - x_{1d}(k+1) - e_1(k) + \Delta\bar{F}_1(k) + \bar{d}_1] \end{aligned} \tag{7.27}$$

For the saturation function $\text{sat}(x_{vd}(k))$, the handling method in Chap. 4 is used. Then, (7.27) can be written as

$$\begin{aligned} s_1(k+1) = s_1(k) + \mathscr{N}' + \mathscr{M}'[\bar{F}_1(k) + \bar{G}_1(k)(e_2(k) + \mathscr{Y}(k) \\ +\lambda\!\!\!^{-}(k)x_{vd}(k)) - x_{1d}(k+1) - e_1(k) + \Delta\bar{F}_1(k) + \bar{d}_1] \end{aligned} \tag{7.28}$$

where $\mathscr{Y}(k)$ is the approximation error, $\lambda\!\!\!^{-}(k)$ is the unknown and bounded diagonal matrix with $\|\lambda\!\!\!^{-}(k)\| \le \bar{\ell}_0$, and $\bar{\ell}_0$ is a positive constant.

To handle the disturbance $\bar{d}_1(k)$ in (7.27), the designed DTDO is employed based on (7.21) and (7.23) as follows:

$$\begin{cases} \hat{S}_{1i}(k+1) = -\sigma_{1i}[\bar{F}_{1i}(k) + \bar{\rho}_{1i}\hat{\theta}_{1i}(k)\phi_{1i}(k) + \bar{u}_{1i}(k)] \\ \sigma_{1i}\hat{\bar{d}}_{1i}(k) = \hat{S}_{1i}(k) + \sigma_{1i}x_{1i}(k) \end{cases} \tag{7.29}$$

where $i = 1, 2, 3$, $x_{1i}(k)$ is the ith element of $x_2(k)$, $\hat{\bar{d}}_{1i}(k)$ is the ith output of DTDO with $\hat{\bar{d}}_1 = [\hat{\bar{d}}_{11}(k), \hat{\bar{d}}_{12}(k), \hat{\bar{d}}_{13}(k)]^{\mathrm{T}}$, $\hat{S}_{1i}(k)$ is the state variable of DTDO, the design parameters $\bar{\rho}_{1i}$ and σ_{1i} satisfy that $\bar{\rho}_{1i} > 0$ and $\sigma_{1i} > 0$. Moreover, the disturbance estimation error is defined as $\tilde{\bar{d}}_{1i}(k) = \bar{d}_{1i}(k) - \hat{\bar{d}}_{1i}(k)$ with $\tilde{\bar{d}}_1 = [\tilde{\bar{d}}_{11}(k), \tilde{\bar{d}}_{12}(k), \tilde{\bar{d}}_{13}(k)]^{\mathrm{T}}$. $\bar{u}_{1i}(k)$ is the ith element of $\bar{u}_1(k)$ with $\bar{u}_1(k) = \bar{G}_2(k)u(k)$,

$\hat{\theta}_{1i}(k)$ is the estimation of the NN optimal weight vector $\theta^*_{1i}(k)$, $\phi_{1i}(k) = \phi_{1i}(Z_1(k))$is the basis function vector, and $Z_1(k) = x_1^{\mathrm{T}}(k)$. According to Assumption 7.2, one can obtain that $|\bar{d}_{1i}(k+1) - \bar{d}_{1i}(k)| \leq \bar{D}_{1i}$, and $\bar{D}_{1i}$ is a positive constant.

According to Lemma 2.2, by using the NN to approximate $(1/\bar{\rho}_{1i})\Delta\bar{F}_{1i}(k)(i = 1, 2, 3)$ with $\Delta\bar{F}_{1i}(k)$ being the ith variable of $\Delta\bar{F}_1(k)$, it yields

$$\begin{aligned} s_1(k+1) = s_1(k) + \mathscr{N}' + \mathscr{M}'[&\bar{F}_1(k) + \bar{G}_1(k)(e_2(k) + \mathscr{Y}(k) \\ &+ \lambda\!\!\!\!\bar{}\,(k)x_{vd}(k)) - x_{1d}(k+1) - e_1(k) + \Phi_1^*(k) + \mathrm{E}_1(k) + \bar{d}_1] \end{aligned} \tag{7.30}$$

where $\Phi_1^*(k) = [\bar{\rho}_{11}\theta_{11}^{*\mathrm{T}}(k)\phi_{11}(k), \bar{\rho}_{12}\theta_{12}^{*\mathrm{T}}(k)\phi_{12}(k), \bar{\rho}_{13}\theta_{13}^{*\mathrm{T}}(k)\phi_{13}(k)]^{\mathrm{T}}$, and the approximation error vector is $\mathrm{E}_1(k) = [\bar{\rho}_{11}\varepsilon_{11}(k), \bar{\rho}_{12}\varepsilon_{12}(k), \bar{\rho}_{13}\varepsilon_{13}(k)]^{\mathrm{T}}$ with $\varepsilon_{1i}(k)$ is the minimum approximation error.

On the basis of the designed method of control signal $x_{vd}(k)$ in Chap. 4, Lemma 2.2 and (5.20), the NN is used to approximate the uncertainty $(1/\bar{\rho}_{0i})\Delta\bar{\mathscr{F}}_{1i}(k)(i = 1, 2, 3)$ with $\Delta\bar{\mathscr{F}}_{1i}(k)$ being the ith variable of $\Delta\bar{\mathscr{F}}_1(k) = -(\lambda\!\!\!\bar{}\,(k)\bar{G}_1(k))^{-1}[\mathscr{M}'^{-1}(s_1(k) + \mathscr{N}') + \bar{F}_1(k) + \hat{\Phi}_1(k) + \hat{\hat{d}}_1(k) + (-\lambda_0 - 1)e_1(k) - x_{1d}(k+1)] + \bar{G}_1^{-1}(k)[\mathscr{M}'^{-1}(s_1(k) + \mathscr{N}') + \bar{F}_1(k) + \hat{\Phi}_1(k) + \hat{\hat{d}}_1(k) + (-\lambda_0 - 1)e_1(k) - x_{1d}(k+1)]$, the control signal $x_{vd}(k)$ can be written as

$$\begin{aligned} x_{vd}(k) = &-\bar{G}_1^{-1}(k)[\mathscr{M}'^{-1}(s_1(k) + \mathscr{N}') + \bar{F}_1(k) + \hat{\Phi}_1(k) + \hat{\hat{d}}_1(k)] \\ &-\bar{G}_1^{-1}(k)[(-\lambda_0 - 1)e_1(k) - x_{1d}(k+1)] + \Phi_0^*(k) + \mathrm{E}_0(k) \end{aligned} \tag{7.31}$$

where $\Phi_0^*(k)=[\bar{\rho}_{01}\theta_{01}^{*\mathrm{T}}(k)\phi_{01}(Z_0(k)), \bar{\rho}_{02}\theta_{02}^{*\mathrm{T}}(k)\phi_{02}(Z_0(k)), \bar{\rho}_{03}\theta_{03}^{*\mathrm{T}}(k)\phi_{03}(Z_0(k))]^{\mathrm{T}}$, and the approximation error vector is $\mathrm{E}_0(k) = [\bar{\rho}_{01}\varepsilon_{01}(k), \bar{\rho}_{02}\varepsilon_{02}(k), \bar{\rho}_{03}\varepsilon_{03}(k)]^{\mathrm{T}}$ with $\varepsilon_{0i}(k)$ is the minimum approximation error, λ_0 is a design positive constant, $\hat{\Phi}_1(k) = [\bar{\rho}_{11}\hat{\theta}_{11}^{\mathrm{T}}(k)\phi_{11}(k), \bar{\rho}_{12}\hat{\theta}_{12}^{\mathrm{T}}(k)\phi_{12}(k), \bar{\rho}_{13}\hat{\theta}_{13}^{\mathrm{T}}(k)\phi_{13}(k)]^{\mathrm{T}}$, $\hat{\Phi}_1(k)$ is the estimation of $\Phi_1(k)$, $Z_0(k) = [x_1^{\mathrm{T}}(k), s_1{}^{\mathrm{T}}(k), e_1{}^{\mathrm{T}}(k), \hat{\Phi}_1^{\mathrm{T}}(k), \hat{\hat{d}}_1^{\mathrm{T}}(k)]^{\mathrm{T}}$, $\theta_{0i}^*(k)$ is the optimal weight vector of the NN, $\varepsilon_{0i}(k)$ is the minimum approximation error, $\phi_{0i}(\cdot)$ is the basis function vector. To facilitate the design of control scheme, define $\phi_{0i}(Z_0(k)) = \phi_{0i}(k)$.

Then, the control signal $x_{vd}(k)$ is designed by

$$\begin{aligned} x_{vd}(k) = &-\bar{G}_1^{-1}(k)[\mathscr{M}'^{-1}(s_1(k) + \mathscr{N}') + \bar{F}_1(k) + \hat{\Phi}_1(k) + \hat{\hat{d}}_1(k)] \\ &-\bar{G}_1^{-1}(k)[(-\lambda_0 - 1)e_1(k) - x_{1d}(k+1)] + \hat{\Phi}_0(k) \end{aligned} \tag{7.32}$$

where $\hat{\Phi}_0(k) = [\bar{\rho}_{01}\hat{\theta}_{01}^{\mathrm{T}}(k)\phi_{01}(k), \bar{\rho}_{02}\hat{\theta}_{02}^{\mathrm{T}}(k)\phi_{02}(k), \bar{\rho}_{03}\hat{\theta}_{03}^{\mathrm{T}}(k)\phi_{03}(k)]^{\mathrm{T}}$.

Moreover, the adaptive law $\hat{\theta}_{1i}(k)$ is chosen as

$$\hat{\theta}_{1i}(k+1) = \kappa_{1i}\phi_{1i}(k)s_{1i}(k) - (\omega_{1i} - 1)\hat{\theta}_{1i}(k) \tag{7.33}$$

where $\kappa_{1i} > 0$ and $\omega_{1i} > 0$ are designed constants.

Substituting (7.32) into (7.28), one has

$$s_1(k+1) = \mathscr{M}'\bar{G}_1(k)e_2(k) + \mathscr{M}'\lambda_0 e_1(k) - \mathscr{M}'\tilde{\Phi}_1(k) - M'\bar{G}_1(k)\lambda(k)\mathrm{E}_0(k)$$
$$+\mathscr{M}'\bar{G}_1(k)\mathscr{Y}(k) + \mathscr{M}'\bar{\bar{d}}_1(k) + M'\bar{G}_1(k)\lambda(k)\tilde{\Phi}_0(k) \quad (7.34)$$

where $\tilde{\Phi}_1(k) = [\bar{\rho}_{11}\tilde{\theta}_{11}^{\mathrm{T}}(k)\phi_{11}(k), \bar{\rho}_{12}\tilde{\theta}_{12}^{\mathrm{T}}(k)\phi_{12}(k), \bar{\rho}_{13}\tilde{\theta}_{13}^{\mathrm{T}}(k)\phi_{13}(k)]^{\mathrm{T}}$, the approximation error vectors are $\tilde{\Phi}_1(k) = \hat{\Phi}_1(k) - \Phi_1^*(k)$ and $\tilde{\Phi}_0(k) = \hat{\Phi}_0(k) - \Phi_0^*(k)$, and $\tilde{\Phi}_0(k) = [\bar{\rho}_{01}\tilde{\theta}_{01}^{\mathrm{T}}(k)\phi_{01}(k), \bar{\rho}_{02}\tilde{\theta}_{02}^{\mathrm{T}}(k)\phi_{02}(k), \bar{\rho}_{03}\tilde{\theta}_{03}^{\mathrm{T}}(k)\phi_{03}(k)]^{\mathrm{T}}$.

According to (7.34), the ith variable of $s_1(k+1)$ can be written as

$$s_{1i}(k+1) = \mathscr{M}_i'E_{1i}(k) + \mathscr{M}_i'\lambda_0 e_{1i}(k) - \mathscr{M}_i'\tilde{\Phi}_{1i}(k)$$
$$+\mathscr{M}_i'\bar{\bar{d}}_{1i}(k) + \mathscr{M}_i'\Delta\Phi_{0i}(k) + \mathscr{M}_i'\mathscr{Y}_i(k) \quad (7.35)$$

where $E_{1i}(k)$ is the ith element of $\bar{G}_1(k)e_2(k)$, $\bar{\mathscr{Y}}_i(k)$ is the ith element of $\bar{G}_1(k)\mathscr{Y}(k) - \bar{G}_1(k)\lambda(k)\mathrm{E}_0(k)$, $s_{1i}(k+1)$ is the ith element of $s_1(k+1)$, $\tilde{\Phi}_{1i}(k)$ is the ith element of $\tilde{\Phi}_1(k)$, $e_{1i}(k)$ the ith element of $e_1(k)$, $\Delta\Phi_{0i}(k)$ is the ith element of $\bar{G}_1(k)\lambda(k)\tilde{\Phi}_0(k)$, and $i = 1, 2, 3$.

From (7.35), one has

$$s_{1i}(k+1) \le \frac{6}{2\rho_{1i}}\mathscr{M}_i'^2 + \frac{1}{2}\rho_{1i}E_{1i}^2(k) + \frac{1}{2}\rho_{1i}\tilde{\Phi}_{1i}^2(k) + \frac{1}{2}\rho_{1i}\bar{\bar{d}}_{1i}^2(k)$$
$$+\frac{1}{2}\rho_{1i}\Delta\Phi_{0i}^2(k) + \frac{1}{2}\rho_{1i}\lambda_0^2 e_{1i}^2(k) + \frac{1}{2}\rho_{1i}\bar{\mathscr{Y}}_i^2(k) \quad (7.36)$$

where ρ_{1i} is a positive constant.

According to (7.33), one can obtain

$$\tilde{\theta}_{1i}(k+1) = \tilde{\theta}_{1i}(k) + \kappa_{1i}\phi_{1i}(k)s_{1i}(k) - \omega_{1i}\hat{\theta}_{1i}(k) \quad (7.37)$$

where $\tilde{\theta}_{1i}(k+1) = \hat{\theta}_{1i}(k+1) - \theta_{1i}^*(k)$ and $\tilde{\theta}_{1i}(k) = \hat{\theta}_{1i}(k) - \theta_{1i}^*(k)$

Based on (7.37), it yields

$$\tilde{\theta}_{1i}^T(k+1)\tilde{\theta}_{1i}(k+1) - \tilde{\theta}_{1i}^T(k)\tilde{\theta}_{1i}(k) = \omega_{1i}^2\left\|\hat{\theta}_{1i}(k)\right\|^2 + 2\kappa_{1i}\tilde{\theta}_{1i}^T(k)\phi_{1i}(k)s_{1i}(k)$$
$$+\kappa_{1i}^2\|\phi_{1i}(k)\|^2 s_{1i}^2(k) - 2\kappa_{1i}\omega_{1i}\phi_{1i}^T(k)\hat{\theta}_{1i}(k)s_{1i}(k) - 2\omega_{1i}\tilde{\theta}_{1i}^T(k)\hat{\theta}_{1i}(k) \quad (7.38)$$

Furthermore, we have

$$-2\omega_{1i}^2\tilde{\theta}_{1i}^T(k)\hat{\theta}_{1i}(k) = -\omega_{1i}^2\left\|\tilde{\theta}_{1i}(k)\right\|^2 - \omega_{1i}^2\left\|\hat{\theta}_{1i}(k)\right\|^2 + \omega_{1i}^2\left\|\theta_{1i}^*(k)\right\|^2 \quad (7.39)$$

Furthermore, the adaptive law $\hat{\theta}_{0i}(k)$ is chosen as

$$\hat{\theta}_{0i}(k+1) = \kappa_{0i}\phi_{0i}(k)s_{1i}(k) - (\omega_{0i} - 1)\hat{\theta}_{0i}(k) \quad (7.40)$$

where $\kappa_{0i} > 0$ and $\omega_{0i} > 0$ are designed constants.

From (7.40), one can obtain

$$\tilde{\theta}_{0i}(k+1) = \tilde{\theta}_{0i}(k) + \kappa_{0i}\phi_{0i}(k)s_{1i}(k) - \omega_{0i}\hat{\theta}_{0i}(k) \tag{7.41}$$

where $\tilde{\theta}_{0i}(k+1) = \hat{\theta}_{0i}(k+1) - \theta^*_{0i}(k)$ and $\tilde{\theta}_{0i}(k) = \hat{\theta}_{0i}(k) - \theta^*_{0i}(k)$

According to (7.41), it yields

$$\begin{aligned}\tilde{\theta}^T_{0i}(k+1)\tilde{\theta}_{0i}(k+1) - \tilde{\theta}^T_{0i}(k)\tilde{\theta}_{0i}(k) &= \omega^2_{0i}\left\|\hat{\theta}_{0i}(k)\right\|^2 + 2\kappa_{0i}\tilde{\theta}^T_{0i}(k)\phi_{0i}(k)s_{1i}(k) \\ &+\kappa^2_{0i}\|\phi_{0i}(k)\|^2 s^2_{1i}(k) - 2\kappa_{0i}\omega_{0i}\phi^T_{0i}(k)\hat{\theta}_{0i}(k)s_{1i}(k) - 2\omega_{0i}\tilde{\theta}^T_{0i}(k)\hat{\theta}_{0i}(k)\end{aligned} \tag{7.42}$$

Moreover, one has

$$-2\omega^2_{0i}\tilde{\theta}^T_{0i}(k)\hat{\theta}_{0i}(k) = -\omega^2_{0i}\left\|\tilde{\theta}_{0i}(k)\right\|^2 - \omega^2_{0i}\left\|\hat{\theta}_{0i}(k)\right\|^2 + \omega^2_{0i}\left\|\theta^*_{0i}(k)\right\|^2 \tag{7.43}$$

The following function candidate is given by:

$$\begin{aligned}V_1(k) &= \sum_{i=1}^{3}|s_{1i}(k)| + \sum_{i=1}^{3}\omega_{1i}\tilde{\theta}^T_{1i}(k)\tilde{\theta}_{1i}(k) \\ &+\sum_{i=1}^{3}\sigma_{1i}\tilde{\bar{d}}^2_{1i}(k) + \sum_{i=1}^{3}\omega_{0i}\tilde{\theta}^T_{0i}(k)\tilde{\theta}_{0i}(k)\end{aligned} \tag{7.44}$$

From (7.44), the first difference is calculated as

$$\begin{aligned}\Delta V_1(k) = &\sum_{i=1}^{3}|s_{1i}(k+1)| + \sum_{i=1}^{3}\omega_{1i}\tilde{\theta}^T_{1i}(k+1)\tilde{\theta}_{1i}(k+1) + \sum_{i=1}^{3}\sigma_{1i}\tilde{\bar{d}}^2_{1i}(k+1) \\ &-\sum_{i=1}^{3}|s_{1i}(k)| - \sum_{i=1}^{3}\omega_{1i}\tilde{\theta}^T_{1i}(k)\tilde{\theta}_{1i}(k) - \sum_{i=1}^{3}\sigma_{1i}\tilde{\bar{d}}^2_{1i}(k) \\ &+\sum_{i=1}^{3}\omega_{0i}\tilde{\theta}^T_{0i}(k+1)\tilde{\theta}_{0i}(k+1) - \sum_{i=1}^{3}\omega_{0i}\tilde{\theta}^T_{0i}(k)\tilde{\theta}_{0i}(k)\end{aligned} \tag{7.45}$$

Combining (7.38) and (7.39), one has

$$\begin{aligned}\omega_{1i}(\tilde{\theta}^T_{1i}(k+1)\tilde{\theta}_{1i}(k+1) - \tilde{\theta}^T_{1i}(k)\tilde{\theta}_{1i}(k)) &\le -(\omega^2_{1i} - \kappa^2_{1i})\left\|\tilde{\theta}_{1i}(k)\right\|^2 \\ +(2\omega^2_{1i}\bar{\phi}^2_{1i} + \omega_{1i}\kappa^2_{1i}\bar{\phi}^2_{1i})s^2_{1i}(k) &- (\omega^2_{1i} - \omega^3_{1i} - \kappa^2_{1i}\omega^2_{1i})\left\|\hat{\theta}_{1i}(k)\right\|^2 + \omega^2_{1i}\bar{\theta}^{*2}_{1i}\end{aligned} \tag{7.46}$$

where $\left\|\theta^*_{1i}(k)\right\| \le \bar{\theta}^*_{1i}$, $\|\phi_{1i}(k)\| \le \bar{\phi}_{1i}$, $\bar{\theta}^*_{1i}$ and $\bar{\phi}_{1i}$ are positive constant.

From (7.42) and (7.43), one can obtain

$$\omega_{0i}(\tilde{\theta}_{0i}^T(k+1)\tilde{\theta}_{0i}(k+1) - \tilde{\theta}_{0i}^T(k)\tilde{\theta}_{0i}(k)) \leq -(\omega_{0i}^2 - \kappa_{0i}^2)\left\|\tilde{\theta}_{0i}(k)\right\|^2$$
$$+(2\omega_{0i}^2\bar{\phi}_{0i}^2 + \omega_{0i}\kappa_{0i}^2\bar{\phi}_{0i}^2)s_{1i}^2(k) - (\omega_{0i}^2 - \omega_{0i}^3 - \kappa_{0i}^2\omega_{0i}^2)\left\|\hat{\theta}_{0i}(k)\right\|^2 + \omega_{0i}^2\bar{\theta}_{0i}^{*2} \tag{7.47}$$

where $\left\|\theta_{0i}^*(k)\right\| \leq \bar{\theta}_{0i}^*$, $\|\phi_{0i}(k)\| \leq \bar{\phi}_{0i}$, $\bar{\theta}_{0i}^*$ and $\bar{\phi}_{0i}$ are positive constant.

According to (7.29), (7.36) and (7.45), one has

$$\Delta V_{1i}(k) \leq -(\omega_{1i}^2 - \kappa_{1i}^2 - 0.5\bar{\phi}_{1i}^2\rho_{1i}\bar{\rho}_{1i}^2)\left\|\tilde{\theta}_{1i}(k)\right\|^2 + \omega_{1i}^2\bar{\theta}_{1i}^{*2} - (\sigma_{1i} - 0.5\rho_{1i})\tilde{\bar{d}}_{1i}^2(k)$$
$$-(\omega_{1i}^2 - \kappa_{1i}^2\omega_{1i}^2 - 3\bar{\phi}_{1i}^2\sigma_{1i}\bar{\rho}_{1i}^2 - \omega_{1i}^3)\left\|\hat{\theta}_{1i}(k)\right\|^2 + 3\sigma_{1i}\bar{D}_{1i}^2$$
$$-|s_{1i}(k)| + 3\sigma_{1i}\bar{\rho}_{1i}^2\bar{\varepsilon}_{1i}^2 + \mathscr{Q}_i(k) - (\omega_{0i}^2 - \kappa_{0i}^2 - 0.5\rho_{1i}\bar{g}_1^2\ell_0^2\bar{\rho}_{0i}^2)\left\|\tilde{\theta}_{0i}(k)\right\|^2$$
$$-(\omega_{0i}^2 - \omega_{0i}^3 - \kappa_{0i}^2\omega_{0i}^2)\left\|\hat{\theta}_{0i}(k)\right\|^2 + \omega_{0i}^2\bar{\theta}_{0i}^{*2} \tag{7.48}$$

where $\Delta V_{1i}(k) = |s_{1i}(k+1)| + \omega_{1i}\tilde{\theta}_{1i}^T(k+1)\tilde{\theta}_{1i}(k+1) + \sigma_{1i}\tilde{\bar{d}}_{1i}^2(k+1) - |s_{1i}(k)| - \omega_{1i}\tilde{\theta}_{1i}^T(k)\tilde{\theta}_{1i}(k) - \sigma_{1i}\tilde{\bar{d}}_{1i}^2(k) + \omega_{0i}\tilde{\theta}_{0i}^T(k+1)\tilde{\theta}_{0i}(k+1) - \omega_{0i}\tilde{\theta}_{0i}^T(k)\tilde{\theta}_{0i}(k)$, $|\varepsilon_{1i}(k)| \leq \bar{\varepsilon}_{1i}$, $\mathscr{Q}_i(k) = 0.5\rho_{1i}\lambda_0^2 e_{1i}^2(k) + 0.5\rho_{1i}E_{1i}^2(k) + 0.5\rho_{1i}\mathscr{Y}_i^2(k) + \frac{6}{\rho_{1i}}\mathscr{M}_i^2 + 0.5\rho_{1i}\Delta\mathscr{F}_{1i}^2(k) + (2\omega_{1i}^2\bar{\phi}_{1i}^2 + \omega_{1i}\kappa_{1i}^2\bar{\phi}_{1i}^2)s_{1i}^2(k) + \omega_{0i}\kappa_{0i}^2\bar{\phi}_{0i}^2)s_{1i}^2(k)$.

Step 2 Considering (7.15) and $e_2(k) = x_2(k) - \mathrm{sat}(x_{vd}(k))$, one has

$$\begin{aligned} e_2(k+1) &= x_2(k+1) - \mathrm{sat}(x_{vd}(k+1)) \\ &= \bar{F}_2(k) + \Delta\bar{F}_2(k) + \bar{G}_2(k)u(k) + \bar{d}_2(k) - v_d(k+1) \end{aligned} \tag{7.49}$$

In order to deal with the disturbance $\bar{d}_2(k)$ in (7.49), the nonlinear DTDO is designed according to (7.21) and (7.23), which is given by

$$\begin{cases} \hat{S}_{2i}(k+1) = -\sigma_{2i}[\bar{F}_{2i}(k) + \bar{\rho}_{2i}\hat{\theta}_{2i}(k)\phi_{2i}(k) + \bar{u}_{2i}(k)] \\ \sigma_{2i}\hat{\bar{d}}_{2i}(k) = \hat{S}_{2i}(k) + \sigma_{2i}x_{2i}(k) \end{cases} \tag{7.50}$$

where $i = 1, 2, 3$, $x_{2i}(k)$ denotes the ith variable of $x_2(k)$, $\hat{\bar{d}}_{2i}(k)$ denotes the ith output of DTDO with $\hat{\bar{d}}_2(k) = [\hat{\bar{d}}_{21}(k), \hat{\bar{d}}_{22}(k), \hat{\bar{d}}_{23}(k)]^{\mathrm{T}}$, $\hat{S}_{2i}(k)$ is the state variable of DTDO. The design constants $\bar{\rho}_{2i}$ and σ_{2i} satisfy that $\bar{\rho}_{2i} > 0$ and $\sigma_{2i} > 0$. Moreover, the disturbance estimation error is defined as $\tilde{\bar{d}}_{2i}(k) = \bar{d}_{2i}(k) - \hat{\bar{d}}_{2i}(k)$ with $\tilde{\bar{d}}_2 = [\tilde{\bar{d}}_{21}(k), \tilde{\bar{d}}_{22}(k), \tilde{\bar{d}}_{23}(k)]^{\mathrm{T}}$. $\bar{u}_{2i}(k)$ denotes the ith variable of $\bar{u}_2(k)$ with $\bar{u}_2(k) = \bar{G}_2(k)u(k)$. $\hat{\theta}_{2i}(k)$ is the estimation of the NN optimal weight vector $\theta_{2i}^*(k)$. $\phi_{2i}(k) = \phi_{2i}(Z_2(k))$ is the basis function vector with $Z_2(k) = [x_1^{\mathrm{T}}(k), x_2^{\mathrm{T}}(k)]^{\mathrm{T}}$. According to Assumption 7.2, we have that $\left|\bar{d}_{2i}(k)\right| \leq \delta_{2i}$ and $\left|\nabla^{\gamma}\bar{d}_{2i}(k)\right| \leq \bar{\delta}_{2i}$, δ_{2i} and $\bar{\delta}_{2i}$ are positive constants. In addition, based on Assumption 7.2, it yields that $\left|\bar{d}_{2i}(k+1) - \bar{d}_{2i}(k)\right| \leq \bar{D}_{2i}$, and $\bar{D}_{2i}$ is a positive constant.

To forecast $v_d(k+1)$, the following discrete-time tracking differentiator based on Lemma 2.3 is employed to estimate $v_d(k+1)$:

$$\begin{cases} \bar{\mathscr{L}}_{11i}(k+1) = \bar{\mathscr{L}}_{11i}(k) + h_{01i}\bar{\mathscr{L}}_{12i}(k) \\ \bar{\mathscr{L}}_{12i}(k+1) = \bar{\mathscr{L}}_{12i}(k) - h_{01i}r_{01i}\bar{\mathscr{L}}_{13i}[\bar{\mathscr{L}}_{14i}(k), \delta_{01i}] \end{cases} \tag{7.51}$$

where $i = 1, 2, 3$, h_{01i} and r_{01i} are design constants, $\bar{\mathscr{L}}_{11i}(k)$ and $\bar{\mathscr{L}}_{12i}(k)$ are the state variables of discrete-time tracking differentiator (7.51).

According to (7.51) and Lemma 2.3, one has

$$v_d(k+1) = \bar{\mathscr{L}}_{11}(k) + h_{01}\bar{\mathscr{L}}_{12}(k) - \tilde{\mathscr{L}}(k) \tag{7.52}$$

where $\bar{\mathscr{L}}_{11}(k) = [\bar{\mathscr{L}}_{111}(k), \bar{\mathscr{L}}_{112}(k), \bar{\mathscr{L}}_{113}(k)]^{\mathrm{T}}$, $\bar{\mathscr{L}}_{12}(k) = [\bar{\mathscr{L}}_{121}(k), \bar{\mathscr{L}}_{122}(k), \bar{\mathscr{L}}_{123}(k)]^{\mathrm{T}}$, $h_{01} = \mathrm{diag}[h_{011}, h_{012}, h_{013}]$, $\mathrm{diag}[\cdot]$ is a diagonal matrix, $r_{01} = \mathrm{diag}[r_{011}, r_{012}, r_{013}]$, $\tilde{\mathscr{L}}(k) = [\tilde{\mathscr{L}}_1(k), \tilde{\mathscr{L}}_2(k), \tilde{\mathscr{L}}_3(k)]^{\mathrm{T}}$ is the estimated error, and $\tilde{\mathscr{L}}(k)$ is bounded. Therefor, according to Definition 2.1, one can obtain that $\left|\nabla^{\gamma}\tilde{\mathscr{L}}_i(k)\right| \le \bar{\mathscr{L}}_i$, where $\bar{\mathscr{L}}_i$ is a positive constant, and $\bar{\mathscr{L}} = [\bar{\mathscr{L}}_1, \bar{\mathscr{L}}_2, \bar{\mathscr{L}}_3]^{\mathrm{T}}$.

Substituting (7.52) into (7.49), one has

$$\begin{aligned} e_2(k+1) = {} & \bar{F}_2(k) + \Delta\bar{F}_2(k) + \bar{G}_2(k)u(k) + \bar{d}_2 \\ & -\bar{\mathscr{L}}_{11}(k) - h_{01}\bar{\mathscr{L}}_{12}(k) + \tilde{\mathscr{L}}(k) \end{aligned} \tag{7.53}$$

The NN-based DTFO controller is designed by

$$\begin{aligned} u(k) = {} & -\bar{G}_2^{-1}(k)(\bar{F}_2(k) - \bar{\mathscr{L}}_{11}(k) + \nabla^{-\gamma}\lambda_1\hat{\bar{d}}_2 + \nabla^{-\gamma}\hat{\Phi}_2(k) - h_{01}\bar{\mathscr{L}}_{12}(k)) \\ & +\bar{G}_2^{-1}(k)[\nabla^{-\gamma}\left(\sum_{n=1}^{k} N_n e_2(k-n)\right) - \lambda_2 e_2(k)] + \bar{G}_2^{-1}(k)\lambda_2\nabla^{-\gamma}N_0 \end{aligned} \tag{7.54}$$

where $N_0 = \sum_{j=0}^{k}(-1)^j\binom{\gamma}{j}e_2(k-j)$, γ is the fractional order, $\lambda_1 > 0$ and $\lambda_2 > 0$ are design constants, $\hat{\Phi}_2(k) = [\bar{\rho}_{21}\hat{\theta}_{21}^{\mathrm{T}}(k)\phi_{21}(k), \bar{\rho}_{22}\hat{\theta}_{22}^{\mathrm{T}}(k)\phi_{22}(k), \bar{\rho}_{23}\hat{\theta}_{23}^{\mathrm{T}}(k)\phi_{23}(k)]^{\mathrm{T}}$, $n = j-1$, $j = 2, \ldots, k+1$ and $N_n = (-1)^{n+1}\binom{\gamma}{n+1}$.

Moreover, the following adaptive law $\hat{\theta}_{2i}(k)$ is chosen as

$$\hat{\theta}_{2i}(k+1) = \kappa_{2i}\phi_{2i}(k)e_{2i}(k) - (\omega_{2i} - 1)\hat{\theta}_{2i}(k) \tag{7.55}$$

where $\kappa_{2i} > 0$ and $\omega_{2i} > 0$ are design constants.

The above design procedure of the NN-based DTFO controller with DTDO for the approximative discrete-time UAV system (7.13) in the presence of system uncertainties and external disturbances can be summarized in the following theorem:

Theorem 7.1 *For the approximative discrete-time UAV system (7.13) with system uncertainties and external disturbances, all initial conditions are in the compact set, the DTFO controller is designed as (7.54), the control signal $x_{vd}(k)$ is designed as (7.32), and the DTDO is designed as (7.21) and (7.23), if the design parameters σ_{2i}, κ_{2i}, ω_{2i}, $\bar{\rho}_{2i}$, λ_0, λ_1, λ_2 and γ are chosen appropriately, then, the proposed NN-based DTFO controller can ensure that all the signals in the closed-loop system are bounded and the tracking errors can satisfy the prescribed performance.*

Proof Combining (7.53) and (7.54), one has

$$
\begin{aligned}
e_2(k+1) = {} & \nabla^{-\gamma}\left(\sum_{n=1}^{k} N_n e_2(k-n)\right) - \nabla^{-\gamma}\lambda_1\hat{\bar{d}}_2 + \bar{d}_2 \\
& + \tilde{\mathscr{L}}(k) - \lambda_2(e_2(k) - \nabla^{-\gamma}N_0) + \Delta\bar{F}_2(k) - \nabla^{-\gamma}\hat{\Phi}_2(k) \quad (7.56)
\end{aligned}
$$

Moreover, (7.56) can be written as

$$
\begin{aligned}
\nabla^{\gamma} e_2(k+1) = {} & \sum_{n=1}^{k} N_n e_2(k-n) - \lambda_1\hat{\bar{d}}_2 + \nabla^{\gamma}\bar{d} + \nabla^{\gamma}\tilde{\mathscr{L}}(k) \\
& - \lambda_2(\nabla^{\gamma} e_2(k) - N_0) + \nabla^{\gamma}\Delta\bar{F}_2(k) - \hat{\Phi}_2(k) \quad (7.57)
\end{aligned}
$$

According to Definition 2.1, one can obtain

$$
\nabla^{\gamma} e_2(k+1) = e_2(k+1) - \gamma e_2(k) + \sum_{n=1}^{k} N_n e_2(k-n) \quad (7.58)
$$

$$
\nabla^{\gamma} e_2(k) = \sum_{j=0}^{k} (-1)^j \binom{\gamma}{j} e_2(k-j) \quad (7.59)
$$

On the basis of (7.57), (7.58) and (7.59), $e_2(k+1)$ can be described as

$$
e_2(k+1) = \gamma e_2(k) - \lambda_1\hat{\bar{d}}_2 + \nabla^{\gamma}\bar{d} + \nabla^{\gamma}\tilde{\mathscr{L}}(k) + \nabla^{\gamma}\Delta\bar{F}_2(k) - \hat{\Phi}_2(k) \quad (7.60)
$$

For the ith element of $e_2(k+1)$ in (7.60), we have

$$
e_{2i}(k+1) = \gamma e_{2i}(k) - \lambda_1\hat{\bar{d}}_{2i} + \nabla^{\gamma}\bar{d}_{2i} + \nabla^{\gamma}\tilde{\mathscr{L}}_i(k) + \nabla^{\gamma}\Delta\bar{F}_{2i}(k) - \hat{\Phi}_{2i}(k) \quad (7.61)
$$

where $e_{2i}(k+1)$ is the ith element of $e_2(k+1)$, $\Delta\bar{F}_{2i}(k)$ is the ith element of $\Delta\bar{F}_2(k)$, $\hat{\Phi}_{2i}(k)$ is the ith estimation of $\hat{\Phi}_2(k)$.

According to $\tilde{\bar{d}}_{2i} = \bar{d}_{2i} - \hat{\bar{d}}_{2i}$ and (7.61), one has

$$\begin{aligned} e_{2i}(k+1) &= \gamma e_{2i}(k) - \lambda_1 \bar{d}_{2i} + \nabla^{\gamma} \bar{d}_{2i} + \nabla^{\gamma} \tilde{\mathscr{L}}_i(k) \\ &\quad + \nabla^{\gamma} \Delta \bar{F}_{2i}(k) - \hat{\Phi}_{2i}(k) + \lambda_1 \bar{\bar{d}}_{2i} \end{aligned} \tag{7.62}$$

Based on $\left|\bar{d}_{2i}(k)\right| \leq \delta_{2i}$, $\left|\nabla^{\gamma} \bar{d}_{2i}(k)\right| \leq \bar{\delta}_{2i}$, $\left|\nabla^{\gamma} \tilde{\mathscr{L}}_i(k)\right| \leq \mathscr{L}_i$ and (7.62), one has

$$\begin{aligned} e_{2i}^2(k+1) &\leq 7\gamma^2 e_{2i}^2(k) + 7\lambda_1^2 \delta_{2i}^2 + 7\bar{\rho}_{2i}^2 \bar{\phi}_{2i}^2 \left\|\hat{\theta}_{2i}(k)\right\|^2 \\ &\quad + 7\bar{\delta}_{2i}^2 + 7\bar{\mathscr{L}}_i^2 + 7(\nabla^{\gamma} \Delta \bar{F}_{2i}(k))^2 + 7\lambda_1^2 \tilde{\bar{d}}_{2i}^2(k) \end{aligned} \tag{7.63}$$

where $\|\phi_{2i}(k)\| \leq \bar{\phi}_{2i}$ and $\bar{\phi}_{2i}$ is a positive constant.

According to (7.55), it yields

$$\tilde{\theta}_{2i}(k+1) = \tilde{\theta}_{2i}(k) + \kappa_{2i}\phi_{2i}(k)e_{2i}(k) - \omega_{2i}\hat{\theta}_{2i}(k) \tag{7.64}$$

where $\tilde{\theta}_{2i}(k+1) = \hat{\theta}_{2i}(k+1) - \theta_{2i}^*(k)$ and $\tilde{\theta}_{2i}(k) = \hat{\theta}_{2i}(k) - \theta_{2i}^*(k)$.

On the basis of (7.64), we have

$$\begin{aligned} \tilde{\theta}_{2i}^{\mathrm{T}}(k+1)\tilde{\theta}_{2i}(k+1) - \tilde{\theta}_{2i}^{\mathrm{T}}(k)\tilde{\theta}_{2i}(k) &= \omega_{2i}^2 \left\|\hat{\theta}_{2i}(k)\right\|^2 + 2\kappa_{2i}\tilde{\theta}_{2i}^{\mathrm{T}}(k)\phi_{2i}(k)e_{2i}(k) \\ + \kappa_{2i}^2 \|\phi_{2i}(k)\|^2 e_{2i}^2(k) - 2\kappa_{2i}\omega_{2i}\phi_{2i}^{\mathrm{T}}(k)\hat{\theta}_{2i}(k)e_{2i}(k) &- 2\omega_{2i}\tilde{\theta}_{2i}^{\mathrm{T}}(k)\hat{\theta}_{2i}(k) \end{aligned} \tag{7.65}$$

In addition, $-2\omega_{2i}^2\tilde{\theta}_{2i}^{\mathrm{T}}(k)\hat{\theta}_{2i}(k)$ can be described

$$-2\omega_{2i}^2\tilde{\theta}_{2i}^{\mathrm{T}}(k)\hat{\theta}_{2i}(k) = -\omega_{2i}^2\left\|\tilde{\theta}_{2i}(k)\right\|^2 - \omega_{2i}^2\left\|\hat{\theta}_{2i}(k)\right\|^2 + \omega_{2i}^2\left\|\theta_{2i}^*(k)\right\|^2 \tag{7.66}$$

According to (7.64) and (7.66), one can obtain

$$\begin{aligned} &\omega_{2i}(\tilde{\theta}_{2i}^{\mathrm{T}}(k+1)\tilde{\theta}_{2i}(k+1) - \tilde{\theta}_{2i}^{\mathrm{T}}(k)\tilde{\theta}_{2i}(k)) \leq \\ &-(\omega_{2i}^2 - \kappa_{2i}^2)\left\|\tilde{\theta}_{2i}(k)\right\|^2 + (2\omega_{2i}^2\bar{\phi}_{2i}^2 + \omega_{2i}\kappa_{2i}^2\bar{\phi}_{2i}^2)e_{2i}^2(k) \\ &-(\omega_{2i}^2 - \omega_{2i}^3 - \kappa_{2i}^2\omega_{2i}^2)\left\|\hat{\theta}_{2i}(k)\right\|^2 + \omega_{2i}^2\bar{\theta}_{2i}^{*2} \end{aligned} \tag{7.67}$$

where $\left\|\theta_{2i}^*(k)\right\| \leq \bar{\theta}_{2i}^*$, and $\bar{\theta}_{2i}^*$ is a positive constant.

The following function candidate is given by:

$$V_2(k) = \sum_{i=1}^{3} e_{2i}^2(k) + \sum_{i=1}^{3} \omega_{2i}\tilde{\theta}_{2i}^{\mathrm{T}}(k)\tilde{\theta}_{2i}(k) + \sum_{i=1}^{3} \sigma_{2i}\tilde{\bar{d}}_{2i}^2(k) + V_1(k) \tag{7.68}$$

Based on (7.68), the first difference is calculated as

$$\Delta V_2(k) = \sum_{i=1}^{3} e_{2i}^2(k+1) + \sum_{i=1}^{3} \omega_{2i}\tilde{\theta}_{2i}^{\mathrm{T}}(k+1)\tilde{\theta}_{2i}(k+1) + \sum_{i=1}^{3} \sigma_{2i}\tilde{\bar{d}}_{2i}^{2}(k+1) - \sum_{i=1}^{3} e_{2i}^2(k) - \sum_{i=1}^{3} \bar{\kappa}_{2i}\tilde{\delta}_{2i}^2(k) - \sum_{i=1}^{3} \sigma_{2i}\tilde{\bar{d}}_{2i}^{2}(k) \tag{7.69}$$

Combining (7.26), (7.48), (7.63) and (7.67), one has

$$\begin{aligned}\Delta V_{2i}(k) \le & -(\omega_{2i}^2 - \kappa_{2i}^2 - 3\bar{\phi}_{2i}^2\sigma_{2i}\bar{\rho}_{2i}^2)\left\|\tilde{\theta}_{2i}(k)\right\|^2 \\ & -(1 - 7\gamma^2 - 2\omega_{2i}^2\bar{\phi}_{2i}^2 - \omega_{2i}\kappa_{2i}^2\bar{\phi}_{2i}^2)e_{2i}^2(k) + V_1(k+1) - V_1(k) \\ & -(\sigma_{2i} - 7\lambda_1^2)\tilde{\bar{d}}_{2i}^{2}(k) + 7\lambda_1^2\delta_{2i}^2 + 7\bar{\delta}_{2i}^2 + 7(\nabla^{\gamma}\Delta\bar{F}_{2i}(k))^2 \\ & -(\omega_{2i}^2 - \omega_{2i}^3 - \kappa_{2i}^2\omega_{2i}^2 - 7\bar{\rho}_{2i}^2\bar{\phi}_{2i}^2)\left\|\hat{\theta}_{2i}(k)\right\|^2 + 7\bar{\mathscr{L}}_i^2 \\ & +\omega_{2i}^2\bar{\theta}_{2i}^{*2} + 3\sigma_{2i}\bar{D}_{2i}^2 + 3\sigma_{2i}\bar{\rho}_{2i}^2\bar{\varepsilon}_{2i}^2 \end{aligned} \tag{7.70}$$

where $\Delta V_{2i}(k) = e_{2i}^2(k+1) + \omega_{2i}\tilde{\theta}_{2i}^{\mathrm{T}}(k+1)\tilde{\theta}_{2i}(k+1) + \sigma_{2i}\tilde{\bar{d}}_{2i}^{2}(k+1) - e_{2i}^2(k) - \omega_{2i}\tilde{\theta}_{2i}^{\mathrm{T}}(k)\tilde{\theta}_{2i}(k) - \sigma_{2i}\tilde{\bar{d}}_{2i}^{2} + V_{1i}(k+1) - V_{1i}(k)$, $V_{1i}(k) = |s_{1i}(k)| + \omega_{1i}\tilde{\theta}_{1i}^T(k)\tilde{\theta}_{1i}(k) + \omega_{0i}\tilde{\theta}_{0i}^T(k)\tilde{\theta}_{0i}(k) + \sigma_{1i}\tilde{\bar{d}}_{1i}^{2}(k)$ and $|\varepsilon_{2i}| \le \bar{\varepsilon}_{2i}$.

According to Lemma 2.2, the NN is employed to approximate the unknown nonlinear function $\mathscr{R}_i(k)$ with $\mathscr{R}_i(k) = \frac{1}{\bar{\rho}_{2i}}7(\nabla^{\gamma}\Delta\bar{F}_{2i}(k))^2 + \frac{1}{\bar{\rho}_{2i}}\mathscr{Q}_i(k)$. Then, $\Delta V_{2i}(k)$ satisfies that

$$\begin{aligned}\Delta V_{2i}(k) \le & -(\omega_{2i}^2 - \kappa_{2i}^2 - 3\bar{\phi}_{2i}^2\sigma_{2i}\bar{\rho}_{2i}^2)\left\|\tilde{\theta}_{2i}(k)\right\|^2 - (\sigma_{2i} - 7\lambda_1^2)\tilde{\bar{d}}_{2i}^{2}(k) \\ & -(1 - 7\gamma^2 - 2\omega_{2i}^2\bar{\phi}_{2i}^2 - \omega_{2i}\kappa_{2i}^2\bar{\phi}_{2i}^2)e_{2i}^2(k) + 7\bar{\delta}_{2i}^2 + \bar{\Phi}_{2i}^* \\ & -(\omega_{2i}^2 - \omega_{2i}^3 - \kappa_{2i}^2\omega_{2i}^2 - 7\bar{\rho}_{2i}^2\bar{\phi}_{2i}^2)\left\|\hat{\theta}_{2i}(k)\right\|^2 + \bar{\Xi}_{2i} + 7\bar{\mathscr{L}}_i^2 + \omega_{2i}^2\bar{\theta}_{2i}^{*2} \\ & +3\sigma_{2i}\bar{D}_{2i}^2 + 3\sigma_{2i}\bar{\rho}_{2i}^2\bar{\varepsilon}_{2i}^2 + 7\lambda_1^2\delta_{2i}^2 - (\omega_{1i}^2 - \kappa_{1i}^2 - 0.5\bar{\phi}_{1i}^2\rho_{1i}\bar{\rho}_{1i}^2)\left\|\tilde{\theta}_{1i}(k)\right\|^2 \\ & -(\sigma_{1i} - 0.5\rho_{1i})\tilde{\bar{d}}_{1i}^{2}(k) - (\omega_{1i}^2 - \kappa_{1i}^2\omega_{1i}^2 - 3\bar{\phi}_{1i}^2\sigma_{1i}\bar{\rho}_{1i}^2 - \omega_{1i}^3)\left\|\hat{\theta}_{1i}(k)\right\|^2 \\ & -|s_{1i}(k)| + 3\sigma_{1i}\bar{\rho}_{1i}^2\bar{\varepsilon}_{1i}^2 + \mathscr{Q}_i(k) - (\omega_{0i}^2 - \kappa_{0i}^2 - 0.5\rho_{1i}\bar{g}_1^2\bar{\ell}_0^2\bar{\rho}_{0i}^2)\left\|\tilde{\theta}_{0i}(k)\right\|^2 \\ & -(\omega_{0i}^2 - \omega_{0i}^3 - \kappa_{0i}^2\omega_{0i}^2)\left\|\hat{\theta}_{0i}(k)\right\|^2 + \omega_{0i}^2\bar{\theta}_{0i}^{*2} + \omega_{1i}^2\bar{\theta}_{1i}^{*2} + 3\sigma_{1i}\bar{D}_{1i}^2 \end{aligned} \tag{7.71}$$

where $\left|\Phi_{2i}^*(k)\right| \le \bar{\Phi}_{2i}^*$, $\bar{\Phi}_{2i}^* = \bar{\rho}_{2i}\theta_{2i}^{*T}(k)\phi_{2i}(k)$, $|\Xi_{2i}(k)| \le \bar{\Xi}_{2i}$, $\Xi_{2i}(k) = \bar{\rho}_{2i}\varepsilon_{2i}(k)$, $\theta_{2i}^*(k)$ is the optimal weight vector, $\varepsilon_{2i}(k)$ is the minimum approximation error, $\phi_{2i}(\cdot)$ is the basis function vector, and $\bar{\Phi}_{2i}^*$ and $\bar{\Xi}_{2i}$ are positive constants.

According to (7.71), one has

$$\begin{aligned}\Delta V_2(k) \leq & -\|s_1(k)\| - \sum_{i=1}^{3} \mathscr{K}_1 \left\|\tilde{\theta}_{1i}(k)\right\|^2 - \sum_{i=1}^{3} \mathscr{K}_2 \left\|\hat{\theta}_{1i}(k)\right\|^2 - \mathscr{K}_3 \left\|\tilde{\bar{d}}_1(k)\right\|^2 \\ & -\mathscr{K}_4 \|e_2(k)\|^2 - \sum_{i=1}^{3} \mathscr{K}_5 \left\|\tilde{\theta}_{2i}(k)\right\|^2 - \sum_{i=1}^{3} \mathscr{K}_6 \left\|\hat{\theta}_{2i}(k)\right\|^2 \\ & -\mathscr{K}_7 \left\|\tilde{\bar{d}}_2(k)\right\|^2 - \sum_{i=1}^{3} \mathscr{K}_8 \left\|\tilde{\theta}_{0i}(k)\right\|^2 - \sum_{i=1}^{3} \mathscr{K}_9 \left\|\hat{\theta}_{0i}(k)\right\|^2 + \mathscr{K}_{10} \end{aligned} \tag{7.72}$$

where $\mathscr{K}_1 = \min(\omega_{1i}^2 - \kappa_{1i}^2 - 0.5\bar{\phi}_{1i}^2 \rho_{1i} \bar{\rho}_{1i}^2)$, $\mathscr{K}_2 = \min(\omega_{1i}^2 - \kappa_{1i}^2 \omega_{1i}^2 - 3\bar{\phi}_{1i}^2 \sigma_{1i} \bar{\rho}_{1i}^2 - \omega_{1i}^3)$, $\mathscr{K}_3 = \min(\sigma_{1i} - 0.5\rho_{1i})$, $\mathscr{K}_4 = \min(1 - 7\gamma^2 - 2\omega_{2i}^2 \bar{\phi}_{2i}^2 - \omega_{2i} \kappa_{2i}^2 \bar{\phi}_{2i}^2)$, $\mathscr{K}_5 = \min(\omega_{2i}^2 - \kappa_{2i}^2 - 3\bar{\phi}_{2i}^2 \sigma_{2i} \bar{\rho}_{2i}^2)$, $\mathscr{K}_6 = \min(\omega_{2i}^2 - \omega_{2i}^3 - \kappa_{2i}^2 \omega_{2i}^2 - 7\bar{\rho}_{2i}^2 \bar{\phi}_{2i}^2)$, $\mathscr{K}_7 = \min(\sigma_{2i} - 7\lambda_1^2)$, $\mathscr{K}_8 = \min(\omega_{0i}^2 - \kappa_{0i}^2 - 0.5\rho_{1i} \bar{g}_1^2 \bar{\ell}_0^2 \bar{\rho}_{0i}^2)$, $\mathscr{K}_9 = \min(\omega_{0i}^2 - \omega_{0i}^3 - \kappa_{0i}^2 \omega_{0i}^2)$, $\mathscr{K}_{10} = \sum_{i=1}^{3} 7\lambda_1^2 \delta_{2i}^2 + \sum_{i=1}^{3} 7\bar{\delta}_{2i}^2 + \sum_{i=1}^{3} 7\bar{\mathscr{L}}_i^2 + \sum_{i=1}^{3} \bar{\Phi}_{2i}^{*2} + \sum_{i=1}^{3} \omega_{2i}^2 \bar{\theta}_{2i}^{*2} + \sum_{i=1}^{3} 3\sigma_{2i} \bar{D}_{2i}^2 + \sum_{i=1}^{3} 3\sigma_{2i} \bar{\rho}_{2i}^2 \bar{\varepsilon}_{2i}^2 + \sum_{i=1}^{3} i\bar{X}i_{2i}^2 + \sum_{i=1}^{3} \omega_{1i}^2 \bar{\theta}_{1i}^{*2} + \sum_{i=1}^{3} 3\sigma_{1i} \bar{D}_{1i}^2 + \sum_{i=1}^{3} \omega_{0i}^2 \bar{\theta}_{0i}^{*2}$, and $\min(\cdot)$ denotes the minimum value.

According to (7.72), if the designed parameters can satisfy that $\mathscr{K}_1 > 0$, $\mathscr{K}_2 > 0$, $\mathscr{K}_3 > 0$, $\mathscr{K}_5 > 0$, $\mathscr{K}_6 > 0$, $\mathscr{K}_7 > 0$, $\mathscr{K}_8 > 0$ and $\mathscr{K}_9 > 0$, we have that $s_1(k)$ is bounded. Hence, the tracking error can satisfy the prescribed performance. Moreover, the conclusion that the disturbance estimation errors $\tilde{\bar{d}}_1(k)$ and $\tilde{\bar{d}}_2(k)$ are bounded from (7.72). This concludes the proof. ◇

7.5 Simulation Study

Numerical simulation results are presented to demonstrate the effectiveness of the proposed DTDO-based DTFO tracking control scheme in this section.

In this numerical simulation, the maximum external wind speed is assumed as V_{Wm}=[10, 5, 4]^{T}m/s, and t_m=3s. The external disturbances are assumed as $\bar{d}_{21}$=0.5 $\Delta T(3\sin(0.2t) - 4\cos(0.3t))$, $\bar{d}_{22} = 0.5\Delta T(4\sin(0.1t) - 4\cos(0.4t))$ and $\bar{d}_{23} = 0.5\Delta T(4\sin(0.5t) - 6\cos(0.2t))$. The DT step $T_0 = 0.002$ and the fractional order $\gamma = 0.001$. The initial values are given by $\beta_0 = 0.2$deg, $\alpha_0 = 0$deg, $\mu_0 = 0$deg, $p_0 = q_0 = r_0 = 0$deg$/s$. The prescribed performance function is chosen as $\varpi_{1i} = (7 - 20^{-1})e^{-t} + 20^{-1} (i = 1, 2, 3)$, and the constant τ_{1i} is chosen as $\tau_{1i} = 1$. The desired attitudes x_d are chosen as $\beta_d = 0$deg, $\alpha_d = (3(\cos(2t) + \sin(t)) + 4)$deg and $\mu_d = (3(\cos(2t) - \cos(3t)) + 4)$deg. The parameters of the disturbance observer are designed as $\sigma_{21} = \sigma_{22} = 0.2$, $\sigma_{23} = 0.1$ and $\sigma_{21} = \sigma_{22} = \sigma_{23} = 0.002$. The parameters of the adaptive laws are chosen as $\kappa_{11} = \kappa_{12} = \kappa_{13} = 0.0001$, $\omega_{11} = \omega_{12} = \omega_{13} = 0.001$, $\kappa_{21} = \kappa_{22} = \kappa_{23} = 0.0001$ and $\omega_{21} = \omega_{22} = \omega_{23} = 0.001$. The parameters are chosen as $\lambda_0 = 0.8$, $\lambda_1 = 0.0001$, $\lambda_2 = 0.2$, $\bar{\rho}_{2i} = 0.001$, $h_{01} = \mathrm{diag}[120, 120, 120]$ and $r_{01} = \mathrm{diag}[20, 20, 20]$. Moreover, we define that $e_\beta = \beta - \beta_d$, $e_\alpha = \alpha - \alpha_d$ and $e_\mu = \mu - \mu_d$.

According to the designed parameters above and the developed DTFO tracking control scheme based on the designed DTDO, the numerical simulation results are presented in Figs. 7.2, 7.3, 7.4, 7.5, 7.6 and 7.7. The tracking control results are shown in Fig. 7.2, which shows that good tracking control performance is achieved. From Fig. 7.2, we can see that the system output can track the desired signal quickly. Figure 7.3 presents that the tracking errors e_β, e_α and e_μ can quickly converge to a bound with small overshoots. Moreover, the tracking errors can satisfy the prescribed performance. According to Figs. 7.4 and 7.6, the output of the designed disturbance observer can estimate the external disturbances well. The responses of disturbance estimation errors are presented in Figs. 7.5 and 7.7. It can be seen from Figs. 7.4, 7.5, 7.6 and 7.7 that the satisfactory estimation performance of the DTDO can be obtained under the proposed NN-based DTDO. Therefore, according to the above simulation results, the satisfactory steady-state and transient performance can be ensured under the proposed DTDO-based DTFO control technique with a prescribed performance for the control of the UAV system.

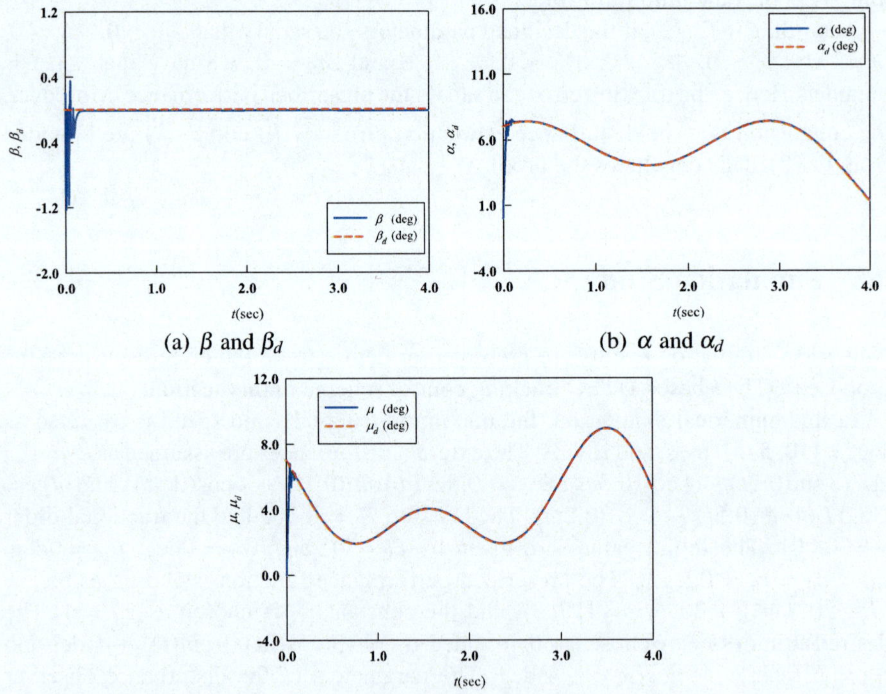

(a) β and β_d (b) α and α_d

(c) μ and μ_d

Fig. 7.2 The responses of the attitude angles under the proposed DTFO control scheme

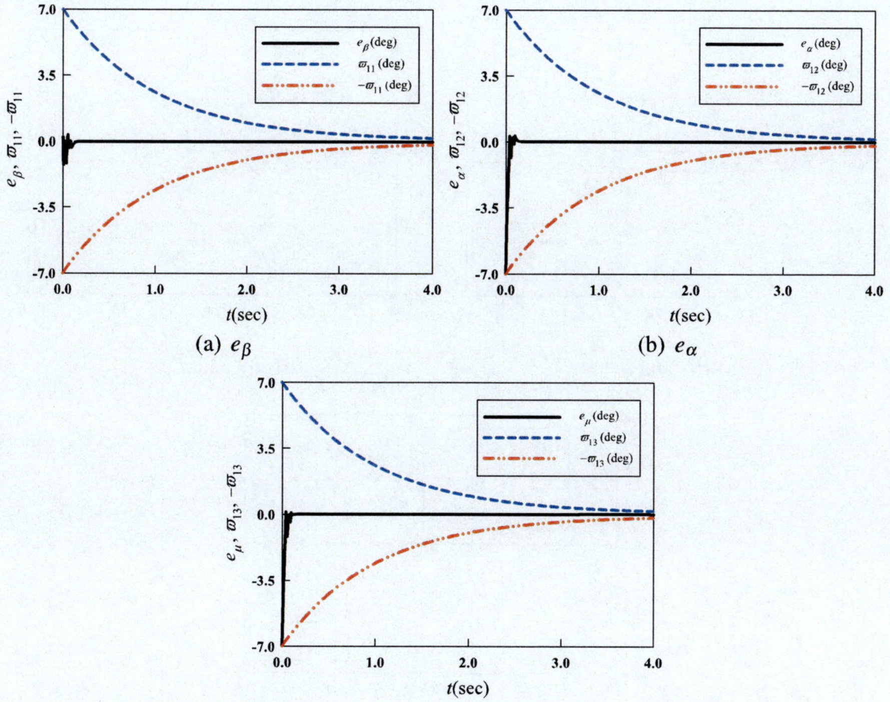

(a) e_β (b) e_α (c) e_μ

Fig. 7.3 The responses of the tracking errors under the proposed DTFO control scheme

To further illustrate the effectiveness of the designed DTDO-based DTFO control scheme, its performance is compared with that of an SMC method and a robust continuous-time integer-order control method [31] via dealing with the tracking control of the uncertain UAV system. The following two cases are given:

Case 1: The designed DO and the NN in the designed controller are replaced using the SM term to handle the uncertainties. Other conditions are the same as those in the above numerical simulation. By properly adjusting control parameter, the good tracking control performance can be obtained. The tracking control results are shown in Fig. 7.8, and the responses of the tracking errors are given in Fig. 7.9. Moreover, for the proposed fractional-order control and the Case 1, the overall results on maximum absolute values of tracking errors and average errors are also summarized in Table 7.1. It can be observed that the desired attitude signals can be followed well. However, the output signals have relatively strong chattering from Figs. 7.8 and 7.9, and the maximum absolute values of tracking errors and average errors in Case 1 are larger than that of the proposed control scheme in this paper.

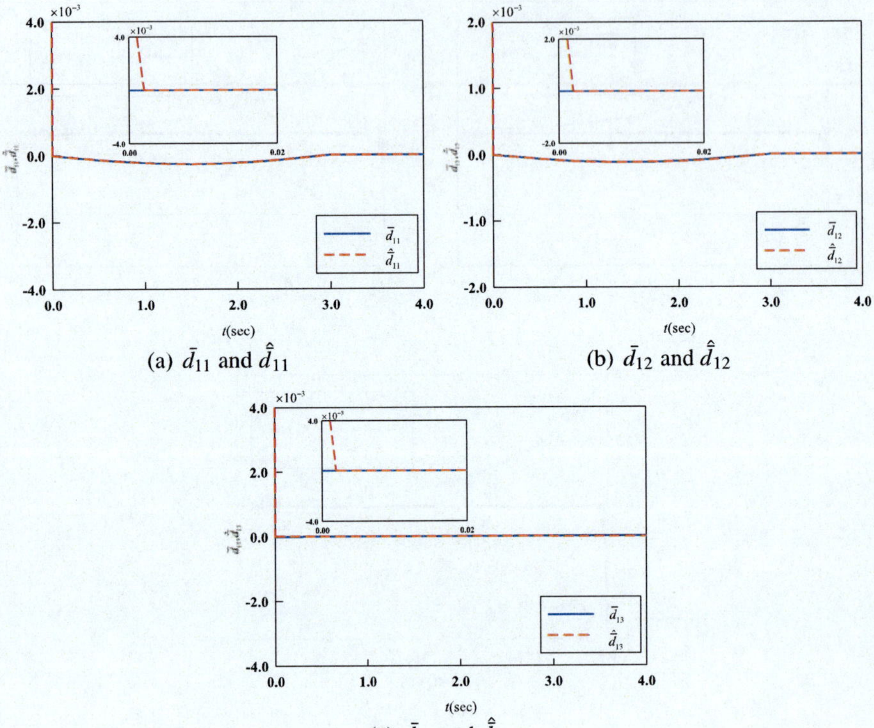

(a) $\bar{d}_{11}$ and $\hat{\bar{d}}_{11}$ (b) $\bar{d}_{12}$ and $\hat{\bar{d}}_{12}$

(c) $\bar{d}_{13}$ and $\hat{\bar{d}}_{13}$

Fig. 7.4 The disturbance estimation performance of the designed DTDO

Fig. 7.5 The responses of the estimation errors of disturbances $\tilde{\bar{d}}_{11}$, $\tilde{\bar{d}}_{12}$ and $\tilde{\bar{d}}_{13}$

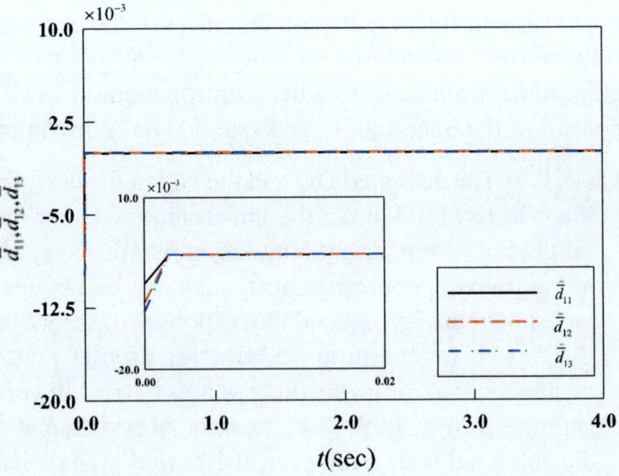

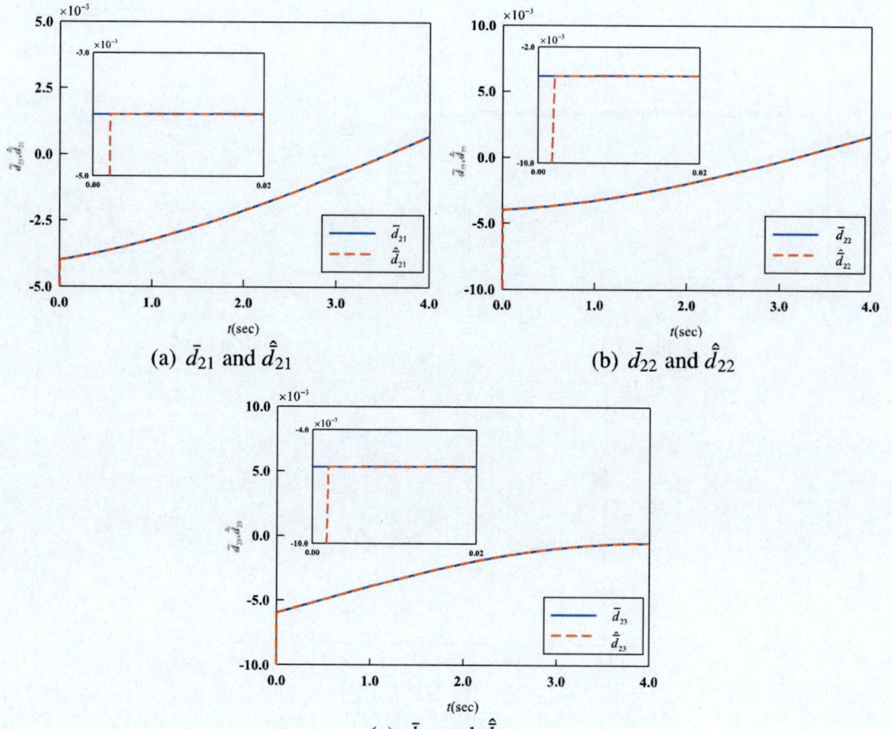

(a) $\bar{d}_{21}$ and $\hat{\bar{d}}_{21}$ (b) $\bar{d}_{22}$ and $\hat{\bar{d}}_{22}$

(c) $\bar{d}_{23}$ and $\hat{\bar{d}}_{23}$

Fig. 7.6 The disturbance estimation performance of the designed DTDO

Fig. 7.7 The responses of the estimation errors of disturbances $\tilde{\bar{d}}_{21}$, $\tilde{\bar{d}}_{22}$ and $\tilde{\bar{d}}_{23}$

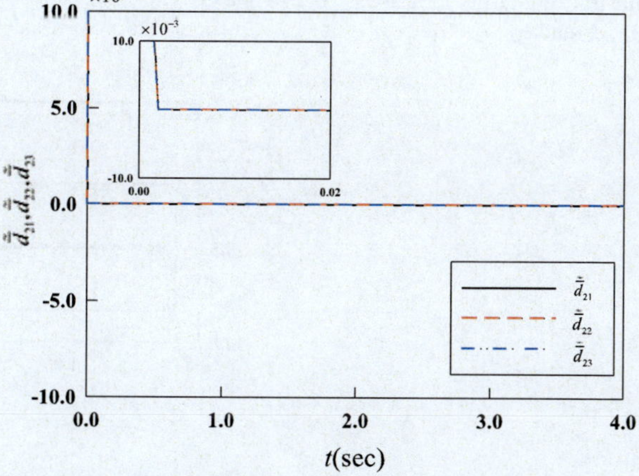

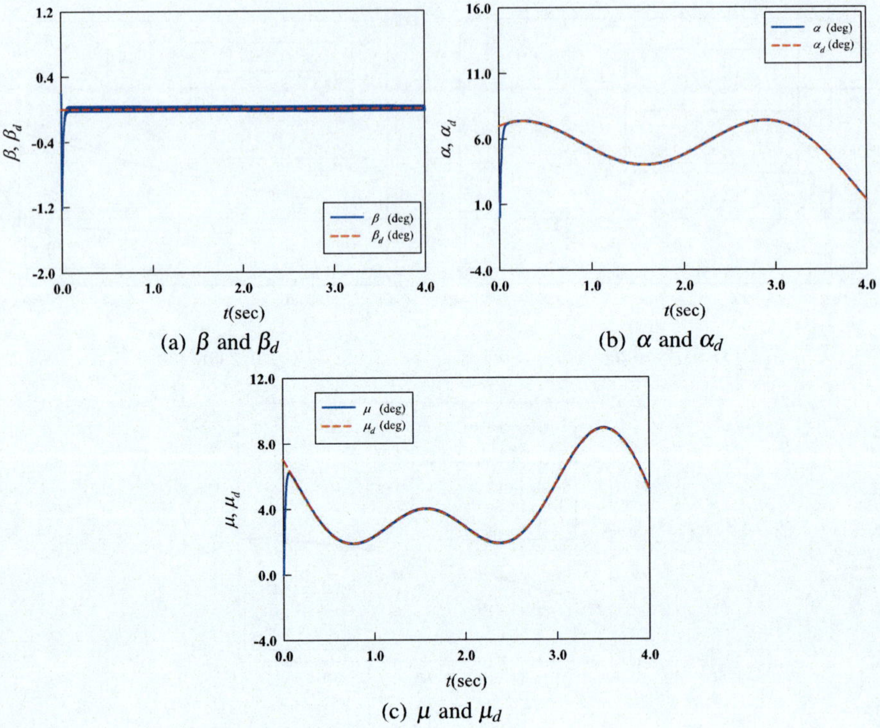

(a) β and β_d

(b) α and α_d

(c) μ and μ_d

Fig. 7.8 The responses of the attitude angles for Case 1

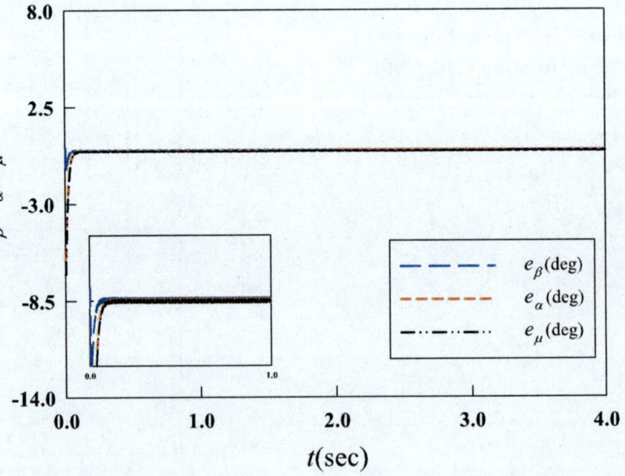

Fig. 7.9 The responses of the tracking errors for Case 1 e_β, e_α and e_μ

Table 7.1 Performance comparison on maximum absolute value of tracking error and average error

Control method		Maximum absolute value of tracking error (deg)	Average error (deg)
FO control	e_β	1.7228	0.0208
	e_α	6.8015	0.036
	e_μ	7	0.0424
SMC	e_β	1.2287	0.0353
	e_α	7.0011	0.0537
	e_μ	7	0.0528
Integer order control	e_β	1.3908	0.2456
	e_α	7.0004	0.5956
	e_μ	7	0.5504

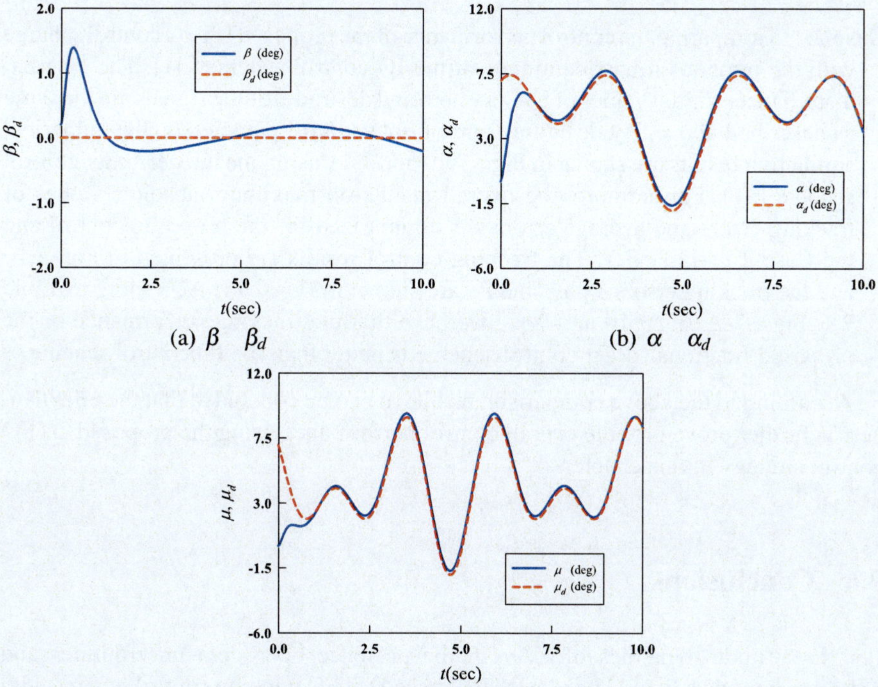

(a) β β_d (b) α α_d
(c) μ μ_d

Fig. 7.10 The responses of the attitude angles for Case 2

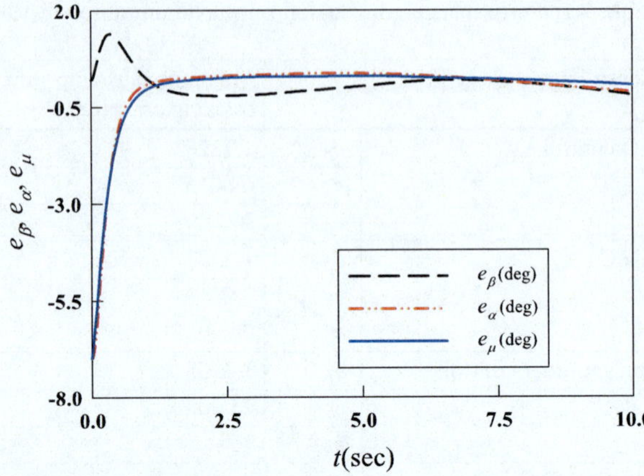

Fig. 7.11 The responses of the tracking errors e_β, e_α and e_μ

Case 2: Comparing the control performance of the proposed DTFO control scheme with the previous robust continuous-time IO control scheme [31]. The external disturbances, initial values of UAV system and desired attitude signals are the same as described above. By designing appropriate control parameters, the numerical simulation results are shown in Figs. 7.10 and 7.11 using the integer-order control scheme [31]. Furthermore, the overall results on maximum absolute values of tracking errors and average errors are summarized for the proposed control and the Case 2 in Table 7.1. The tracking control results are presented in Fig. 7.10 and the tracking errors e_β, e_α and e_μ are shown in Fig. 7.11. According to Table 7.1, Figs. 7.2, 7.3, 7.10 and 7.11, we have that the tracking performance of the proposed fractional-order control scheme is better than the IO control scheme.

According to the above comparison results, it can be concluded that the effectiveness is further proven for the overall control performance using the proposed DTFO control strategy in this chapter.

7.6 Conclusions

For the attitude dynamics of UAV in the presence of system uncertainties and unknown bounded disturbances, a DTDO-based DTFO tracking control scheme with prescribed performance has been proposed. The NN approximation has first been employed to approximate system uncertainties. Then, a DTDO has been designed to counteract the adverse effects of disturbances based on NN. Next, a DTFO control scheme with prescribed performance has been developed using the NN, the designed DTDO and the theory of DTFO. Finally, numerical simulation results have been given to illustrate the effectiveness of the proposed adaptive neural control scheme. Meanwhile, the stability of the closed-loop system has been achieved under the proposed DTFO control technology.

References

1. Wang, Y., Hu, J.: Improved prescribed performance control for air-breathing hypersonic vehicles with unknown deadzone input nonlinearity. ISA Trans. **79**, 95–107 (2018)
2. Yang, Q., Chen, M.: Adaptive neural prescribed performance tracking control for near space vehicles with input nonlinearity. Neurocomputing **174**, 780–789 (2016)
3. Wang, J., Shao, C., Chen, Y.-Q.: Fractional order sliding mode control via disturbance observer for a class of fractional order systems with mismatched disturbance. Mechatronics **53**, 8–19 (2018)
4. Freeborn, T.J.: A survey of fractional-order circuit models for biology and biomedicine. IEEE J. Emerg. Sel. Top. Circuits Syst. **3**(3), 416–424 (2013)
5. Ionescu, C., Lopes, A., Copot, D., et al.: The role of fractional calculus in modeling biological phenomena: a review. Commun. Nonlinear Sci. Numer. Simul. **51**, 141–159 (2017)
6. Sopasakis, P., Sarimveis, H., Macheras, P., et al.: Fractional calculus in pharmacokinetics. J. Pharmacokinet. Pharmacodyn. **45**(1), 107–125 (2018)
7. Shivanian, E., Jafarabadi, A.: Applications of Fractional Calculus in Physics. World Scientific, New Jersey (2000)
8. Chen, W.-C.: Nonlinear dynamics and chaos in a fractional-order financial system. Chaos Solitons Fractals **36**(5), 1305–1314 (2008)
9. Wang, Z., Huang, X., Shi, G.: Analysis of nonlinear dynamics and chaos in a fractional order financial system with time delay. Comput. Math. Appl. **62**(3), 1531–1539 (2011)
10. Tacha, O., Volos, C.K., Kyprianidis, I.M., et al.: Analysis, adaptive control and circuit simulation of a novel nonlinear finance system. Appl. Math. Comput. **276**, 200–217 (2016)
11. Škovránek, T., Podlubny, I., Petráš, I.: Modeling of the national economies in state-space: a fractional calculus approach. Econ. Model. **29**(4), 1322–1327 (2012)
12. Flores-Tlacuahuac, A., Biegler, L.T.: Optimization of fractional order dynamic chemical processing systems. Ind. Eng. Chem. Res. **53**(13), 5110–5127 (2014)
13. Yadav, V.K., Das, S., Bhadauria, B.S., et al.: Stability analysis, chaos control of a fractional order chaotic chemical reactor system and its function projective synchronization with parametric uncertainties. Chin. J. Phy. **55**(3), 594–605 (2017)
14. Yu, W., Luo, Y., Chen, Y., et al.: Frequency domain modelling and control of fractional-order system for permanent magnet synchronous motor velocity servo system. IET Control Theory Appl. **10**(2), 136–143 (2016)
15. Gutierrez, R.E., Rosario, J.M., Tenreiro Machado, J.: Fractional order calculus: basic concepts and engineering applications. Math. Problems Eng. **2010**, Article ID 375858, 19p (2010)
16. Abbas, I.A.: A problem on functional graded material under fractional order theory of thermoelasticity. Theoret. Appl. Fract. Mech. **74**, 18–22 (2014)
17. Tripathy, M.C., Mondal, D., Biswas, K., et al.: Experimental studies on realization of fractional inductors and fractional-order bandpass filters. Int. J. Circuit Theory Appl. **43**(9), 1183–1196 (2015)
18. Ebaid, A.: Analysis of projectile motion in view of fractional calculus. Appl. Math. Model. **35**(3), 1231–1239 (2011)
19. Miller, K.S., Ross, B.: An Introduction to the Fractional Calculus and Fractional Differential Equations. Wiley, New York (1993)
20. Kilbas, A.A., Srivastava, H.M., Trujillo, J.J.: Theory and Applications of Fractional Differential Equations. Elsevier, Netherlands (2006)
21. Manabe, S.: A suggestion of fractional-order controller for flexible spacecraft attitude control. Nonlinear Dyn. **29**(1–4), 251–268 (2002)
22. Aboelela, M.A., Ahmed, M.F., Dorrah, H.T.: Design of aerospace control systems using fractional PID controller. J. Adv. Res. **3**(3), 225–232 (2012)
23. Seyedtabaii, S.: New flat phase margin fractional order PID design: perturbed UAV roll control study. Robot. Auton. Syst. **96**, 58–64 (2017)
24. Chao, H., Luo, Y., Di, L., et al.: Roll-channel fractional order controller design for a small fixed-wing unmanned aerial vehicle. Control Eng. Pract. **18**(7), 761–772 (2010)

25. Luo, Y., Chao, H., Di, L., et al.: Lateral directional fractional order $(PI)^{\alpha}$ control of a small fixed-wing unmanned aerial vehicles: controller designs and flight tests. IET Control Theory Appl. **5**(18), 2156–2167 (2011)
26. Han, J., Di, L., Coopmans, C., et al.: Pitch loop control of a VTOL UAV using fractional order controller. J. Intell. Robot. Syst. **73**(1–4), 187–195 (2014)
27. Kostarigka, A.K., Rovithakis, G.A.: Adaptive dynamic output feedback neural network control of uncertain MIMO nonlinear systems with prescribed performance. IEEE Trans. Neural Netw. Learn. Syst. **23**(1), 138–149 (2012)
28. Meng, W., Yang, Q., Ying, Y., et al.: Adaptive power capture control of variable-speed wind energy conversion systems with guaranteed transient and steady-state performance. IEEE Trans. Energy Convers. **28**(3), 716–725 (2013)
29. Mareels, I.M., Penfold, H., Evans, R.J.: Controlling nonlinear time-varying systems via Euler approximations. Automatica **28**(4), 681–696 (1992)
30. Franklin, G.F., Powell, J.D., Workman, M.L.: Digital Control of Dynamic Systems. Addison-Wesley, Menlo Park, CA (1998)
31. Chen, M., Jiang, B.: Robust attitude control of near space vehicles with time-varying disturbances. Int. J. Control Autom. Syst. **11**(1), 182–187 (2013)

Chapter 8
DTFO Control for UAV with External Disturbances

8.1 Introduction

The FOC has a long history of development, and the FOCS have been widely used in the fields of biology, physics, chemistry, economics, materials science and engineering [1–15]. With the development of the FOC in practical applications, the FO controllers have been successfully applied in aircraft control design [16–21]. In addition, The studies of FOC with discrete-time form have also received extensive attention [22], and some design methods of DTFO controllers were studied and some discrete-time control schemes based on DTFO controllers were proposed [23–29]. Although the above mentioned results have studied the design methods and applications of DTFO controllers, the research on flight control methods based on DTFO controllers is relatively rare. Therefore, for the UAV system with wind disturbances, the discrete-time flight control methods based on DTFO controllers need to be further studies. According to the above described DTFO control problem, based on the research in Chap. 6, a DTDO-based DTFO control scheme is proposed for the trajectory control system model with external wind disturbances and the UAV attitude dynamic model with wind disturbances and input saturation.

8.2 Problem Formulation and Preliminaries

8.2.1 Model of Longitudinal Flight Control System with External Disturbances

According to (6.4) and (6.24) in Chap. 6, the approximate discrete-time trajectory control system model with only external disturbance can be obtained, and the expression can be described as

$$\begin{cases} Z(k+1) = F_z(k) + G_z(k)u_z(k) + \bar{d}_z(k) \\ H(k+1) = F_H(k) + G_H(k)u_H(k) + \bar{d}_H(k) \end{cases} \tag{8.1}$$

S. Shao et al., *Robust Discrete-Time Flight Control of UAV with External Disturbances*, Studies in Systems, Decision and Control 317,
https://doi.org/10.1007/978-3-030-57957-9_8

where $Z(k+1) = \bar{z}_g(k+1)$, $F_z(k) = \bar{z}_g(k)$, $G_z(k) = -\Delta T \bar{V}_A$, ΔT is the sampling period, and $u_z(k) = \sin\gamma(k)$.

8.2.2 *Attitude Dynamics System Model with External Disturbance*

To facilitate the design of the DTFO control scheme, this chapter firstly converts the continuous time UAV attitude dynamics model (2.41) into the discrete-time UAV attitude dynamics model with wind disturbances by using Euler approximation method [30], and the approximate expression can be written as:

$$\begin{cases} x_1(k+1) = x_1(k) + \Delta T(F_1(x_1(k)) + G_1(x_1(k))x_2(k)) + \bar{d}_1(k) \\ x_2(k+1) = x_2(k) + \Delta T(F_2(x(k)) + G_2(x(k))u(k)) + \bar{d}_2(k) \\ y(k) = x_1(k) \end{cases} \tag{8.2}$$

where $\bar{d}_1 \in \Re^3$ and $\bar{d}_2 \in \Re^3$ are unknown wind disturbances.

To simplify the formulas, the following definitions are given as $\bar{F}_1(k) = \Delta T F_1(x_1(k))$, $\bar{F}_2(k) = \Delta T F_2(x(k))$, $\bar{G}_1(k) = \Delta T G_1(x_1(k))$ and $\bar{G}_2(k) = \Delta T G_1(x_1(k))$. Therefore, (8.2) can be written as

$$\begin{cases} x_1(k+1) = x_1(k) + \bar{F}_1(k) + \bar{G}_1(k)x_2(k) + \bar{d}_1(k) \\ x_2(k+1) = x_2(k) + \bar{F}_2(k) + \bar{G}_2(k)u(k) + \bar{d}_2(k) \\ y(k) = x_1(k) \end{cases} \tag{8.3}$$

In this chapter, a DTDO-based DTFO control scheme is proposed for the discrete-time UAV system with wind disturbances using the backstepping technology. The control objective aims at designing an adaptive DTFO tracking controller such that: (1) the reference signals $\bar{z}_{gd}(k)$, $\bar{V}_{Ad}(k)$, $\gamma_d(k)$ and $x_{1d}(k) = [\beta_d, \alpha_d, \mu_d]^T$ can be followed by the system output ${}^g\bar{z}(k)$, $\bar{V}_A(k)$, $\gamma(k)$ and $y(k)$ respectively in the presence of disturbances in a bounded compact set, where the reference signal $\bar{z}_{gd}(k)$, $\bar{V}_{Ad}(k)$, $\gamma_d(k)$ and $x_{1d}(k)$ are known function; (2) all the signals in the closed-loop system are bounded. On the basis of the design thread of the FO control scheme, the developed control method is presented by a block diagram in Fig. 8.1.

To proceed with the design of the adaptive DTFO control scheme for the UAV systems (8.1) and (8.3) subject to wind disturbances, the following assumptions are introduced:

Assumption 8.1 For the considered UAV system (2.41), the velocity is known. Thus, the known control coefficient matrix $\bar{G}_1(k)$, there exist positive constants $\underline{g}_1 > 0$ and $\bar{g}_1 > 0$ such that $\underline{g}_1 \leq \|\bar{G}_1(k)\| \leq \bar{g}_1$.

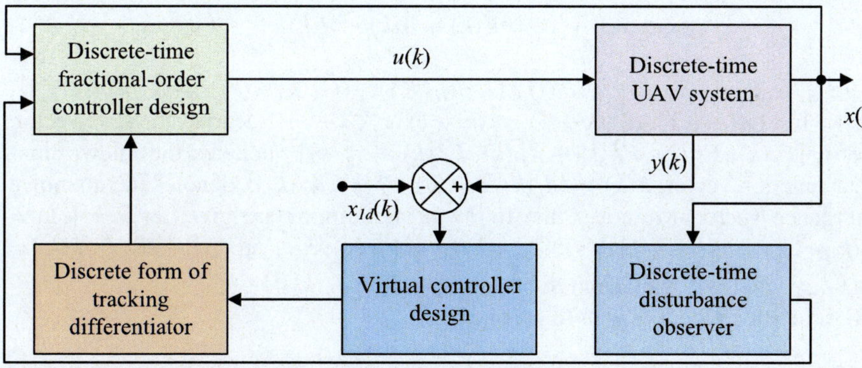

Fig. 8.1 The block diagram of the developed control strategy

Assumption 8.2 The unknown disturbance $d_{i_0}(k)$ is slowly time-varying, $\Delta d_{i_0}(k)$ is bounded and $\Delta d_{i_0}(k) = d_{i_0}(k+1) - d_{i_0}(k)$ with $i_0 = 1, 2$. According to Assumption 6.2, $\Delta \bar{d}_z(k)$, $\Delta \bar{d}_H(k)$ and $\Delta d_{i_0}(k)$ are bounded, and $\Delta \bar{d}_z(k) = \bar{d}_z(k+1) - \bar{d}_z(k)$, $\Delta \bar{d}_H(k) = \bar{d}_H(k+1) - \bar{d}_H(k)$, $\Delta \bar{d}_{i_0}(k) = \bar{d}_{i_0}(k+1) - \bar{d}_{i_0}(k)$, $i_0 = 1, 2$.

Remark 8.1 Based on the Definitions 2.1 and 2.2, for the function vector $\psi(k) = [\psi_1(k), \psi_2(k), \psi_3(k)]^{\mathrm{T}}$, the following definition is given by:

$$\nabla^{\bar{\gamma}} \psi(k) = \left[\sum_{j=0}^{k} (-1)^j \binom{\bar{\gamma}}{j} \bar{\psi}_1, \sum_{j=0}^{k} (-1)^j \binom{\bar{\gamma}}{j} \bar{\psi}_2, \sum_{j=0}^{k} (-1)^j \binom{\bar{\gamma}}{j} \bar{\psi}_3 \right]^{\mathrm{T}},$$

where $\bar{\gamma}$ is the fractional order and $\bar{\psi}_i = \psi_i(k-j)$. Furthermore, we define that

$$\nabla^{-\bar{\gamma}} \psi(k) = \left[\sum_{j=0}^{k} \binom{\bar{\gamma}}{k-j} \psi_1(j), \sum_{j=0}^{k} \binom{\bar{\gamma}}{k-j} \psi_2(j), \sum_{j=0}^{k} \binom{\bar{\gamma}}{k-j} \psi_3(j) \right]^{\mathrm{T}}.$$

8.3 DTFO Control Based on DTDO

In this section, the DTDO-based adaptive discrete-time fractional-order control schemes are developed for the discrete-time trajectory control system model with wind disturbances (8.1) and the attitude dynamics model of UAV (8.3). The detailed design procedure of the DTDO-based DTFO controller is presented by using backstepping technology.

8.3.1 *Design of Nonlinear DTDO*

On the basis of (8.1) and (8.3), we have

$$\bar{x}(k+1) = \bar{F}(k) + \bar{u}(k) + d(k) \tag{8.4}$$

where $\bar{u}(k) = [G_Z(k)u_z(k), (G_H(K)u_H(K))^{\mathrm{T}}, (\bar{G}_1(k)x_2(k))^{\mathrm{T}}, (\bar{G}_2(k)u(k))^{\mathrm{T}}]^{\mathrm{T}}$, $\bar{x}(k+1) = [Z(k+1), H^{\mathrm{T}}(k+1), x_1^{\mathrm{T}}(k+1), x_2^{\mathrm{T}}(k+1)]^{\mathrm{T}}$ denotes the state vector, $\bar{F}(k) = [F_z(k), F_H^{\mathrm{T}}(k), \bar{F}_1^{\mathrm{T}}(k) + x_1^{\mathrm{T}}(k), \bar{F}_2^{\mathrm{T}}(k) + x_2^{\mathrm{T}}(k)]^{\mathrm{T}}$ denotes the known nonlinear function vector, $d(k) = [\bar{d}_z(k), \bar{d}_H^{\mathrm{T}}(K), \bar{d}_1^{\mathrm{T}}(k), \bar{d}_2^{\mathrm{T}}(k)]^{\mathrm{T}}$ denotes the unknown disturbance vector. Moreover, the following definitions are given as $\bar{x}(k+1) = [\bar{x}_1(k+1), \ldots, \bar{x}_9(k+1)]^{\mathrm{T}} \in \Re^9$, $\bar{F}(k) = [\bar{f}_1(k), \ldots, \bar{f}_9(k)]^{\mathrm{T}} \in \Re^9$, $\bar{u}(k) = [\bar{u}_1(k), \ldots, \bar{u}_9(k)]^{\mathrm{T}} \in \Re^9$, and $d(k) = [\bar{\bar{d}}_1(k), \ldots, \bar{\bar{d}}_9(k)]^{\mathrm{T}} \in \Re^9$.

For the $\bar{i}$th element of $\bar{x}$ in (8.4), one has

$$\bar{x}_{\bar{i}}(k+1) = \bar{f}_{\bar{i}}(k) + \bar{u}_{\bar{i}}(k) + \bar{\bar{d}}_{\bar{i}}(k) \tag{8.5}$$

where $\bar{i} = 1, 2, \ldots, 9$.

To design the nonlinear DTDO, an auxiliary variable is introduced as follows:

$$\mathscr{M}_{\bar{i}}(k) = \zeta_{\bar{i}}\bar{\bar{d}}_{\bar{i}}(k) - \zeta_{\bar{i}}\bar{x}_{\bar{i}}(k) \tag{8.6}$$

where $\zeta_{\bar{i}}$ denotes a design positive constant.

According to (8.5) and (8.6), $\mathscr{M}_{\bar{i}}(k+1)$ is given by

$$\mathscr{M}_{\bar{i}}(k+1) = \zeta_{\bar{i}}\bar{\bar{d}}_{\bar{i}}(k+1) - \zeta_{\bar{i}}\bar{f}_{\bar{i}}(k) - \zeta_{\bar{i}}\bar{\bar{d}}_{\bar{i}}(k) - \zeta_{\bar{i}}\bar{u}_{\bar{i}}(k) \tag{8.7}$$

To estimate the disturbance, the estimation of auxiliary variable $\mathscr{M}_{\bar{i}}(k)$ is defined as

$$\hat{\mathscr{M}}_{\bar{i}}(k+1) = -\zeta_{\bar{i}}\bar{f}_{\bar{i}}(k) - \zeta_{\bar{i}}\bar{u}_{\bar{i}}(k) \tag{8.8}$$

where $\hat{\mathscr{M}}_{\bar{i}}(k)$ denotes the estimation of $\mathscr{M}_{\bar{i}}(k)$.

From (8.6), the disturbance $\bar{\bar{d}}_{\bar{i}}(k)$ can be estimated by

$$\hat{\bar{\bar{d}}}_{\bar{i}}(k) = \frac{1}{\zeta_{\bar{i}}}\hat{\mathscr{M}}_{\bar{i}}(k) + \bar{x}_{\bar{i}}(k) \tag{8.9}$$

where $\hat{\bar{\bar{d}}}_{\bar{i}}(k)$ denotes the estimation of $\bar{\bar{d}}_{\bar{i}}(k)$.

Defining $\tilde{\bar{\bar{d}}}_{\bar{i}}(k) = \bar{\bar{d}}_{\bar{i}}(k) - \hat{\bar{\bar{d}}}_{\bar{i}}(k)$ and $\tilde{\mathscr{M}}_{\bar{i}}(k) = \mathscr{M}_{\bar{i}}(k) - \hat{\mathscr{M}}_{\bar{i}}(k)$, and considering (8.6) and (8.9), one has

$$\tilde{\mathscr{M}}_{\bar{i}}(k) = \zeta_{\bar{i}}\tilde{\bar{\bar{d}}}_{\bar{i}}(k) \tag{8.10}$$

Combining (8.7) and (8.8), one can obtain

$$\tilde{\mathcal{M}}_{\bar{i}}(k+1) = \zeta_{\bar{i}} \Delta \bar{\bar{d}}_{\bar{i}} \tag{8.11}$$

where $\Delta \bar{\bar{d}}_{\bar{i}} = \bar{\bar{d}}_{\bar{i}}(k+1) - \bar{\bar{d}}_{\bar{i}}(k)$.

The above design procedure of the DTDO can be summarized in the following theorem:

Theorem 8.1 *Considering the discrete-time UAV system (8.1) and (8.3) in the presence of disturbances, a discrete-time nonlinear DO is designed as (8.8) and (8.9). The designed DTDO can guarantee that the disturbance estimation errors between the outputs of DTDO and the disturbances are bounded.*

Proof To analyze the boundedness of disturbance estimate error $\tilde{\bar{\bar{d}}}_{\bar{i}}(k)$, a Lyapunov function candidate is chosen as

$$V_{\bar{d}_{\bar{i}}}(k) = \zeta_{\bar{i}} \tilde{\bar{\bar{d}}}_{\bar{i}}^2(k) = \frac{1}{\zeta_{\bar{i}}} \tilde{\mathcal{M}}_{\bar{i}}^2(k) \tag{8.12}$$

Based on (8.11), $\Delta V_{\bar{d}_{\bar{i}}}(k)$ can be written as

$$\begin{aligned} \Delta V_{\bar{d}_{\bar{i}}}(k) &= \frac{1}{\zeta_{\bar{i}}} \tilde{\mathcal{M}}_{\bar{i}}^2(k+1) - \frac{1}{\zeta_{\bar{i}}} \tilde{\mathcal{M}}_{\bar{i}}^2(k) \\ &= -\frac{1}{\zeta_{\bar{i}}} \tilde{\mathcal{M}}_{\bar{i}}^2(k) + \zeta_{\bar{i}} \Delta \bar{\bar{d}}_{\bar{i}}^2 \end{aligned} \tag{8.13}$$

According to Assumption 8.2, one has

$$\left| \Delta \bar{\bar{d}}_{\bar{i}} \right| \leq D_{\bar{i}} \tag{8.14}$$

where $D_{\bar{i}}$ denotes a small positive constant.

Combining (8.13) and (8.14), it yields

$$\Delta V_{\bar{d}_{\bar{i}}}(k) \leq -\frac{1}{\zeta_{\bar{i}}} \tilde{\mathcal{M}}_{\bar{i}}^2(k) + \zeta_{\bar{i}} D_{\bar{i}}^2 \tag{8.15}$$

To ensure the boundedness of disturbance estimate error $\tilde{\bar{\bar{d}}}_{\bar{i}}(k)$, the disturbance observer gain $\zeta_{\bar{i}}$ should be chosen to ensure that $\zeta_{\bar{i}} > 0$. The conclusion that the signals $\tilde{\bar{\bar{d}}}_{\bar{i}}(k)$ and $\tilde{\mathcal{M}}_{\bar{i}}(k)$ are bounded from (8.8) and (8.9). Furthermore, $\Delta V_{\bar{d}_{\bar{i}}}(k) \leq 0$, once the error $\tilde{\mathcal{M}}_{\bar{i}}(k)$ is larger than $\sqrt{\zeta_{\bar{i}}^2 D_{\bar{i}}^2}$. Hence the upper bound of the approximation error $\tilde{\bar{\bar{d}}}_{\bar{i}}(k)$ is $D_{\bar{i}}$. This concludes the proof. ◇

8.3.2 DTFO Controller Design and Stability Analysis

In this section, the DTFO controller will be designed to ensure that the output signals $\bar{z}_g(k)$, $\bar{V}_A(k)$, $\gamma(k)$ and $y(k)$ can track the reference signals $\bar{z}_{gd}(k)$, $\bar{V}_{Ad}(k)$, $\gamma_d(k)$ and $x_{1d}(k)$ to the bounded compact set, respectively. The detailed design procedure of the DTDO-based DTFO tracking controllers is developed by backstepping technology as follows:

8.3.2.1 A DTFO Trajectory Control Scheme Based on DTDO

This section is for a trajectory control system model with external disturbances (8.1), a DTFO control scheme is proposed based on a DTDO. Firstly, the height tracking error is defined as $e_z(k) = Z(k) - \bar{z}_{gd}(k)$. According to (8.1), one has

$$\begin{aligned} e_z(k+1) &= Z(k+1) - \bar{z}_{gd}(k+1) \\ &= F_z(k) + G_z(k)u_z(k) + \bar{d}_z(k) - \bar{z}_{gd}(k+1) \end{aligned} \tag{8.16}$$

In order to compensate the negative affects of external disturbance $\bar{d}_z(k)$ on the control performance of the system (8.16), a DIDO is designed based on (8.8) and (8.9), which can be described as

$$\begin{cases} \hat{\mathscr{M}}_z(k+1) = -\zeta_z F_z(k) - \zeta_z \mathscr{U}_z(k) \\ \zeta_z \hat{\bar{d}}_z(k) = \hat{\mathscr{M}}_z(k) + \zeta_z Z(k) \end{cases} \tag{8.17}$$

where $\hat{\bar{d}}_z(k)$ is the output of DTDO, $\hat{\mathscr{M}}_z(k)$ is the state variable of DTDO, the design parameter ζ_z satisfies that $\zeta_z > 0$. Moreover, the disturbance estimation error is defined as $\tilde{\bar{d}}_z(k) = \bar{d}_z(k) - \hat{\bar{d}}_z(k)$, and $\mathscr{U}_z(k) = G_z(k)u_z(k)$. According to Assumption 8.2, one can obtain that $\left|\bar{d}_z(k+1) - \bar{d}_z(k)\right| \leq D_z$, and D_z is a positive constant.

The DTFO height control law is given as follows

$$\begin{aligned} u_z(k) = &-\bar{G}_z^{-1}(k)(F_z(k) + \lambda_z \nabla^{-\bar{\gamma}_z}\hat{\bar{d}}_z(k) - \lambda_{1z}(\nabla^{-\bar{\gamma}_z} M_{0z} - e_z(k))) \\ &+ \bar{G}_z^{-1}(k)\nabla^{-\bar{\gamma}_z}\left(\sum_{nz=1}^{k} M_{nz} e_z(k-nz)\right) + \bar{G}_z^{-1}(k)\bar{z}_{gd}(k+1) \end{aligned} \tag{8.18}$$

where $M_{0z} = \sum_{j=0}^{k} (-1)^j \binom{\bar{\gamma}_z}{j} e_z(k-j)$, $\bar{\gamma}_z$ is the fractional order, λ_z and λ_{1z} are design constants, $nz = j-1$, $j = 2, \ldots, k+1$ and $M_{nz} = (-1)^{n+1}\binom{\bar{\gamma}_z}{n+1}$.

Combining (8.16) and (8.18), one has

$$e_z(k+1) = \nabla^{-\bar{\gamma}_z}\left(\sum_{nz=1}^{k} M_{nz}e_z(k-nz)\right) - \lambda_z\nabla^{-\bar{\gamma}_z}\hat{\bar{d}}_z(k)$$
$$+\bar{d}_z(k) - \lambda_{1z}(e_z(k) - \nabla^{-\bar{\gamma}_z}M_{0z}) \tag{8.19}$$

Moreover, (8.19) can be written as

$$\nabla^{\bar{\gamma}_z}e_z(k+1) = \sum_{nz=1}^{k} M_{nz}e_z(k-nz) - \lambda_z\hat{\bar{d}}_z(k) + \nabla^{\bar{\gamma}_z}\bar{d}_z(k)$$
$$-\lambda_{1z}(\nabla^{\bar{\gamma}_z}e_z(k) - M_{0z}) \tag{8.20}$$

According to Definition 2.1, one can obtain

$$\nabla^{\bar{\gamma}_z}e_z(k+1) = e_z(k+1) - \bar{\gamma}_z e_z(k) + \sum_{nz=1}^{k} M_{nz}e_z(k-nz) \tag{8.21}$$

$$\nabla^{\bar{\gamma}_z}e_z(k) = \sum_{j=0}^{k}(-1)^j\binom{\bar{\gamma}_z}{j}e_z(k-j) \tag{8.22}$$

On the basis of (8.20), (8.21) and (8.22), $e_z(k+1)$ can be described as

$$e_z(k+1) = \bar{\gamma}_z e_z(k) - \lambda_z\hat{\bar{d}}_z(k) + \nabla^{\bar{\gamma}_z}\bar{d}_z(k) \tag{8.23}$$

According to (8.23), one has

$$e_z^2(k+1) \le 4\bar{\gamma}_z^2 e_z^2(k) + 4\lambda_z^2\delta_z^2 + 4\bar{\delta}_z^2 + 4\lambda_z^2\tilde{\bar{d}}_z^2(k) \tag{8.24}$$

where $|\bar{d}_z(k)| \le \delta_z$, $|\nabla^{\bar{\gamma}_z}\bar{d}_z(k)| \le \bar{\delta}_z$, δ_z and $\bar{\delta}_z$ are positive constants.

To prove the effectiveness of nonlinear DTDO (8.17) and height controller $u_z(k)$ (8.18), the following Lyapunov function is chosen as

$$V_z(k) = e_z^2(k) + \zeta_z\tilde{\bar{d}}_z^2(k) \tag{8.25}$$

Based on (8.25), the first order difference of Lyapunov function $V_z(k)$ is described as

$$\begin{aligned}\Delta V_z(k) &= V_z(k+1) - V_z(k)\\ &= e_z^2(k+1) + \zeta_z\tilde{\bar{d}}_z^2(k+1) - e_z^2(k) - \zeta_z\tilde{\bar{d}}_z^2(k)\end{aligned} \tag{8.26}$$

From (8.15), one has

$$\zeta_z(\tilde{\bar{d}}_z^2(k+1) - \tilde{\bar{d}}_z^2(k)) \leq -\zeta_{2i}\tilde{\bar{d}}_z^2(k) + \zeta_z D_z^2 \tag{8.27}$$

Combining (8.24) and (8.27), it yields

$$\Delta V_z(k) \leq -\chi_{z1} e_z^2(k) - \chi_{z2}\tilde{\bar{d}}_z^2(k) + \chi_{z3} \tag{8.28}$$

where $\chi_{z3} = \zeta_z D_z^2 + 4\lambda_z^2\delta_z^2 + 4\bar{\delta}_z^2$, $\chi_{z1} = 1 - 4\bar{\gamma}_z^2$ and $\chi_{z2} = \zeta_z - 4\bar{\gamma}_z^2$.

The following control analysis is conducted for the velocity $\bar{V}_A$ and the track angle γ. The following error is defined as $e_H(k) = H(k) - H_d(k)$, and $H_d(k) = [\bar{V}_{Ad}(k), \gamma_d(k)]^{\mathrm{T}}$. Therefore, according to (8.1), one can obtain

$$\begin{aligned} e_H(k+1) &= H(k+1) - H_d(k+1) \\ &= F_H(k) + G_H(k)u_H(k) + \bar{d}_H(k) - H_d(k+1) \end{aligned} \tag{8.29}$$

In order to deal with the disturbance $\bar{d}_H(k)$ in (8.29), a DTDO is designed based on (8.8) and (8.9). The following DTDO is described as

$$\begin{cases} \hat{\mathscr{M}}_{Hi_0}(k+1) = -\zeta_{Hi} F_{Hi_0}(k) - \zeta_{Hi_0}\mathscr{U}_{Hi_0}(k) \\ \zeta_{Hi}\hat{\bar{d}}_{Hi_0}(k) = \hat{\mathscr{M}}_{Hi_0}(k) + \zeta_{Hi_0} H_{i_0}(k) \end{cases} \tag{8.30}$$

where $i_0 = 1, 2$, $H_{i_0}(k)$ denotes the i_0th variable of $H(k)$, $F_{Hi_0}(k)$ denotes the i_0th variable of $F_H(k)$, $\hat{\bar{d}}_{Hi_0}(k)$ denotes the i_0th output of DTDO, and $\hat{\bar{d}}_H(k) = [\hat{\bar{d}}_{H1}(k), \hat{\bar{d}}_{H2}(k)]^{\mathrm{T}}$, $\hat{\mathscr{M}}_{Hi_0}(k)$ is the state variable of DTDO, the control parameter ζ_{Hi_0} satisfies that $\zeta_{Hi_0} > 0$. Furthermore, the disturbance estimation error is defined as $\tilde{\bar{d}}_{Hi_0}(k) = \bar{d}_{Hi_0}(k) - \hat{\bar{d}}_{Hi_0}(k)$, and $\bar{d}_H(k) = [\bar{d}_{H1}(k), \bar{d}_{H2}(k)]^{\mathrm{T}}$. $\mathscr{U}_{Hi_0}(k)$ denotes the i_0th variable of $\mathscr{U}_H(k)$, and $\mathscr{U}_H(k) = G_H(k)u_H(k)$. According to Assumption 8.2, one can obtain that $\left|\bar{d}_{Hi_0}(k+1) - \bar{d}_{Hi_0}(k)\right| \leq D_{Hi_0}$, and D_{Hi_0} is a positive constant. In addition, $\tilde{\bar{d}}_H(k) = [\tilde{\bar{d}}_{H1}(k), \tilde{\bar{d}}_{H2}(k)]^{\mathrm{T}}$ is defined.

Moreover, the control law u_H is designed by

$$\begin{aligned} u_H(k) = &-\bar{G}_H^{-1}(k)(F_H(k) + \lambda_H \nabla^{-\bar{\gamma}_H}\hat{\bar{d}}_H(k) - \lambda_{1H}(\nabla^{-\bar{\gamma}_H} M_{0H} - e_H(k))) \\ &+\bar{G}_H^{-1}(k)\nabla^{-\bar{\gamma}_H}\left(\sum_{nH=1}^{k} M_{nH} e_H(k - nH)\right) + \bar{G}_H^{-1}(k) H_d(k+1) \end{aligned} \tag{8.31}$$

where $M_{0H} = \sum_{j=0}^{k}(-1)^j\binom{\bar{\gamma}_H}{j} e_H(k-j)$, $\bar{\gamma}_H$ is the fractional order, λ_H and λ_{Hz} are design contestants, $nH = j-1$, $j = 2, \ldots, k+1$, and $M_{nH} = (-1)^{n+1}\binom{\bar{\gamma}_H}{n+1}$.

Combining (8.1) and (8.31), one has

$$e_H(k+1) = \nabla^{-\bar{\gamma}_H}\left(\sum_{nH=1}^{k} M_{nH} e_H(k-nH)\right) - \lambda_H \nabla^{-\bar{\gamma}_H} \hat{\bar{d}}_H(k)$$
$$+\bar{d}_H(k) - \lambda_{1H}(e_H(k) - \nabla^{-\bar{\gamma}_H} M_{0H}) \tag{8.32}$$

Furthermore, (8.32) can be written as

$$\nabla^{\bar{\gamma}_H} e_H(k+1) = \sum_{nH=1}^{k} M_{nH} e_H(k-nH) - \lambda_H \hat{\bar{d}}_H(k) + \nabla^{\bar{\gamma}_H} \bar{d}_H(k)$$
$$-\lambda_{1H}(\nabla^{\bar{\gamma}_H} e_H(k) - M_{0H}) \tag{8.33}$$

According to Definition 2.1, one has

$$\nabla^{\bar{\gamma}_H} e_H(k+1) = e_H(k+1) - \bar{\gamma}_H e_H(k) + \sum_{nH=1}^{k} M_{nH} e_H(k-nH) \tag{8.34}$$

$$\nabla^{\bar{\gamma}_H} e_H(k) = \sum_{j=0}^{k} (-1)^j \binom{\bar{\gamma}_H}{j} e_H(k-j) \tag{8.35}$$

One the basis of (8.33), (8.34) and (8.35), $e_H(k+1)$ can be written as

$$e_H(k+1) = \bar{\gamma}_H e_H(k) - \lambda \hat{\bar{d}}_H(k) + \nabla^{\bar{\gamma}_H} \bar{d}_H(k) \tag{8.36}$$

Furthermore, the i_0th variable of $e_H(k+1)$ can be described by

$$e_{Hi_0}(k+1) = \bar{\gamma}_H e_{Hi_0}(k) + \lambda_H \tilde{\bar{d}}_{Hi_0}(k) - \lambda_H \bar{d}_{Hi_0}(k) + \nabla^{\bar{\gamma}_H} \bar{d}_{Hi_0}(k) \tag{8.37}$$

where $e_{Hi_0}(k+1)$ is the i_0th variable of $e_H(k+1)$.

According to (8.37), one has

$$e^2_{Hi_0}(k+1) \leq 4\bar{\gamma}^2_H e^2_{Hi_0}(k) + 4\lambda^2_H \delta^2_{Hi_0} + 4\bar{\delta}^2_{Hi_0} + 4\lambda^2_H \tilde{\bar{d}}^2_{Hi_0}(k) \tag{8.38}$$

where $\left|\bar{d}_{Hi_0}(k)\right| \leq \delta_{Hi_0}$, $\left|\nabla^{\bar{\gamma}_H} \bar{d}_{Hi_0}(k)\right| \leq \bar{\delta}_{Hi_0}$, δ_{Hi_0} and $\bar{\delta}_{Hi_0}$ are positive constants.

To prove the effectiveness of nonlinear DTDO (8.30) and the controller $u_H(k)$ (8.31), the following Lyapunov function is chosen as:

$$V_H(k) = \sum_{i_0=1}^{2} e^2_{Hi_0}(k) + \sum_{i_0=1}^{2} \zeta_{Hi_0} \tilde{\bar{d}}^2_{Hi_0}(k) \tag{8.39}$$

According to (8.39), the first order difference of Lyapunov function $V_H(k)$ can be written as

$$\begin{aligned}\Delta V_H(k) &= V_H(k{+}1) - V_H(k)\\ &= \sum_{i_0=1}^{2} e_{Hi_0}^2(k+1) + \sum_{i_0=1}^{2} \zeta_{Hi_0}\tilde{\bar{d}}_{Hi_0}^2(k+1)\\ &\quad - \sum_{i_0=1}^{2} e_{Hi_0}^2(k) - \sum_{i_0=1}^{2} \zeta_{Hi_0}\tilde{\bar{d}}_{Hi_0}^2(k)\end{aligned} \tag{8.40}$$

From (8.15), one has

$$\sum_{i_0=1}^{2} \zeta_{Hi_0}(\tilde{\bar{d}}_{Hi_0}^2(k+1) - \tilde{\bar{d}}_{Hi_0}^2(k)) \le -\sum_{i_0=1}^{2} \zeta_{Hi_0}\tilde{\bar{d}}_{Hi_0}^2(k) + \sum_{i_0=1}^{2} \zeta_{Hi_0}D_{Hi_0}^2 \tag{8.41}$$

Combining (8.38) and (8.41), one can obtain

$$\Delta V_H(k) \le -\chi_{H1}\|e_H(k)\|^2 - \chi_{H2}\left\|\tilde{\bar{d}}_H(k)\right\|^2 + \chi_{H3} \tag{8.42}$$

where $\chi_{H3} = \sum_{i_0=1}^{2} (\zeta_{Hi_0}D_{Hi_0}^2 + 4\lambda_H^2\delta_{Hi_0}^2 + 4\bar{\delta}_{Hi_0}^2)$, $\chi_{H1} = \min(1 - 4\bar{\gamma}_H^2)$, $\min(\cdot)$ denotes the minimum value, and $\chi_{H2} = \min(\zeta_{Hi_0} - 4\bar{\gamma}_H^2)$.

8.3.2.2 DTFO Attitude Control Scheme Based on DTDO

Step 1: In the 1st step, the error variables as $e_1(k) = x_1(k) - x_{1d}(k)$ and $e_2(k) = x_2(k) - \mathrm{sat}(x_{vd}(k))$, $x_{vd}(k)$ is the control signal and will be designed. From (8.3), one has

$$\begin{aligned}e_1(k+1) &= x_1(k+1) - x_{1d}(k+1)\\ &= x_1(k) + \bar{F}_1(k) + \bar{G}_1(k)(e_2(k) + \mathrm{sat}(x_{vd}(k)))\\ &\quad + \bar{d}_1(k) - x_{1d}(k+1)\end{aligned} \tag{8.43}$$

where $x_{1d}(k)$ is tracking signal which can be obtained based on the analysis in Sect. 8.3.2.

For the saturation function $\mathrm{sat}(x_{vd}(k))$, the handling method in Chap. 4 is used. Then, (8.43) can be written as

$$\begin{aligned}e_1(k+1) &= x_1(k) + \bar{F}_1(k) + \bar{G}_1(k)(e_2(k) + \mathscr{Y}(k)\\ &\quad + \bar{\lambda}(k)x_{vd}(k)) + \bar{d}_1(k) - x_{1d}(k+1)\end{aligned} \tag{8.44}$$

where $\mathscr{Y}(k)$ is the approximation error, $\bar{\lambda}(k)$ is the unknown and bounded diagonal matrix with $\|\bar{\lambda}(k)\| \leq \bar{\ell}_0$, and $\bar{\ell}_0$ is a positive constant.

For the variable $x_{1d}(k+1)$, $x_{1d}(k+1)$ is predicted by the discrete-time tracking differentiator given in Lemma 2.3. The discrete-time tracking differentiator can be written as

$$\begin{cases} \bar{\mathscr{I}}_{01i}(k+1) = \bar{\mathscr{I}}_{01i}(k) + h_{11i}\bar{\mathscr{I}}_{02i}(k) \\ \bar{\mathscr{I}}_{02i}(k+1) = \bar{\mathscr{I}}_{02i}(k) - h_{11i}r_{11i}\bar{\mathscr{I}}_{03i}(\bar{\mathscr{I}}_{04i}(k), \delta_{11i}) \end{cases} \tag{8.45}$$

where $i = 1, 2, 3$, h_{11i} and r_{11i} are design constants, $\bar{\mathscr{I}}_{01i}(k)$ and $\bar{\mathscr{I}}_{02i}(k)$ are state variables of the discrete-time tracking differentiator.

According to (8.45) and Lemma 2.3, one has

$$x_{1d}(k+1) = \bar{\mathscr{I}}_{01}(k) + h_{11}\bar{\mathscr{I}}_{02}(k) - \tilde{\mathscr{I}}_{01}(k) \tag{8.46}$$

where $\bar{\mathscr{I}}_{01}(k) = [\bar{\mathscr{I}}_{011}(k), \bar{\mathscr{I}}_{012}(k), \bar{\mathscr{I}}_{013}(k)]^{\mathrm{T}}$, $\bar{\mathscr{I}}_{02}(k) = [\bar{\mathscr{I}}_{021}(k), \bar{\mathscr{I}}_{022}(k), \bar{\mathscr{I}}_{023}(k)]^{\mathrm{T}}$, $h_{11} = \mathrm{diag}[h_{111}, h_{112}, h_{113}]$, $r_{11} = \mathrm{diag}[r_{111}, r_{112}, r_{113}]$, the estimated error $\tilde{\mathscr{I}}_{01}(k) = [\tilde{\mathscr{I}}_{011}(k), \tilde{\mathscr{I}}_{012}(k), \tilde{\mathscr{I}}_{013}(k)]^{\mathrm{T}}$, and $\tilde{\mathscr{I}}_{01}(k)$ is bounded.

Substituting (8.46) into (8.44), one has

$$\begin{aligned} e_1(k+1) = {} & x_1(k) + \bar{F}_1(k) + \bar{G}_1(k)(e_2(k) + \mathscr{Y}(k) + \bar{\lambda}(k)x_{vd}(k)) \\ & + \Upsilon_1(k) + \bar{d}_1(k) - \bar{\mathscr{I}}_{01}(k) - h_{11}\bar{\mathscr{I}}_{02}(k) + \tilde{\mathscr{I}}_{01}(k) \end{aligned} \tag{8.47}$$

In order to tackle the disturbance $\bar{d}_1(k)$ in (8.44), the designed DTDO is introduced based on (8.8) and (8.9) as follows:

$$\begin{cases} \hat{\mathscr{M}}_{1i}(k+1) = -\zeta_{1i}\bar{F}_{1i}(k) - \zeta_{1i}\mathscr{U}_{1i}(k) \\ \zeta_{1i}\hat{\bar{d}}_{1i}(k) = \hat{\mathscr{M}}_{1i}(k) + \zeta_{1i}x_{1i}(k) \end{cases} \tag{8.48}$$

where $i = 1, 2, 3$, $x_{1i}(k)$ is the ith element of $x_1(k)$, $\bar{F}_{1i}(k)$ is the ith element of $\bar{F}_1(k) + x_1(k)$, $\hat{\bar{d}}_{1i}(k)$ is the ith output of DTDO, and $\hat{\bar{d}}_1(k) = [\hat{\bar{d}}_{11}(k), \hat{\bar{d}}_{12}(k), \hat{\bar{d}}_{13}(k)]^{\mathrm{T}}$. $\hat{\mathscr{M}}_{1i}(k)$ is the state variable of DTDO. The control parameter ζ_{1i} satisfies that $\zeta_{1i} > 0$. Furthermore, The disturbance estimation error is defined as $\tilde{\bar{d}}_{1i}(k) = \bar{d}_{1i}(k) - \hat{\bar{d}}_{1i}(k)$, and $\bar{d}_1(k) = [\bar{d}_{11}(k), \bar{d}_{12}(k), \bar{d}_{13}(k)]^{\mathrm{T}}$ $\mathscr{U}_{1i}(k)$ is the ith element of $\mathscr{U}_1(k)$, and $\mathscr{U}_1(k) = \bar{G}_1(k)x_2(k)$. According to Assumption 8.2, we have that $\left|\bar{d}_{1i}(k+1) - \bar{d}_{1i}(k)\right| \leq D_{1i}$, and D_{1i} is a positive constant. Furthermore, $\tilde{\bar{d}}_1(k) = [\tilde{\bar{d}}_{11}(k), \tilde{\bar{d}}_{12}(k), \tilde{\bar{d}}_{13}(k)]^{\mathrm{T}}$ is defined.

According to the designed method of control signal $x_{vd}(k)$ in Chap. 4, Lemma 2.2 and (8.47), the NN is used to approximate the uncertainty $(1/\tau_1)\Delta\bar{\mathscr{F}}_{1\bar{j}}(k)(\bar{j} = 1, 2, 3)$ with $\Delta\bar{\mathscr{F}}_{1i}(k)$ being the ith variable of $\Delta\bar{\mathscr{F}}_1(k) = -(\bar{\lambda}(k)\bar{G}_1(k))^{-1}(x_1(k) + \bar{F}_1(k) - x_{1d}(k+1) + \hat{\bar{d}}_1(k)) + \mathscr{N}_1 e_1(k) - \bar{\mathscr{I}}_{01}(k) - h_{11}\bar{\mathscr{I}}_{02}(k)) + \bar{G}_1^{-1}(k)(x_1(k) + \bar{F}_1(k) - x_{1d}(k+1) + \hat{\bar{d}}_1(k)) + \mathscr{N}_1 e_1(k) - \bar{\mathscr{I}}_{01}(k) - h_{11}\bar{\mathscr{I}}_{02}(k))$, the control signal

$x_{vd}(k)$ can be written as

$$\begin{aligned} x_{vd}(k) &= -\bar{G}_1^{-1}(k)(x_1(k) + \bar{F}_1(k) - x_{1d}(k+1) + \hat{\bar{d}}_1(k)) \\ &\quad -\bar{G}_1^{-1}(k)(\mathcal{N}_1 e_1(k) - \bar{\mathcal{I}}_{01}(k) - h_{11}\bar{\mathcal{I}}_{02}(k)) + \bar{\Theta}_1(k) + \Upsilon_1(k) \end{aligned} \tag{8.49}$$

where $\Theta_1(k) = [\frac{1}{\tau_1}\zeta_{11}^{*\mathrm{T}}(k)\varphi_{11}(z_1), \frac{1}{\tau_1}\zeta_{12}^{*\mathrm{T}}(k)\varphi_{12}(z_1), \frac{1}{\tau_1}\zeta_1^{*\mathrm{T}}(k)\varphi_{13}(z_1)]^{\mathrm{T}}$, $\tau_1 > 0$ is a design constant, $\Upsilon_1(k) = [\frac{1}{\tau_1}\varepsilon_{11}(k), \frac{1}{\tau_1}\varepsilon_{12}(k), \frac{1}{\tau_1}\varepsilon_{13}(k)]^{\mathrm{T}}$, $\zeta_{1i}^*(k)$ is the optimal weight vector of the NN, $\varepsilon_{1i}(k)$ is the minimum approximation error, $\varphi_{1i}(\cdot)$ is the basis function vector, $i = 1, 2, 3$, and $z_1 = [x_1^{\mathrm{T}}(k), e_1^{\mathrm{T}}(k), \bar{\mathcal{I}}_{01}^{\mathrm{T}}(k), \bar{\mathcal{I}}_{02}^{\mathrm{T}}(k), \hat{\bar{d}}_1^{\mathrm{T}}(k)]^{\mathrm{T}}$. To facilitate the design of control scheme, define $\varphi_{1i}(z_1) = \varphi_{1i}(k)$.

Then, the control signal $x_{vd}(k)$ is designed as

$$\begin{aligned} x_{vd}(k) &= -\bar{G}_1^{-1}(k)(x_1(k) + \bar{F}_1(k) - x_{1d}(k+1) + \hat{\bar{d}}_1(k)) \\ &\quad -\bar{G}_1^{-1}(k)(\mathcal{N}_1 e_1(k) - \bar{\mathcal{I}}_{01}(k) - h_{11}\bar{\mathcal{I}}_{02}(k)) + \hat{\Theta}_1(k) \end{aligned} \tag{8.50}$$

where $\hat{\Theta}_1(k) = [\frac{1}{\tau_1}\hat{\zeta}_{11}^{\mathrm{T}}(k)\varphi_{11}(z_1), \frac{1}{\tau_1}\hat{\zeta}_{12}^{\mathrm{T}}(k)\varphi_{12}(z_1), \frac{1}{\tau_1}\hat{\zeta}_1^{\mathrm{T}}(k)\varphi_{13}(z_1)]^{\mathrm{T}}$, $\hat{\bar{d}}_1(k)$ is the estimation of $\bar{d}_1(k)$, $\hat{\Theta}_1(k)$ is the estimation of $\Theta(k)$, the diagonal matrix $\mathcal{N}_1 = \mathrm{diag}[\mathcal{N}_{11}, \mathcal{N}_{12}, \mathcal{N}_{13}]$ and $\mathcal{N}_{1i}$ is a designed constant.

Substituting (8.50) into (8.47), we obtain

$$\begin{aligned} e_1(k+1) &= \tilde{\bar{d}}_1(k) + \bar{G}_1(k)e_2(k) - \mathcal{N}_1 e_1(k) + \tilde{\mathcal{I}}_{01}(k) + \bar{G}_1(k)\mathcal{Y}(k) \\ &\quad + \bar{\lambda}(k)\bar{G}_1(k)\tilde{\Theta}_1(k) - \bar{\lambda}(k)\bar{G}_1(k)\Upsilon_1(k) \end{aligned} \tag{8.51}$$

where $\tilde{\Theta}_1(k) = [\frac{1}{\tau_1}\tilde{\zeta}_{11}^{\mathrm{T}}(k)\varphi_{11}(z_1), \frac{1}{\tau_1}\tilde{\zeta}_{12}^{\mathrm{T}}(k)\varphi_{12}(z_1), \frac{1}{\tau_1}\tilde{\zeta}_1^{\mathrm{T}}(k)\varphi_{13}(z_1)]^{\mathrm{T}}$ and $\tilde{\Theta}_1(k) = \hat{\Theta}_1(k) - \Theta_1(k)$.

According to (8.51), the ith element of $e_1(k+1)$ can be written as

$$e_{1i}(k+1) = \tilde{\bar{d}}_{1i}(k) + E_{1i}(k) - \mathcal{N}_{1i}e_{1i}(k) + \tilde{\mathcal{I}}_{01i}(k) + \Delta\tilde{\Theta}_{1i}(k) + \bar{\mathcal{Y}}_i(k) \tag{8.52}$$

where $e_{1i}(k+1)$ is the ith element of $e_1(k+1)$, $E_{1i}(k)$ is the ith element of $\bar{G}_1(k)e_2(k)$, $\Delta\tilde{\Theta}_{1i}(k)$ is the ith element of $\bar{\lambda}(k)\bar{G}_1(k)\tilde{\Theta}_1(k)$, and $\bar{\mathcal{Y}}_i(k)$ is the ith element of $\bar{G}_1(k)\mathcal{Y}(k) - \bar{\lambda}(k)\bar{G}_1(k)\Upsilon_1(k)$.

From (8.52), one has

$$\begin{aligned} \zeta_{1i}^2 e_{1i}^2(k+1) &\le 5\zeta_{1i}^2\tilde{\bar{d}}_{1i}^2(k) + 5\zeta_{1i}^2\mathcal{N}_{1i}^2 e_{1i}^2(k) + 5\zeta_{1i}^2 E_{1i}^2(k) \\ &\quad + 5\zeta_{1i}^2\bar{\mathcal{I}}_{2i}^2 + 5\zeta_{1i}^2\Delta\tilde{\Theta}_{1i}^2(k) \end{aligned} \tag{8.53}$$

where $\left|\tilde{\mathcal{I}}_{01i}(k) + \bar{\mathcal{Y}}_i(k)\right| \le \bar{\mathcal{I}}_{2i}$ and $\bar{\mathcal{I}}_{2i}$ is a positive constant.

In addition, the adaptive law $\hat{\zeta}_{1\bar{j}}(k)$ is designed by

$$\hat{\zeta}_{1i}(k+1) = \kappa_{1i}\varphi_{1i}(k)e_{1i}(k) - (\omega_{1i} - 1)\hat{\zeta}_{1i}(k) \tag{8.54}$$

where $\kappa_{1i} > 0$ and $\omega_{1i} > 0$ are design constants, $e_{1i}(k)$ is the ith variables of $e_1(k)$.

According to (8.54), one has

$$\tilde{\zeta}_{1i}(k+1) = \tilde{\zeta}_{1i}(k) + \kappa_{1i}\varphi_{1i}(k)e_{1i}(k) - \omega_{1i}\hat{\zeta}_{1i}(k) \tag{8.55}$$

where $\tilde{\zeta}_{1i}(k+1) = \hat{\zeta}_{1i}(k+1) - \zeta_{1i}^*(k+1)$, and $\tilde{\zeta}_{1i}(k) = \hat{\zeta}_{1i}(k) - \zeta_{1i}^*(k)$.

From (8.55), it yields

$$\begin{aligned}\left\|\tilde{\zeta}_{1i}(k+1)\right\|^2 - \left\|\tilde{\zeta}_{1i}(k)\right\|^2 &= \kappa_{1i}^2\|\varphi_{1i}(k)\|^2 e_{1i}^2(k) \\ &+\omega_{1i}^2\left\|\hat{\zeta}_{1i}(k)\right\|^2 - 2\kappa_{1i}\omega_{1i}\varphi_{1i}^{\mathrm{T}}(k)\hat{\zeta}_{1i}(k)e_{1i}(k) \\ &+2\kappa_{1i}\tilde{\zeta}_{1i}^{\mathrm{T}}(k)\varphi_{1i}(k)e_{1i}(k) - 2\omega_{1i}\tilde{\zeta}_{1i}^{\mathrm{T}}(k)\hat{\zeta}_{1i}(k)\end{aligned} \tag{8.56}$$

Furthermore, one can obtain

$$-2\omega_{1i}^2\tilde{\zeta}_{1i}^{\mathrm{T}}(k)\hat{\zeta}_{1i}(k) = -\omega_{1i}^2\left\|\tilde{\zeta}_{1i}(k)\right\|^2 - \omega_{1i}^2\left\|\hat{\zeta}_{1i}(k)\right\|^2 + \omega_{1i}^2\|\zeta_{1i}^*(k)\|^2 \tag{8.57}$$

Combining (8.56) and (8.57), one has

$$\begin{aligned}\omega_{1i}(\left\|\tilde{\zeta}_{1i}(k+1)\right\|^2 - \left\|\tilde{\zeta}_{1i}(k)\right\|^2) &\leq -(\omega_{1i}^2 - \omega_{1i}^3 - \kappa_{1i}^2\omega_{1i}^2)\left\|\hat{\zeta}_{1i}(k)\right\|^2 \\ &+(2\omega_{1i}^2\bar{\varphi}_1 + \omega_{1i}\kappa_{1i}^2\bar{\varphi}_1)e_{1i}^2(k) - (\omega_{1i}^2 - \kappa_{1i}^2)\left\|\tilde{\zeta}_{1i}(k)\right\|^2 + \omega_{1i}^2\bar{\zeta}_1^*\end{aligned} \tag{8.58}$$

where $\|\varphi_{1i}(k)\|^2 \leq \bar{\varphi}_1$, $\left\|\zeta_{1i}^*(k)\right\|^2 \leq \bar{\zeta}_1^*$, $\bar{\varphi}_1 > 0$ and $\bar{\zeta}_1^* > 0$ are constants.

Consider the following Lyapunov function candidate:

$$V_1(k) = \sum_{i=1}^{3}\varsigma_{1i}^2 e_{1i}^2(k) + \sum_{i=1}^{3}\varsigma_{1i}\tilde{\bar{d}}_{1i}^2(k) + \sum_{i=1}^{3}\omega_{1i}\left\|\tilde{\zeta}_{1i}(k)\right\|^2 \tag{8.59}$$

Then, the first difference of $V_1(k)$ is calculated as

$$\begin{aligned}\Delta V_1(k) &= V_1(k+1) - V_1(k) \\ &= \sum_{i=1}^{3}\varsigma_{1i}^2 e_{1i}^2(k+1) + \sum_{i=1}^{3}\varsigma_{1i}\tilde{\bar{d}}_{1i}^2(k+1) + \sum_{i=1}^{3}\omega_{1i}\left\|\tilde{\zeta}_{1i}(k+1)\right\|^2 \\ &-\sum_{i=1}^{3}\varsigma_{1i}^2 e_{1i}^2(k) - \sum_{i=1}^{3}\varsigma_{1i}\tilde{\bar{d}}_{1i}^2(k) - \sum_{i=1}^{3}\omega_{1i}\left\|\tilde{\zeta}_{1i}(k)\right\|^2\end{aligned} \tag{8.60}$$

Based on (8.53), one has

$$\sum_{i=1}^{3}\zeta_{1i}^2(e_{1i}^2(k+1)-e_{1i}^2(k)) \le \sum_{i=1}^{3}5\zeta_{1i}^2\tilde{\bar{d}}_{1i}^2(k)+\sum_{i=1}^{3}5\zeta_{1i}^2\mathscr{I}_{2i}^2$$
$$-\sum_{i=1}^{3}(\zeta_{1i}^2(1-5\mathscr{N}_{1i}^2)e_{1i}^2(k)+5\zeta_{1i}^2E_{1i}^2(k))+\sum_{i=1}^{3}\frac{5\zeta_{1i}^2\bar{\varphi}_1\bar{g}_1^2\bar{\ell}_0^2}{\tau_1^2}\left\|\tilde{\zeta}_{1i}(k)\right\|^2 \quad (8.61)$$

According to (8.15), one can obtain

$$\sum_{i=1}^{3}\zeta_{1i}(\tilde{\bar{d}}_{1i}^2(k+1)-\tilde{\bar{d}}_{1i}^2(k)) \le \sum_{i=1}^{3}(-\zeta_{1i}\tilde{\bar{d}}_{1i}^2(k)+\zeta_{1i}D_{1i}^2) \quad (8.62)$$

Combining (8.61) and (8.62), $\Delta V_1(k)$ can be written as

$$\Delta V_1(k) \le -\chi_{11}\|e_1(k)\|^2-\chi_{12}\left\|\tilde{\bar{d}}_1(k)\right\|^2+\chi_{13}\|e_2(k)\|^2$$
$$-\sum_{i=1}^{3}\chi_{14}\left\|\tilde{\zeta}_{1i}(k)\right\|^2-\sum_{i=1}^{3}\chi_{15}\left\|\hat{\zeta}_{1i}(k)\right\|^2+\chi_{16} \quad (8.63)$$

where $\chi_{11}=\min(\zeta_{1i}^2-\zeta_{1i}^2 5\mathscr{N}_{1i}^2-2\omega_{1i}^2\bar{\varphi}_1-\omega_{1i}\kappa_{1i}^2\bar{\varphi}_1)$, $\chi_{12}=\min(\zeta_{1i}-4\zeta_{1i}^2)$, $\chi_{13}=\max(4\zeta_{1i}^2\bar{g}_1^2)$, $\chi_{14}=\omega_{1i}^2-\kappa_{1i}^2-\frac{5\zeta_{1i}^2\bar{\varphi}_1\bar{g}_1^2\bar{\ell}_0^2}{\tau_1^2}$, $\chi_{15}=\omega_{1i}^2-\omega_{1i}^3-\kappa_{1i}^2\omega_{1i}^2$, $\max(\cdot)$ denotes the maximum value, and $\chi_{16}=\sum_{i=1}^{3}\zeta_{1i}D_{1i}^2+\sum_{i=1}^{3}5\zeta_{1i}^2\mathscr{I}_{2i}^2+\sum_{i=1}^{3}\omega_{1i}^2\bar{\zeta}_1^*$.

Step 2: Considering (8.3) and $e_2(k)=x_2(k)-\text{sat}(x_{vd}(k))$, one has

$$\begin{aligned} e_2(k+1) &= x_2(k+1)-\text{sat}(x_{vd}(k+1)) \\ &= x_2(k)+\bar{F}_2(k)+\bar{G}_2(k)u(k)+\bar{d}_2(k)-v_d(k+1) \end{aligned} \quad (8.64)$$

In order to tackle the disturbance $\bar{d}_2(k)$ in (8.64), the designed DTDO is introduced based on (8.8) and (8.9) as follows:

$$\begin{cases} \hat{\mathscr{M}}_{2i}(k+1)=-\zeta_{2i}\bar{F}_{2i}(k)-\zeta_{2i}\mathscr{U}_{2i}(k) \\ \zeta_{2i}\hat{\bar{d}}_{2i}(k)=\hat{\mathscr{M}}_{2i}(k)+\zeta_{2i}x_{2i}(k) \end{cases} \quad (8.65)$$

where $i=1,2,3$, $x_{2i}(k)$ is the ith element of $x_2(k)$, $\bar{F}_{2i}(k)$ is the ith element of $\bar{F}_2(k)+x_2(k)$, $\hat{\bar{d}}_{2i}(k)$ is the ith output of DTDO, and $\hat{\bar{d}}_2(k)=[\hat{\bar{d}}_{21}(k),\hat{\bar{d}}_{22}(k),\hat{\bar{d}}_{23}(k)]^{\mathrm{T}}$. $\hat{\mathscr{M}}_{2i}(k)$ is the state variable of DTDO. The designed control parameter ζ_{2i} satisfies that $\zeta_{2i}>0$. Moreover, the disturbance estimation error is defined as $\tilde{\bar{d}}_{2i}(k)=\bar{d}_{2i}(k)-\hat{\bar{d}}_{2i}(k)$, and $\bar{d}_2(k)=[\bar{d}_{21}(k),\bar{d}_{22}(k),\bar{d}_{23}(k)]^{\mathrm{T}}$. $\mathscr{U}_{2i}(k)$ is the ith element of $\mathscr{U}_2(k)$, and $\mathscr{U}_2(k)=\bar{G}_2(k)u(k)$. According to Assumption 8.2, one can obtain that $|\bar{d}_{2i}(k+1)-\bar{d}_{2i}(k)|\le D_{2i}$, and D_{2i} is a positive constant. In addition, $\tilde{\bar{d}}_2(k)=[\tilde{\bar{d}}_{21}(k),\tilde{\bar{d}}_{22}(k),\tilde{\bar{d}}_{23}(k)]^{\mathrm{T}}$ is introduced.

To forecast $v_d(k+1)$, the following discrete form of tracking differentiator based on Lemma 2.3 is employed to estimate $v_d(k+1)$:

$$\begin{cases} \bar{\mathscr{I}}_{11i}(k+1) = \bar{\mathscr{I}}_{11i}(k) + h_{01i}\bar{\mathscr{I}}_{12i}(k) \\ \bar{\mathscr{I}}_{12i}(k+1) = \bar{\mathscr{I}}_{12i}(k) - h_{01i}r_{01i}\bar{\mathscr{I}}_{13i}(\bar{\mathscr{I}}_{14i}(k), \delta_{01i}) \end{cases} \tag{8.66}$$

where $i = 1, 2, 3$, h_{01i} and r_{01i} denote design constants, $\bar{\mathscr{I}}_{11i}(k)$ and $\bar{\mathscr{I}}_{12i}(k)$ state variables of discrete-time tracking differentiator.

According to (8.66) and Lemma 2.3, one has

$$x_{vd}(k+1) = \bar{\mathscr{I}}_{11}(k) + h_{01}\bar{\mathscr{I}}_{12}(k) - \tilde{\mathscr{I}}(k) \tag{8.67}$$

where $\bar{\mathscr{I}}_{11}(k) = [\bar{\mathscr{I}}_{111}(k), \bar{\mathscr{I}}_{112}(k), \bar{\mathscr{I}}_{113}(k)]^{\mathrm{T}}$, $\bar{\mathscr{I}}_{12}(k) = [\bar{\mathscr{I}}_{121}(k), \bar{\mathscr{I}}_{122}(k), \bar{\mathscr{I}}_{123}(k)]^{\mathrm{T}}$, $h_{01} = \mathrm{diag}[h_{011}, h_{012}, h_{013}]$, $r_{01} = \mathrm{diag}[r_{011}, r_{012}, r_{013}]$, $\tilde{\mathscr{I}}(k) = [\tilde{\mathscr{I}}_1(k), \tilde{\mathscr{I}}_2(k), \tilde{\mathscr{I}}_3(k)]^{\mathrm{T}}$ is the estimation error vector, and $\tilde{\mathscr{I}}(k)$ is bounded. Therefore, according to Definition 2.1, we have that $\left|\nabla^{\bar{\gamma}}\tilde{\mathscr{I}}_i(k)\right| \le \bar{\mathscr{I}}_i$, and $\bar{\mathscr{I}}_i$ is a positive constant with $\bar{\mathscr{I}} = [\bar{\mathscr{I}}_1, \bar{\mathscr{I}}_2, \bar{\mathscr{I}}_3]^{\mathrm{T}}$.

Substituting (8.67) into (8.64), one has

$$\begin{aligned} e_2(k+1) = {} & x_2(k) + \bar{F}_2(k) + \bar{G}_2(k)u(k) + \bar{d}_2(k) - \bar{\mathscr{I}}_{11}(k) \\ & -h_{01}\bar{\mathscr{I}}_{12}(k) + \tilde{\mathscr{I}}(k) \end{aligned} \tag{8.68}$$

The DTFO controller is designed as

$$\begin{aligned} u(k) = {} & -\bar{G}_2^{-1}(k)(x_2(k) + \bar{F}_2(k) - \bar{\mathscr{I}}_{11}(k) + \lambda\nabla^{-\bar{\gamma}}\hat{\bar{d}}_2(k)) \\ & +\bar{G}_2^{-1}(k)(h_{01}\bar{\mathscr{I}}_{12}(k) + \lambda_1(\nabla^{-\bar{\gamma}}M_0 - e_2(k))) \\ & +\bar{G}_2^{-1}(k)\nabla^{-\bar{\gamma}}\left(\sum_{n=1}^{k} M_n e_2(k-n)\right) \end{aligned} \tag{8.69}$$

where $M_0 = \sum_{j=0}^{k}(-1)^j\binom{\bar{\gamma}}{j}e_2(k-j)$, $\bar{\gamma}$ is the fractional order, λ and λ_1 are two designed constants, $n = j-1$, $j = 2, \ldots, k+1$ and $M_n = (-1)^{n+1}\binom{\bar{\gamma}}{n+1}$.

Combining (8.68) and (8.69), one has

$$\begin{aligned} e_2(k+1) = {} & \nabla^{-\bar{\gamma}}\left(\sum_{n=1}^{k} M_n e_2(k-n)\right) - \lambda\nabla^{-\bar{\gamma}}\hat{\bar{d}}_2(k) \\ & +d_2(k) + \tilde{\mathscr{I}}(k) - \lambda_1(e_2(k) - \nabla^{-\bar{\gamma}}M_0) \end{aligned} \tag{8.70}$$

Furthermore, (8.70) can be written as

$$\nabla^{\bar{\gamma}} e_2(k+1) = \sum_{n=1}^{k} M_n e_2(k-n) - \lambda \hat{\bar{d}}_2(k) + \nabla^{\bar{\gamma}} \bar{d}(k)$$
$$+ \nabla^{\bar{\gamma}} \tilde{\mathscr{I}}(k) - \lambda_1 (\nabla^{\bar{\gamma}} e_2(k) - M_0) \tag{8.71}$$

From Definition 2.1, we obtain

$$\nabla^{\bar{\gamma}} e_2(k+1) = e_2(k+1) - \bar{\gamma} e_2(k) + \sum_{n=1}^{k} M_n e_2(k-n) \tag{8.72}$$

$$\nabla^{\bar{\gamma}} e_2(k) = \sum_{j=0}^{k} (-1)^j \binom{\bar{\gamma}}{j} e_2(k-j) \tag{8.73}$$

On the basis of (8.71), (8.72) and (8.73), $e_2(k+1)$ can be written as

$$e_2(k+1) = \bar{\gamma} e_2(k) - \lambda \hat{\bar{d}}_2(k) + \nabla^{\bar{\gamma}} \bar{d}(k) + \nabla^{\bar{\gamma}} \tilde{\mathscr{I}}(k) \tag{8.74}$$

For the ith element of $e_2(k+1)$, we have

$$e_{2i}(k+1) = \bar{\gamma} e_{2i}(k) + \lambda \tilde{\bar{d}}_{2i}(k) - \lambda \bar{d}_{2i}(k) + \nabla^{\bar{\gamma}} \bar{d}_{2i}(k) + \nabla^{\bar{\gamma}} \tilde{\mathscr{I}}_i(k) \tag{8.75}$$

where $e_{2i}(k+1)$ is the ith element of $e_2(k+1)$.

According to (8.75), one has

$$e_{2i}^2(k+1) \leq 5\bar{\gamma}^2 e_{2i}^2(k) + 5\lambda^2 \delta_{2i}^2 + 5\bar{\delta}_{2i}^2 + 5\bar{\mathscr{I}}_i^2 + 5\lambda^2 \tilde{\bar{d}}_{2i}^2(k) \tag{8.76}$$

where $|d_{2i}(k)| \leq \delta_{2i}$, $\left|\nabla^{\bar{\gamma}} d_{2i}(k)\right| \leq \bar{\delta}_{2i}$, δ_{2i} and $\bar{\delta}_{2i}$ are positive constants.

Consider the following Lyapunov function candidate:

$$V_2(k) = \sum_{i=1}^{3} e_{2i}^2(k) + \sum_{i=1}^{3} \zeta_{2i} \tilde{\bar{d}}_{2i}^2(k) \tag{8.77}$$

Based on (8.77), the first difference of Lyapunov $V_2(k)$ is calculated as

$$\begin{aligned} \Delta V_2(k) &= V_2(k+1) - V_2(k) \\ &= \sum_{i=1}^{3} e_{2i}^2(k+1) + \sum_{i=1}^{3} \zeta_{2i} \tilde{\bar{d}}_{2i}^2(k+1) - \sum_{i=1}^{3} e_{2i}^2(k) - \sum_{i=1}^{3} \zeta_{2i} \tilde{\bar{d}}_{2i}^2(k) \end{aligned} \tag{8.78}$$

Referring (8.15), it yields

$$\sum_{i=1}^{3} \zeta_{2i}(\tilde{\bar{d}}_{2i}^2(k+1) - \tilde{\bar{d}}_{2i}^2(k)) \leq -\sum_{i=1}^{3} \zeta_{2i}\tilde{\bar{d}}_{2i}^2(k) + \sum_{i=1}^{3} \zeta_{2i} D_{2i}^2 \tag{8.79}$$

Combining (8.76) and (8.79), one has

$$\Delta V_2(k) \leq -\chi_{21}\|e_2(k)\|^2 - \chi_{22}\left\|\tilde{\bar{d}}_2(k)\right\|^2 + \chi_{23} \tag{8.80}$$

where $\chi_{23} = \sum_{i=1}^{3}(\zeta_{2i} D_{2i}^2 + 5\lambda^2\delta_{2i}^2 + 5\bar{\delta}_{2i}^2 + 5\bar{\mathscr{I}}_i^2)$, $\chi_{21} = \min(1 - 5\bar{\gamma}^2)$ and $\chi_{22} = \min(\zeta_{2i} - 5\bar{\gamma}^2)$.

The above design procedure of DTFO controller based on DTDO for the trajectory control system model (8.1) and the attitude dynamics system (8.3) with disturbances can be summarized in the following theorem:

Theorem 8.2 *Consider the discrete-time UAV system with disturbances (8.1) and (8.3), if the design parameters ζ_z, ζ_{Hi_0}, ζ_{1i}, ζ_{2i}, λ_z, λ_{1z}, λ_H, λ_{1H}, λ, λ_1, $\bar{\gamma}_z$, $\bar{\gamma}_H$, $\bar{\gamma}$ and $\mathscr{N}_{1i}$ are chosen appropriately, by designing the height controller (8.18), the velocity and track control law (8.31), the control signal $x_{vd}k$ (8.50), the DTFO attitude controller (8.69) and the DTDOs (8.8) and (8.9), then, the proposed DTFO controller can guarantee that all the signals in the closed-loop system are bounded and the tracking errors can converge to a bounded compact set.*

Proof For the whole system, the following Lyapunov function is given by

$$V(k) = \sum_{i_0=1}^{2} V_{i_0}(k) + V_z(k) + V_H(k) \tag{8.81}$$

According to (8.28), (8.42), (8.63) and (8.80), one has

$$\begin{aligned}
\Delta V(k) &= \sum_{i_0=1}^{2} V_{i_0}(k+1) - \sum_{i_0=1}^{2} V_{i_0}(k) + V_z(k+1) + V_H(k+1) - V_z(k) - V_H(k) \\
&\leq -\chi_{11}\|e_1(k)\|^2 - \chi_{12}\left\|\tilde{\bar{d}}_1(k)\right\|^2 + \chi_{13}\|e_2(k)\|^2 - \chi_{z1}e_z^2(k) \\
&\quad -\chi_{z2}\tilde{\bar{d}}_z^2(k) + \chi_{z3} - \chi_{21}\|e_2(k)\|^2 - \chi_{22}\left\|\tilde{\bar{d}}_2(k)\right\|^2 + \chi_{16} + \chi_{23} \\
&\quad -\chi_{H1}\|e_H(k)\|^2 - \chi_{H2}\left\|\tilde{\bar{d}}_H(k)\right\|^2 + \chi_{H3} - \sum_{i=1}^{3} \chi_{14}\left\|\tilde{\zeta}_{1i}(k)\right\|^2 \\
&\quad -\sum_{i=1}^{3} \chi_{15}\left\|\hat{\zeta}_{1i}(k)\right\|^2
\end{aligned} \tag{8.82}$$

From (8.82), if $\chi_{11} > 0$, $\chi_{12} > 0$, $\chi_{14} > 0$, $\chi_{15} > 0$, $\chi_{z1} > 0$, $\chi_{z2} > 0$, $\chi_{H1} > 0$, $\chi_{H2} > 0$, $\chi_{21} - \chi_{13} > 0$ and $\chi_{22} > 0$ are satisfied by choosing appropriate parameters, we can obtain that the tracking error $e_z(k)$, $e_H(k)$ and $e_1(k)$ are bounded. Furthermore, the conclusion that the estimation errors $\tilde{\bar{d}}_z(k)$, $\tilde{\bar{d}}_H(k)$, $\tilde{d}_1(k)$ and $\tilde{d}_2(k)$ are bounded from (8.82). This concludes the proof. ◇

8.4 Simulation Study

In this section, the simulation results are given for the trajectory control system model (8.1) with wind disturbances and the attitude dynamic model of the UAV system with wind disturbances (8.2) to illustrate the effectiveness of DTDO-based DTFO tracking control scheme.

In the simulation, the effects of gust on the control of UAV system are considered, and the specific form of gust is shown in (2.40). In this numerical simulation, the maximum external wind speed is assumed as $V_{Wm} = [10, 5, 4]^{\mathrm{T}}$m/s, and $t_m = 3$s. Furthermore, The external harmonic disturbances are set as $\bar{d}_2 = \frac{\Delta T}{2}[3\sin(0.2t) - 4\cos(0.3t), 4\sin(0.1t) - 4\cos(0.4t), 4\sin(0.5t) - 6\cos(0.2t)]^T$. The sampling period ΔT is selected as $\Delta T = 0.002$. The fractional order $\bar{\gamma}_z$, $\bar{\gamma}_H$ and $\bar{\gamma}$ are chosen as $\bar{\gamma} = \bar{\gamma}_z = \bar{\gamma}_H = 0.001$. The initial values of UAV system are chosen as $\beta_0 = 0$deg, $\alpha_0 = 0.2$deg, $\mu_0 = 0$deg and $p_0 = q_0 = r_0 = 0$deg/s. The flight altitude is 1005m and the flight velocity is 90m/s. The desired tracking signals β_d and μ_d are given as $\beta_d = 0$deg and $\mu_d = 0$deg, and the desired signals α_d and γ_d can be obtained by calculating $u_H(k)$ and $u_z(k)$, respectively. The tracking signals of velocity and altitude are selected as $100\mathrm{erf}(0.2t) + 100$ and $300\mathrm{erf}(0.2t) + 1000$. The control parameters in the DTDO are chosen as $\zeta_z = 0.2$, $\zeta_{H1} = \zeta_{H2} = 0.2$, $\zeta_{11} = \zeta_{12} = 0.2$, $\zeta_{13} = 0.1$ and $\zeta_{21} = \zeta_{22} = \zeta_{23} = 0.002$. Set the control parameters as $\lambda_z = 0.0001$, $\lambda_{1z} = -1$, $\lambda_z = 0.0001$, $\lambda_{1H} = -1$, $\lambda = 0.0001$, $\lambda_1 = 0.2$, $\mathcal{N}_1 = \mathrm{diag}[-0.31, -0.31, -0.31]$, $h_{01} = \mathrm{diag}[6.5, 6.5, 6.5]$ and $r_{01} = \mathrm{diag}[20, 20, 20]$. Furthermore, the following definitions are that $\hat{\bar{d}}_v$ and $\hat{\bar{d}}_\gamma$ are the estimations of $\bar{d}_v$ and $\bar{d}_\gamma$, respectively. In addition, the disturbance estimation errors are defined as $\tilde{\bar{d}}_v = \bar{d}_v - \hat{\bar{d}}_v$ and $\tilde{\bar{d}}_\gamma = \bar{d}_\gamma - \hat{\bar{d}}_\gamma$, and the tracking errors are defined as $e_z = \bar{z}_g - \bar{z}_{gd}$, $e_v = \bar{V}_A - \bar{V}_{Ad}$, $e_\gamma = \gamma - \gamma_d$, $e_\beta = \beta - \beta_d$, $e_\alpha = \alpha - \alpha_d$ and $e_\mu = \mu - \mu_d$.

According to the above designed parameters, on the basis of the proposed DTDO-based DTFO tracking control approach, the simulation results are shown in Figs. 8.2, 8.3, 8.4, 8.5, 8.6, 8.7, 8.8, 8.9, 8.10, 8.11, 8.12 and 8.13. According to the tracking result in Fig. 8.2, the UAV can fly at the desired altitude. The flight velocity control performance of the UAV is presented in Fig. 8.3. It can also show that the velocity of UAV under discrete time control can fly according to the desired velocity. According to Fig. 8.4 and Fig. 8.5, the track angle γ of the UAV can keep up with the desired signal, and the thrust $\bar{T}$ changes over time. The tracking control results and tracking errors e_β, e_α and e_μ for the discrete-time attitude dynamics of the UAV system with external disturbances (8.2) are shown in Fig. 8.6 and Fig. 8.7, respectively. According

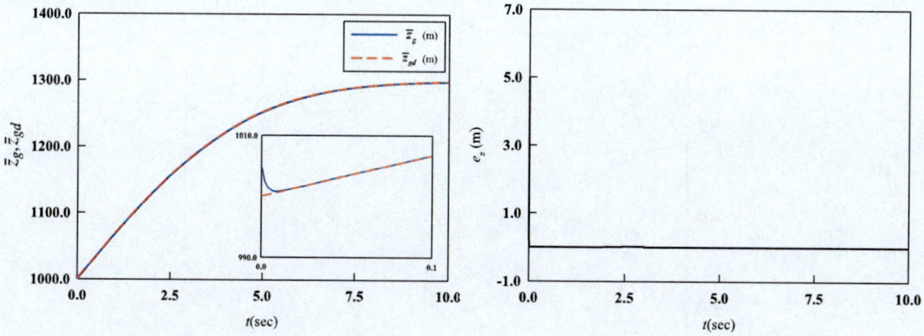

(a) Height tracking signals $\bar{z}_g$ and $\bar{z}_{gd}$ (b) Height tracking error e_z

Fig. 8.2 Tracking control results of UAV flight height

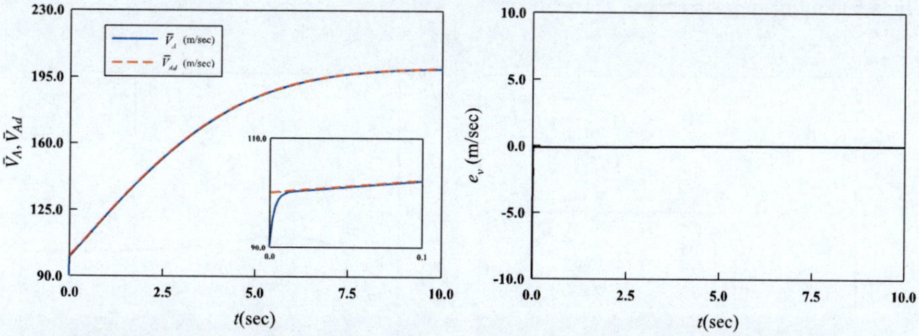

(a) Velocity tracking response $\bar{V}_A$ and $\bar{V}_{Ad}$ (b) Velocity tracking error e_v

Fig. 8.3 Tracking control results of flight velocity

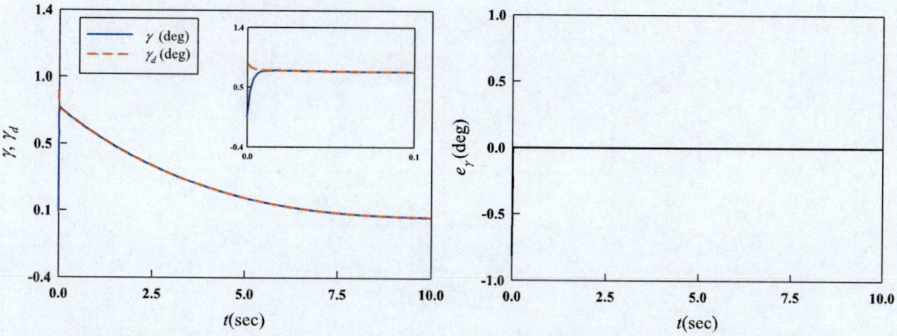

(a) Track angle response γ and γ_d (b) Track angle tracking error e_γ

Fig. 8.4 Tracking control results of track angle

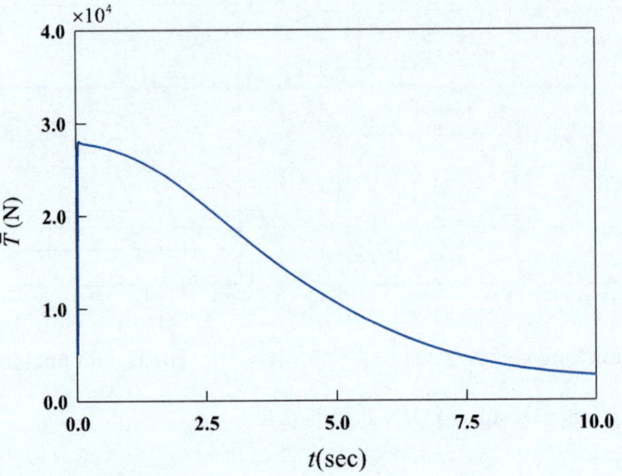

Fig. 8.5 Engine thrust response $\bar{T}$

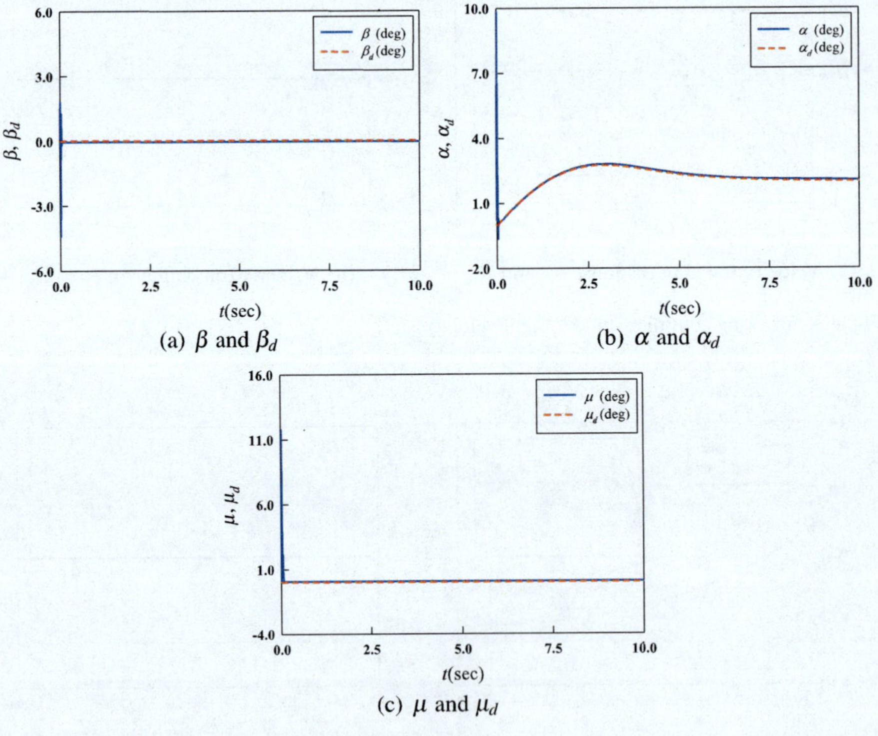

Fig. 8.6 The tracking results of attitude angles

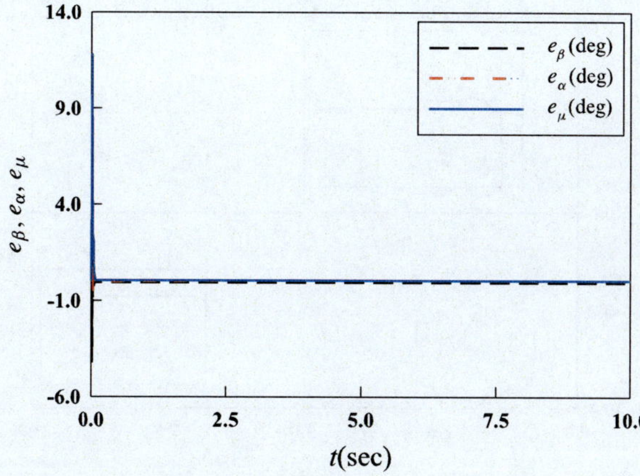

Fig. 8.7 The attitude angle tracking errors e_β, e_α and e_μ

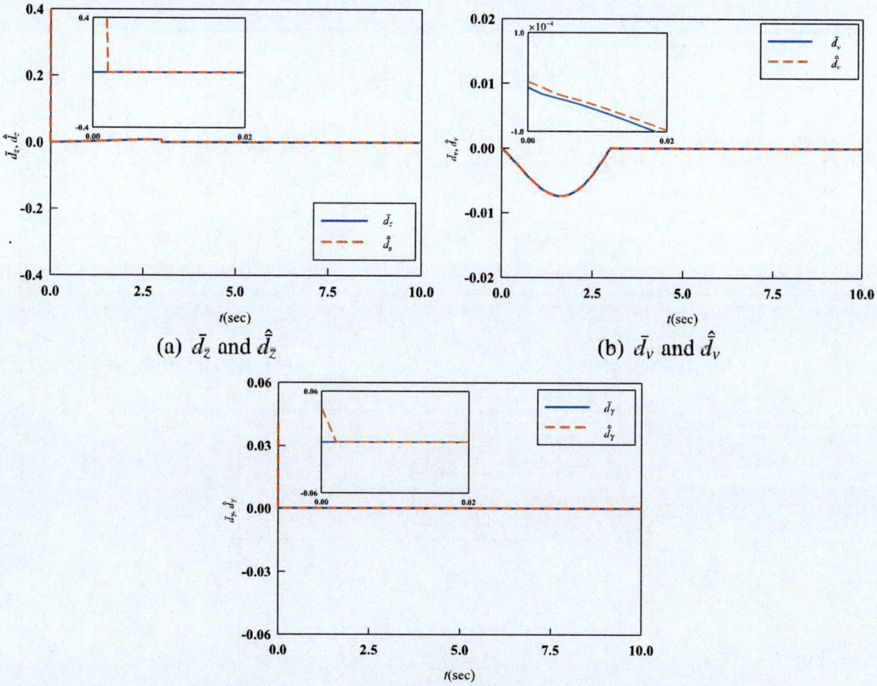

(a) $\bar{d}_z$ and $\hat{\bar{d}}_z$

(b) $\bar{d}_v$ and $\hat{\bar{d}}_v$

(c) $\bar{d}_\gamma$ and $\hat{\bar{d}}_\gamma$

Fig. 8.8 The estimation performance of the designed SMDO

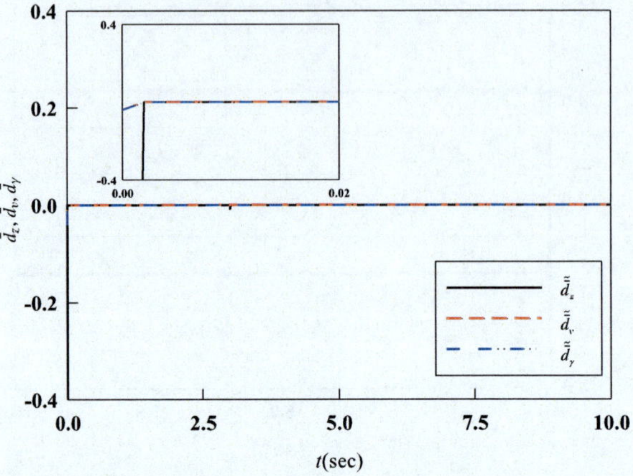

Fig. 8.9 Estimation errors of the disturbance observer $\tilde{\bar{d}}_z$, $\tilde{\bar{d}}_v$ and $\tilde{\bar{d}}_\gamma$

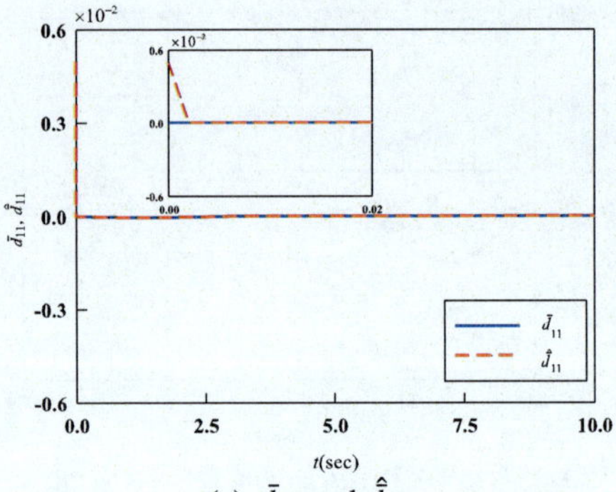

(a) $\bar{d}_{11}$ and $\hat{\bar{d}}_{11}$

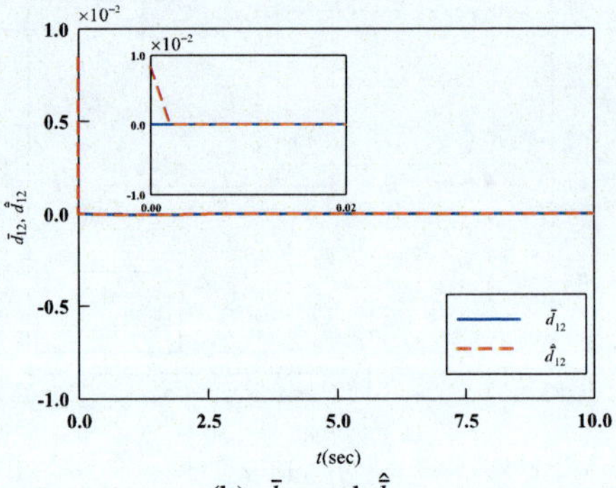

(b) $\bar{d}_{12}$ and $\hat{\bar{d}}_{12}$

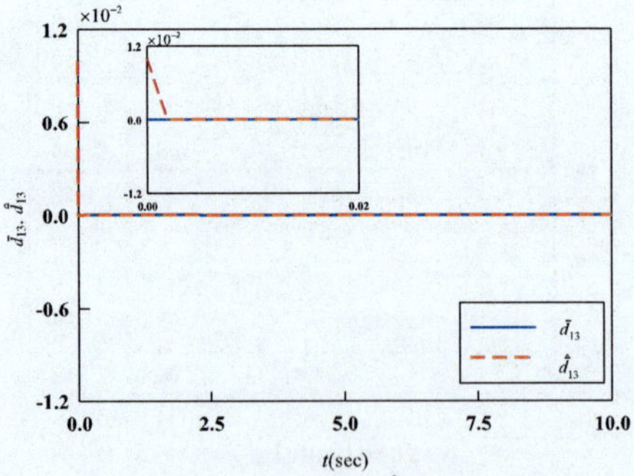

(c) $\bar{d}_{13}$ and $\hat{\bar{d}}_{13}$

Fig. 8.10 The estimation performance of the designed DTDO

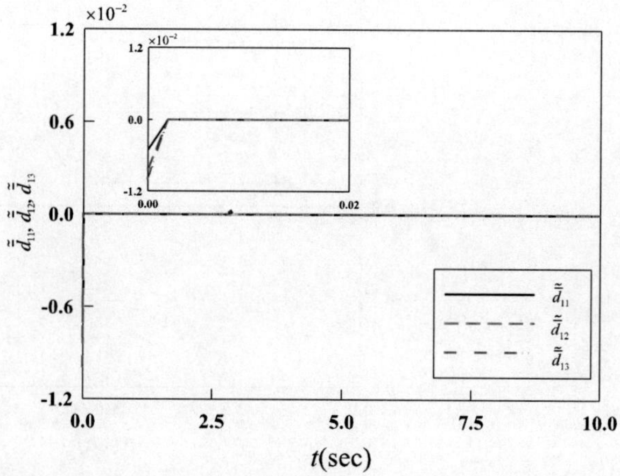

Fig. 8.11 Estimation errors of the disturbance observer $\tilde{\bar{d}}_{11}$, $\tilde{\bar{d}}_{12}$ and $\tilde{\bar{d}}_{13}$

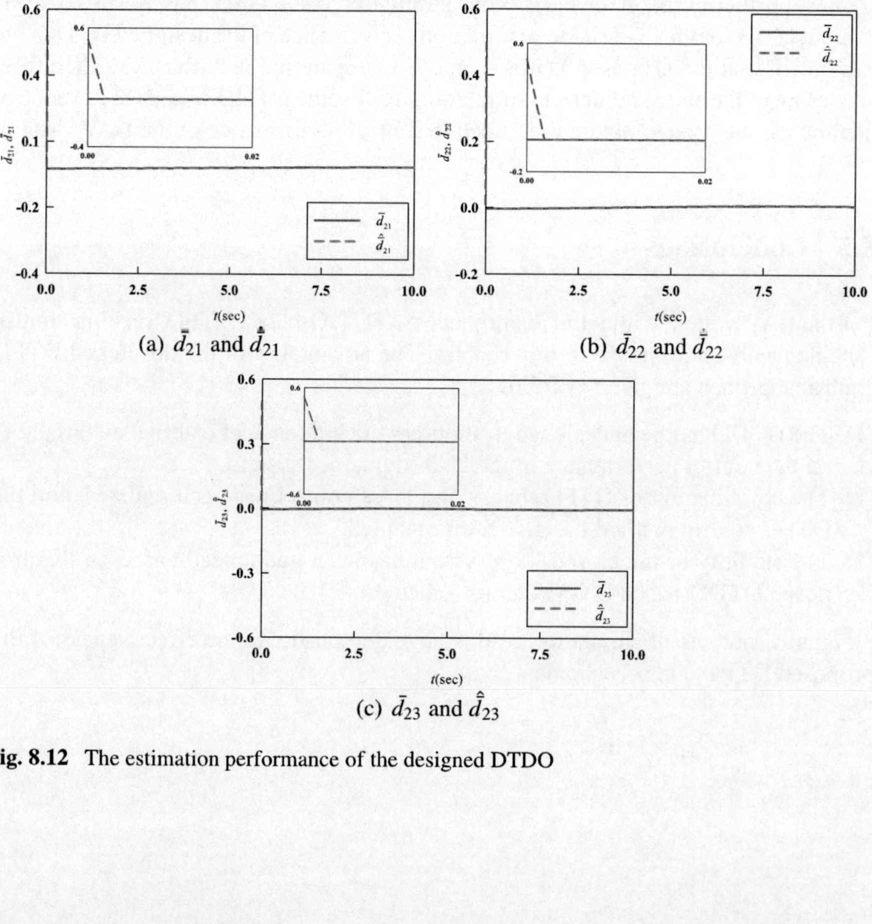

(a) $\bar{d}_{21}$ and $\hat{\bar{d}}_{21}$ (b) $\bar{d}_{22}$ and $\hat{\bar{d}}_{22}$

(c) $\bar{d}_{23}$ and $\hat{\bar{d}}_{23}$

Fig. 8.12 The estimation performance of the designed DTDO

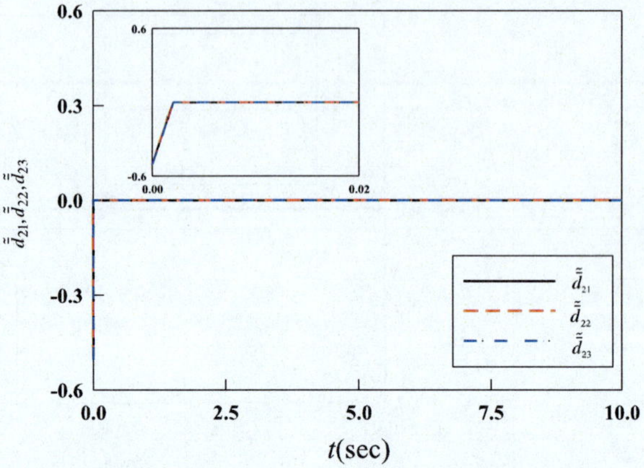

Fig. 8.13 Estimation errors of the disturbance observer $\tilde{\tilde{d}}_{21}$, $\tilde{\tilde{d}}_{22}$ and $\tilde{\tilde{d}}_{23}$

to Figs. 8.6 and 8.7, the output signals of UAV attitude system can track the desired signals. Furthermore, on the basis of the simulation results Figs. 8.8, 8.9, 8.10, 8.11, 8.12 and 8.13 for the disturbance estimation performance of the designed DTDO, we can obtain that the proposed DTDO can estimate external disturbances. Therefore, according to the above numerical simulation results, the DTDO-based DTFO control method can be used to obtain the effective control performance of the UAV system.

8.5 Conclusions

For the UAV system with wind disturbances, a DTDO-based DTFO tracking control scheme has been proposed in this chapter. The advantages of the developed DTFO control approach are given as follows:

(1) The DTDO has been designed to suppress the influence of external disturbances on the control performance of the UAV dynamics system.
(2) The combination of DTFO theory and UAV control has been realized, and the DTFO control method has also been obtained.
(3) The stability of the closed-loop system has been guaranteed based on the proposed DTDO-based DTFO control scheme.

Finally, numerical simulation results have demonstrated the effectiveness of the proposed DTFO control scheme.

References

1. Freeborn, T.J.: A survey of fractional-order circuit models for biology and biomedicine. IEEE J. Emerg. Sel. Top. Circuits Syst. **3**(3), 416–424 (2013)
2. Ionescu, C., Lopes, A., Copot, D., et al.: The role of fractional calculus in modeling biological phenomena: a review. Commun. Nonlinear Sci. Numer. Simul. **51**, 141–159 (2017)
3. Sopasakis, P., Sarimveis, H., Macheras, P., et al.: Fractional calculus in pharmacokinetics. J. Pharmacokinet. Pharmacodyn. **45**(1), 107–125 (2018)
4. Shivanian, E., Jafarabadi, A.: Applications of Fractional Calculus in Physics. World Scientific, New Jersey (2000)
5. Chen, W.-C.: Nonlinear dynamics and chaos in a fractional-order financial system. Chaos Solitons Fractals **36**(5), 1305–1314 (2008)
6. Wang, Z., Huang, X., Shi, G.: Analysis of nonlinear dynamics and chaos in a fractional order financial system with time delay. Comput. Math. Appl. **62**(3), 1531–1539 (2011)
7. Tacha, O., Volos, C.K., Kyprianidis, I.M., et al.: Analysis, adaptive control and circuit simulation of a novel nonlinear finance system. Appl. Math. Comput. **276**, 200–217 (2016)
8. Škovránek, T., Podlubny, I., Petráš, I.: Modeling of the national economies in state-space: a fractional calculus approach. Econ. Model. **29**(4), 1322–1327 (2012)
9. Flores-Tlacuahuac, A., Biegler, L.T.: Optimization of fractional order dynamic chemical processing systems. Ind. Eng. Chem. Res. **53**(13), 5110–5127 (2014)
10. Yadav, V.K., Das, S., Bhadauria, B.S., et al.: Stability analysis, chaos control of a fractional order chaotic chemical reactor system and its function projective synchronization with parametric uncertainties. Chin. J. Phys. **55**(3), 594–605 (2017)
11. Yu, W., Luo, Y., Chen, Y., et al.: Frequency domain modelling and control of fractional-order system for permanent magnet synchronous motor velocity servo system. IET Control Theory Appl. **10**(2), 136–143 (2016)
12. Gutierrez, R.E., Rosario, J.M., Tenreiro Machado, J.: Fractional order calculus: basic concepts and engineering applications. Math. Problems Eng. **2010**, Article ID 375858, 19p (2010)
13. Abbas, I.A.: A problem on functional graded material under fractional order theory of thermoelasticity. Theoret. Appl. Fract. Mech. **74**, 18–22 (2014)
14. Tripathy, M.C., Mondal, D., Biswas, K., et al.: Experimental studies on realization of fractional inductors and fractional order bandpass filters. Int. J. Circuit Theory Appl. **43**(9), 1183–1196 (2015)
15. Ebaid, A.: Analysis of projectile motion in view of fractional calculus. Appl. Math. Model. **35**(3), 1231–1239 (2011)
16. Manabe, S.: A suggestion of fractional-order controller for flexible spacecraft attitude control. Nonlinear Dyn. **29**(1–4), 251–268 (2002)
17. Aboelela, M.A., Ahmed, M.F., Dorrah, H.T.: Design of aerospace control systems using fractional PID controller. J. Adv. Res. **3**(3), 225–232 (2012)
18. Seyedtabaii, S.: New flat phase margin fractional order PID design: perturbed UAV roll control study. Robot. Auton. Syst. **96**, 58–64 (2017)
19. Chao, H., Luo, Y., Di, L., et al.: Roll-channel fractional order controller design for a small fixed-wing unmanned aerial vehicle. Control Eng. Pract. **18**(7), 761–772 (2010)
20. Luo, Y., Chao, H., Di, L., et al.: Lateral directional fractional order $(PI)^{\alpha}$ control of a small fixed-wing unmanned aerial vehicles: controller designs and flight tests. IET Control Theory Appl. **5**(18), 2156–2167 (2011)
21. Han, J., Di, L., Coopmans, C., et al.: Pitch loop control of a VTOL UAV using fractional order controller. J. Intell. Robot. Syst. **73**(1–4), 187–195 (2014)
22. Goodrich, C., Peterson, A.C.: Discrete Fractional Calculus. Springer, Berlin (2015)
23. Machado, J.: Discrete-time fractional-order controllers. Fract. Calcu. Appl. Anal. **4**, 47–66 (2001)
24. Ma, C., Hori, Y.: The time-scaled trapezoidal integration rule for discrete fractional order controllers. Nonlinear Dyn. **38**(1–4), 171–180 (2004)

25. Barbosa, R.S., Machado, J.T.: Implementation of discrete-time fractional-order controllers based on LS approximations. Acta Polytech. Hung. **3**(4), 5–22 (2006)
26. Das, S., Pan, I., Halder, K., et al.: LQR based improved discrete PID controller design via optimum selection of weighting matrices using fractional order integral performance index. Appl. Math. Model. **37**(6), 4253–4268 (2013)
27. Merrikh-Bayat, F., Mirebrahimi, N., Khalili, M.R.: Discrete-time fractional-order PID controller: definition, tuning, digital realization and some applications. Int. J. Control Autom. Syst. **13**(1), 81–90 (2015)
28. Huang, L., Wang, L., Shi, D.: Discrete fractional order chaotic systems synchronization based on the variable structure control with a new discrete reaching-law. IEEE/CAA J. Autom. Sin. (2016). https://doi.org/10.1109/JAS.2016.7510148
29. Sun, G., Ma, Z., Yu, J.: Discrete-time fractional order terminal sliding mode tracking control for linear motor. IEEE Trans. Ind. Electron. **65**(4), 3386–3394 (2018)
30. Mareels, I.M., Penfold, H., Evans, R.J.: Controlling nonlinear time-varying systems via Euler approximations. Automatica **28**(4), 681–696 (1992)

Chapter 9
Summary and Scope

9.1 Summary of Full Text

UAVs have been widely used in civilian and military fields due to their outstanding combat capability in remote areas or dangerous environments. However, the structures of UAVs are relatively complex, so there will be modeling errors between the accurate model and the constructed model, leading to the system uncertainties in UAV systems. At the same time, the flight environment of UAVs is always changeable, so the stability of UAVs flight control system is vulnerable to external disturbances. The above mentioned system uncertainties and external disturbances will not only reduce the control performance of the flight control system, but also lead to the instability of the UAV systems and bring insecurity to the UAVs flight. In addition, with the wide use of digital computers and microprocessors in control applications, the studies of discrete-time control methods have attracted more and more attention. Therefore, studying the discrete-time control methods with strong robustness and suppressing the adverse effects of system uncertainties and external disturbances on the flight control performance of UAVs, so as to improve the closed-loop performance of UAV systems, has become the key issues to be considered in the research of discrete-time flight control schemes for UAV systems. In addition, the deflection angle of the rudder surface for fixed-wing UAV is limited to a limited range [1], and the input saturation problems are less considered in the traditional design of flight control schemes for UAV systems. However, when the deflection angle of the aerodynamic rudder reaches the limit, it will affect the control performance of the UAV systems and even destroy the stability of the UAV systems. Therefore, for the research on the discrete-time flight control methods of UAV systems, robust discrete-time constrained control is also one of the important problems in the design of UAVs control methods. In addition, studies on transient performance of fixed-wing UAV systems such as overshooting and convergence speed are still relatively few. Excessive overshooting may cause the actuator of UAVs to exceed the physical limit, thus leading to the instability of closed-loop UAV systems. Therefore, when designing

S. Shao et al., *Robust Discrete-Time Flight Control of UAV with External Disturbances*, Studies in Systems, Decision and Control 317,
https://doi.org/10.1007/978-3-030-57957-9_9

a robust discrete-time flight controller, the steady-state performance and transient performance of the closed-loop system should be guaranteed, which leads to the problem of prescribed performance control.

This book studies the external disturbances, system uncertainties and input saturation in the fixed-wing UAV system, and studies the robust discrete-time flight control technologies under external disturbances. The main research results of this book are as follows:

(1) This book introduces the research background and significance of UAVs control, the research status of advanced flight control methods, the research status of discrete-time flight control methods, the research status of flight control based on DO, the research status of input nonlinearities and system uncertainties and the research status of flight control based on FO theory.
(2) Firstly, the dynamic mathematical model of fixed-wing UAV six degrees of freedom with wind influence is derived, and the nonlinear affine model of attitude dynamics of UAV is given, which provides the model basis for the design of discrete-time flight control schemes. Secondly, the input saturation, the dead-zone nonlinearity and the non-symmetric saturation and dead-zone nonlinearity are introduced, which provides the basis of the input nonlinearity model for the further study of the input restricted control of UAV. Finally, some lemmas and definitions are introduced, including Grünwald-Letnikow difference operator definition of DTFO form and RBFNNs approximation theory of discrete-time form, which provide theoretical basis for the robust design of discrete-time schemes of UAV nonlinear models under external disturbances.
(3) The methods of the discrete-time BC and the discrete-time SMC are employed for the flight control of the fixed-wing UAV system without wind disturbances. According to the BC technology, the discrete-time controllers are designed, and the stability theory of Lyapunov in the form of discrete-time is introduced to prove that the discrete-time controller designed can ensure the boundedness of the closed-loop system signals. Finally, the simulation results are given to show the effectiveness of the discrete-time control schemes.
(4) A discrete-time flight control scheme based on the DTDO and the NN is proposed for attitude dynamics model of UAV with external wind disturbances and system uncertainties. Firstly, the system uncertainties in the attitude dynamics model are approximated by using the NN. Secondly, a DTDO based on NN is designed, and the designed DTDO is used to estimate the external disturbances. Then, according to the tracking differentiator in the form of discrete-time, the nonlinear DTDO based on NN, and the technique of BC, a discrete-time controller based on NN is designed. Finally, the attitude dynamics model of UAV with external wind disturbances and system uncertainties is simulated and analyzed, and the simulation results further illustrate the effectiveness of the proposed discrete-time flight control scheme based on the DTDO and the NN.
(5) On the basis of previous studies, a NN tracking control scheme based on the DTDO and an auxiliary system are proposed for the attitude dynamics model of uncertain UAV with external wind disturbances and input saturation. Firstly,

the system uncertainties in the attitude dynamics model of discrete UAV are dealt with by the NN approximation principle. Secondly, in order to suppress the adverse effects of external disturbances on the system, a DTDO based on NN is designed, and the output of DTDO is applied to the design of the controller. In addition, the auxiliary system is used to compensate the negative influence of input saturation on the control of UAV attitude dynamics model. Then, according to the DTDO, the auxiliary system, the discrete-time tracking differentiator, and the BC technique, an adaptive discrete-time controller based on NN is designed, and the boundedness of the closed-loop system signals is proved by the stability theory of Lyapunov. Finally, by selecting appropriate control parameters, numerical simulation results show the effectiveness of the proposed discrete-time flight control scheme based on the NN, the DTDO and an auxiliary system.

(6) Based on the previous research on the input saturation problem, a trajectory control scheme of discrete-time based on the SMDO is proposed, and an adaptive attitude control scheme based on a discrete-time SMDO, an auxiliary system and the NN is proposed. Firstly, the approximate discrete-time trajectory control system model with wind disturbances and system uncertainties and the attitude dynamics model with wind disturbances and input saturation are obtained based on the Euler approximation method. Secondly, a discrete-time SMDO is designed, and the output of SMDO is used to compensate the negative influence of external disturbances. Then, according to the auxiliary system, the discrete-time tracking differentiator, the NN-based SMDO and the BC technology, an adaptive attitude controller of discrete-time is designed, and the boundedness of the closed-loop system signals is proved by the stability theory of Lyapunov. Finally, numerical simulation results show the effectiveness of the discrete-time flight control scheme based on the SMDO, the NN and the auxiliary system.

(7) According to the design of the continuous time control schemes, an adaptive DTFO control scheme based on the NN, the DTDO and the prescribed performance control method is proposed for the attitude dynamics model of the UAV with external disturbances and system uncertainties. Firstly, the system uncertainties in attitude dynamics model of UAV are tackled by using NN approximation. Secondly, a DTDO based on NN is designed to compensate the adverse effect of external disturbances on the control performance. Then, based on the NN, the designed DTDO, the prescribed performance control method and the BC technology, a DTFO controller is designed based on the DTDO and the NN, and the stability theory of Lyapunov is used to prove the boundedness of closed-loop system signals. Finally, numerical simulation results show the effectiveness of the proposed prescribed performance adaptive control scheme based on the NN.

(8) From the previous discrete-time FOC research, a DTFO scheme based on the DTDO is proposed for the trajectory control system model with external wind disturbances and the attitude dynamics model of UAV with wind disturbances and input saturation. Firstly, a DTDO is designed to compensate for the adverse effects of external disturbances on the attitude dynamics system of the UAV. Secondly, by using the output of DTDO, the DTFO theory and the BC technology, a control method of DTFO based on the DTDO is designed, and the boundedness

of closed-loop system is proved by Lyapunov stability theory. Finally, the simulation analysis is made on the UAV trajectory control system model and attitude dynamics model with external disturbances, and the simulation results show the effectiveness of the proposed DTFO control scheme based on the designed DTDO.

9.2 Future Research Prospect

In this book, focusing on the control problem of dynamics model of fixed-wing UAV with external disturbances, the discrete-time flight control of UAV has been studied, and some research results have been obtained. However, due to the limitations of our time and ability level, the research on the discrete-time flight control method of UAV still has much to be explored and perfected:

(1) For the discrete-time flight control of fixed-wing UAVs six degrees of freedom dynamic model, although the robust discrete-time flight control of the attitude dynamics model of fixed-wing UAV with external disturbances has been studied in this book and some theoretical results have been obtained, the research on all states of fixed-wing UAV has not been discussed. As the flight environment of fixed-wing UAVs is often changeable during the mission, the flight speed and altitude of UAVs often need to be adjusted according to the actual situation during the flight of UAVs. Therefore, when studying the control of UAVs, it is necessary to consider the problem of speed and height control in the design of the discrete-time flight control of UAVs, which can further improve the ability for UAVs to perform a variety of tasks.

(2) Robust discrete-time flight control for fixed-wing UAV systems with multiple actuator nonlinearities. Due to the time constraints, only input saturation nonlinear characteristics are considered in the design of the discrete-time flight control when studying the discrete-time input restricted control of the UAV attitude dynamics system. However, practical UAV systems often have a variety of actuator nonlinearities, such as hysteresis and dead-zone nonlinearities, which will not only affect the control performance of the UAV system but also cause the instability of the flight control system. In addition, the design of the discrete-time flight control of fixed-wing UAV with a variety of actuator nonlinear characteristics will become more complex. Therefore, it is necessary to consider a variety of nonlinear characteristics in the subsequent study of UAVs control to enhance the robustness of the performance of discrete-time flight control system.

(3) The discrete-time flight control methods are designed by using a variety of discrete-time approximation methods. In this book, the attitude dynamics system of continuous time fixed-wing UAVs is transformed into the discrete-time attitude dynamics system by using the Euler approximation method to design the discrete-time flight control schemes and some research conclusions are obtained. However, the research on designing discrete-time flight control schemes by using

other discrete-time approximation methods has not been successful. Due to the different control precision of the discrete-time flight control designed by different discrete-time approximation methods, the flight control performance of the UAV will be different.

(4) Study on finite time discrete flight control problem. For the studies of discrete-time flight control in this book, the finite time discrete-time flight control method of fixed wing UAV system has not been studied. For the control time optimization of the UAV system, the control convergence time of the UAV closed-loop system can be optimized by using the finite time discrete-time flight control method. In addition, when studying the design of the finite time discrete flight control based on DO, the finite time DTDO should be designed firstly, and then the finite time discrete flight control method should be studied in combination with the designed DTDO. However, the above problems have certain complexity. Therefore, in the future design of the UAVs control, the finite time discrete flight control method can be further studied.

Reference

1. Sonneveldt, L.: Nonlinear F-16 model description, Tech. rep., Delft University of Technology, Netherlands (2006)